17|95

St. Olaf College

JAN 5 1981

Science Library

THE
LOGARITHMIC POTENTIAL
AND OTHER MONOGRAPHS

THE LOGARITHMIC POTENTIAL DISCONTINUOUS DIRICHLET AND NEUMANN PROBLEMS
BY GRIFFITH CONRAD EVANS

FUNDAMENTAL EXISTENCE THEOREMS
BY GILBERT AMES BLISS

DIFFERENTIAL-GEOMETRIC ASPECTS OF DYNAMICS
BY EDWARD KASNER

CHELSEA PUBLISHING COMPANY, INC.
NEW YORK, N. Y.

Published at New York, N.Y., 1980
Printed on 'long-life' acid-free paper
International Standard Book Number 0-8284-0305-8
Library of Congress Catalog Card Number 79-57336
Printed in the United States of America

PREFACE

The present work contains three monographs on mathematics, as follows:

THE LOGARITHMIC POTENTIAL, by Griffith Conrad Evans, originally published in 1927 as *American Mathematical Society Colloquium Publications, Volume VI* and here re-issued with revisions and corrections by the author.

FUNDAMENTAL EXISTENCE THEOREMS, by Gilbert Ames Bliss of the University of Chicago, originally published in 1913 as the first of the two monographs comprising *The Princeton Colloquium*. It is here re-issued unabridged and without textual alteration.

DIFFERENTIAL GEOMETRIC ASPECTS OF DYNAMICS, by Edward Kasner of Columbia University, originally published in 1913 as the second of the two monographs comprising *The Princeton Colloquium*. It is here re-issued unabridged and without textual alteration.

All three of these monographs were originally published by The American Mathematical Society at New York, N.Y. The original pagination has been retained, except for certain pages of *The Logarithmic Potential*, where the late Professor Evans has added to the text.

04623

THE LOGARITHMIC POTENTIAL

DISCONTINUOUS DIRICHLET AND NEUMANN PROBLEMS

BY

GRIFFITH CONRAD EVANS, PH. D.
PROFESSOR OF PURE MATHEMATICS
THE RICE INSTITUTE

TO

VITO VOLTERRA

PREFACE

This small treatise is an outgrowth of a study of Stieltjes integrals and potential theory which the author published in the 1920 volume of the Rice Institute Pamphlet, and a needed revision and development of the last part of that essay in the direction indicated by three notes which appeared in 1923, in the Comptes rendus des séances de l'Academie des Sciences. Two of these were written in conjunction with my colleague, Professor H. E. Bray. The work gives a unified treatment of the basis of the theory of Laplace's equation in two dimensions, suitable, it is hoped, for graduate students of a moderate degree of advancement, and is intended to be of service in the development of the theory of partial differential equations of elliptic type. These developments are generating a compound of two of the most important elements of modern analysis—the concepts of Lebesgue on the one hand, and of Volterra on the other.

An earlier form of part of the treatise was given in lectures at the Rice Institute in the academic year 1924-25, in connection with a course in the theory of functions of a real variable, and at the University of Chicago during the Summer Quarter of 1925. Chapter VII furnished the substance of an invited discourse at the meeting of the Southwestern Section of the American Mathematical Society in November, 1926.

The author is much indebted to Professor O. D. Kellogg, who has seen a large portion of the manuscript, and aided with kindly criticism, to Professor Bray, who has read the proof sheets, and, finally, to the American Mathematical Society, through whose generosity the publication is possible.

HOUSTON, TEXAS.
June, 1927.

GRIFFITH C. EVANS.

TABLE OF CONTENTS

PAGES

I. PRELIMINARY CONCEPTS. STIELTJES INTEGRALS AND FOURIER SERIES 1
  1. Functions of limited variation. 2. Continuation of the preceding. 3. Integrals with respect to a function of limited variation. 4. Note on the second law of the mean. 5. Classical theorems on integrals and limits of integrals. 6. The limits of Stieltjes integrals. 7. Note on Lebesgue integrals. 8. Convergence of Fourier series. 9. Summability of series.

II. FUNCTIONS HARMONIC WITHIN A CIRCLE 26
  10. Preliminary theorems. 11. A preliminary result. 12. Note on integral identities. 13. Digression: Functions of points and of point sets. 14. Properties of the Poisson-Stieltjes integral. 15. Continuation of the preceding: Behaviour of $u(r,\theta)$ in the neighborhood of the boundary. 16. The Poisson integral: $F(\varphi)$ absolutely continuous.

III. NECESSARY AND SUFFICIENT CONDITIONS. THE DIRICHLET PROBLEMS FOR THE CIRCLE 46
  17. Fundamental theorem and lemma. 18. Proof of fundamental theorem. 19. Special cases of the Poisson integral. 20. The Dirichlet problem and its extension.

IV. POTENTIALS OF A SINGLE LAYER AND THE NEUMANN PROBLEM 55
  21. The Stieltjes integral for potentials of a single layer. 22. Necessary and sufficient conditions. 23. Further properties. 24. The Neumann problems. 25. General points of view. 26. Digression: Physical interpretation of a general distribution of mass. 27. Cauchy's integral formula.

	PAGES
V. GENERAL SIMPLY CONNECTED PLANE REGIONS AND THE ORDER OF THEIR BOUNDARY POINTS	71

28. Conformal transformations and general regions. 29. Invariant forms of conditions (i), (ii) etc. 30. Invariant forms of conclusions. 31. Order of boundary points. 32. Integrals on the boundary and the Dirichlet problems. 33. Special cases of the condition (ii). The continuous boundary value problem. 34. A new continuous boundary value problem. 35. The generalized Neumann problem in the general region.

VI. PLANE REGIONS OF FINITE CONNECTIVITY	86

36. Functions harmonic outside a circle. 37. The multiply connected region bounded by $n+1$ distinct circles. 38. Representation in terms of the Green's function. 39. Boundary integrals and Stieltjes integral equations. 40. General regions of finite connectivity. Isolated point boundaries. 41. Annular regions. Determination of the functions $F_0(\theta)$ and $F_1(\theta)$. 42. Uniqueness of the representation of Theorem 3 for S.

VII. RELATED PROBLEMS	121

43. A simple discontinuous boundary value problem. 44. Continuous boundary value problems. 45. Regions with continuous boundaries. 46. Regions with rectifiable boundaries. 47. Regions of infinite connectivity. 48. Remarks on necessary and sufficient conditions. 49. Convergence in the mean of positive order less than one. 50. Integro-differential equations of Bôcher type.

CHAPTER I

PRELIMINARY CONCEPTS
STIELTJES INTEGRALS AND FOURIER SERIES

1. Functions of limited variation. In the chapters which follow, the Dirichlet and Neumann problems are recast under general points of view which derive, as did earlier formulations of the problems, from physical considerations. These problems are to be discussed in relation to conformal transformations and the most general distributions of matter, simply or in doublets, on the boundary of a general simply or finitely connected open region bounded by circles. In this way new classes of boundary conditions arise, and the appropriate boundary value problems may be solved. Thus for the old problems some new results, as well as some familiar ones, appear as special cases.

The principal instrument for the investigation of harmonic functions, from this point of view, will be the *Stieltjes integral*. This integral refers fundamentally to functions *of limited variation*. For the convenience of the reader some important properties of the relation between these two concepts will be briefly summarized, and one or two theorems obtained.

Let $f(x)$ be defined for every x in the closed interval (a, b), and let this interval be divided into a finite number n of subintervals by points $x_1, x_2, \cdots, x_{n-1}$, writing $x_0 = a$, $x_n = b$ for convenience. If the quantity

$$\sum_{i=0}^{n-1} |f(x_{i+1}) - f(x_i)|, \qquad x_0 < x_1 < x_2 \cdots < x_n,$$

is bounded, $\leq N$, for all such positions of the x_i in (a, b) and all n, the function is said to be *of limited variation* in (a, b). The least value of N which will satisfy this condition is said to be the *total variation* of $f(x)$ in (a, b), and is designated by T.

In any closed interval (a, x), $a < x \leq b$, let us choose a set of non-overlapping subintervals $(x_1', x_1''), (x_2', x_2''), \cdots,$

finite in number, which add together into the whole or less than the whole of (a, x). The upper bound, for all such choices, of the quantity

$$\sum_k (f(x_k'') - f(x_k'))$$

exists, and we designate it by $\varphi(x)$ or φ_{ax}; that is to say, $\varphi(x)$ is the smallest number which is not exceeded by any possible value of the given sum. Let us definitely admit a single point as a special case of an interval, namely one whose end points coincide. We see then directly that $\varphi(x) \geq 0$, and again that $\varphi(x)$ is a non-decreasing function of x. It is called the *positive variation function* of $f(x)$.

EXERCISE. Show that $\varphi(x)$ has the same value and the same properties even if we do not admit intervals of zero length. These properties may be demonstrated by proving that given ε arbitrarily small we can find in any interval (x_1, x_2) some subinterval (x', x'') for which $|f(x'') - f(x')| < \varepsilon$.

In order to complete the definition of $\varphi(x)$ we must assign it a value at $x = a$; we assign it there the value 0.

It is interesting to note that if X is some value intermediate between a and x we have

$$\varphi_{ax} = \varphi_{aX} + \varphi_{Xx}$$

or

$$\varphi(x) = \varphi(X) + \varphi_{Xx}.$$

For we can form a sum for the interval (a, x) as near to φ_{ax} as we desire; if X is an interior point of a subinterval the sum is not changed if we insert X as the end of one interval and the beginning of the next; thus in any case we can form partial sums which relate to φ_{aX} and φ_{Xx} respectively, yet cannot exceed them. Hence $\varphi_{ax} \leq \varphi_{aX} + \varphi_{Xx}$. Also we can form partial sums for (a, X) and (X, x) as near respectively to φ_{aX} and φ_{Xx} as we desire; the sum total will be $\leq \varphi_{ax}$. Hence $\varphi_{ax} \geq \varphi_{aX} + \varphi_{Xx}$, from which the conclusion follows.

We define another function of x, which is a non-increasing function:

$$\psi(x) = f(x) - f(a) - \varphi(x).$$

PRELIMINARY CONCEPTS

In fact, if x_1, x_2 are two values of x, with $x_2 > x_1$, we have

$$\psi(x_2) - \psi(x_1) = f(x_2) - f(x_1) - \{\varphi(x_2) - \varphi(x_1)\}.$$

But $\varphi(x_2) - \varphi(x_1) = \varphi_{x_1 x_2} \geq f(x_2) - f(x_1)$. Hence $\psi(x_2) - \psi(x_1) \leq 0$, and $\psi(x)$ is a non-increasing function of x.

The function $\psi(x)$ is called the *negative variation function* of $f(x)$. Moreover by the definition of $\psi(x)$ we have

$$f(x) = f(a) + \varphi(x) + \psi(x).$$

In other words, a function of limited variation in the closed interval (a, b) can be written as the difference of two non-decreasing functions of x. The converse of this statement is obviously true.

The function $t(x) = \varphi(x) - \psi(x)$ is again a non-decreasing function of x, being the sum of two such functions, and is called the *total variation function* of $f(x)$.

EXERCISE. Show that definitions of $\psi(x)$ and $t(x)$ analogous to that of $\varphi(x)$ may be given, and that $t(b) = T$.

A non-decreasing function approaches a limiting value as x approaches a value α from the left, and also as x approaches α from the right. Hence the same property holds for the difference between two such functions, that is, for any function of limited variation. The two values, the limit from the left and the limit from the right, need not be equal, or equal to the value $f(\alpha)$, since $f(x)$ need not be continuous at α. It is therefore convenient to have symbols for these values and write $f(\alpha + 0)$ for the quantity $\lim\limits_{\substack{x = \alpha \\ x > \alpha}} f(x)$, and $f(\alpha - 0)$ for the quantity $\lim\limits_{\substack{x = \alpha \\ x < \alpha}} f(x)$, respectively. If $\alpha = 0$, the symbols $f(0+)$ and $f(0-)$ are used. The statement we have just made then amounts to saying that if $f(x)$ is of limited variation, $f(x+0)$ and $f(x-0)$ exist if $a < x < b$; and also $f(a+0)$ and $f(b-0)$ exist.

An important property of a function of limited variation is that the aggregate of its points of discontinuity must be denumerable. To prove this fact it is sufficient to consider a non-decreasing function, since the discontinuities of a function of limited variation will also be discontinuities of its total variation function. But for such a function the number of points x where $f(x+0)-f(x)$ or $f(x)-f(x-0)$ is $\geq T/2$ is finite; also the number of points such that these jumps are $< T/2$ but $\geq T/4$; also the number of points for which these jumps are $< T/4$ but $\geq T/8$, etc. In such a classification however every point of discontinuity is ultimately included, since a point where both $f(x+0)-f(x)$ and $f(x)-f(x-0)$ are both zero is a point of continuity.

2. Continuation. We have discussed the function of limited variation with respect to a closed interval (a, b). It is convenient however to be able to consider the same sort of situation with respect to an open interval. That is we say that $f(x)$ is of limited variation if

$$\sum |f(x_k'')-f(x_k')| \leq N,$$

N being some constant, no matter how the subintervals (x_k', x_k'') are chosen in the open interval (a, b). We define the positive variation function as before; it is again a non-decreasing function of x, with $\varphi(a+0)=0$; moreover $\varphi(x)=\varphi(X)+\varphi_{Xx}$ if $a < X < x < b$. Further let $\psi_1(x)=f(x)-\varphi(x)$; this is a non-increasing function. Hence $\psi_1(a+0)$ exists. We therefore define the negative variation function as

$$\psi(x) = \psi_1(x)-\psi_1(a+0),$$

so that $\psi(a+0)=0$. We define the total variation function as

$$t(x) = \varphi(x)-\psi(x),$$

and we have also $t(a+0)=0$. Whether or not for completeness we define $\varphi(a)$, $\psi(a)$ and $t(a)$ as 0 is immaterial, since a is outside the open interval. Finally, we have

$$f(x) = \varphi(x)+\psi(x)+\psi_1(a+0),$$

so that $f(a+0)=\psi_1(a+0)$. The quantities $f(a+0)$ and $f(b-0)$ are thus seen to be determinate, although $f(a)$ and $f(b)$ are not defined. The quantity $t(b-0)-t(a+0)$ is seen to be the least value possible of the N above, and may be spoken of as the total variation T' of $f(x)$ for the open interval (a, b).

It is obvious that a function which is of limited variation on an open interval may be extended so that it will be of limited variation on the corresponding closed interval. This is a particular case of the following theorem.

THEOREM 1. *Let E be a set dense in (a, b), not necessarily including the end points a, b, and let $f(x)$ be of limited variation on E. Then $f(x)$ may be defined on the complementary set CE so as to be of limited variation on the closed interval (a, b).*

Our hypothesis about E is that every subinterval of (a, b) not of zero length contains at least one point of E. Let x be any value $a < x \leq b$. Consider a finite number of non-overlapping intervals (x_i', x_i''), $a \leq x_i' \leq x_i'' \leq x$, of which the ends belong to E, and consider the sums $\sum |f(x_i'') - f(x_i')|$, $\sum \{f(x_i'') - f(x_i')\}$. If the former sum has an upper bound T, when $x = b$, $f(x)$ is said to be of limited variation on E. In this case, given x, the latter sum has an upper bound $\varphi(x) = \varphi_{ax}$ which is not negative. The definition of $\varphi(x)$ is completed by writing $\varphi(a) = 0$.

We notice that $\varphi(x)$ is a non-decreasing function of x, and that if x is a point of CE we have $\varphi(x) = \varphi(x-0)$. Moreover if $x > X$ and X is in E, we have $\varphi(x) = \varphi(X) + \varphi_{Xx}$.

EXERCISE. The reader may prove that if X is in CE we have $\varphi(x) = \varphi(X+0) + \varphi_{Xx}$.

We define also $\psi(x)$ on the points of E,

$$\psi(x) = f(x) - \varphi(x) + \text{const.}$$

where the constant is to be assigned. The reasoning previously employed shows that on E the function $\psi(x)$ is a non-increasing function. Hence the upper bound of $\psi(x)$ on E is $\psi(a)$ if a is in E; otherwise it is $\psi(a+0)$. But in either case we may choose the arbitrary constant so that this upper bound is zero.

We may now proceed to extend the definition of $\psi(x)$ by defining a function $\beta(x)$ for the function $-\psi(x)$ in the same

way as before we defined $\varphi(x)$ for $f(x)$. But since on the points of E, $-\psi(x)$ is a non-decreasing function with lower bound zero, it follows that $\beta(x) = -\psi(x)$ on E. In order to complete the definition of $\psi(x)$ we write for the points of CE
$$\psi(x) = -\beta(x).$$

In particular, since $\varphi(x)$ and $\psi(x)$ are monotonic, the quantities $\varphi(a+0)$, $\psi(a+0)$, $\varphi(b-0)$, $\psi(b-0)$ and the quantities $\varphi(x\pm 0)$, $\psi(x\pm 0)$ for $a < x < b$ are all defined and equal to the values which are obtained for them by considering merely the points of E; moreover for points of CE we have $\varphi(x-0) = \varphi(x)$, $\psi(x-0) = \psi(x)$.

It follows then that $f(a+0)$ is defined by means of the points of E, and if a is not a point of E we are at liberty to define $f(a) = f(a+0)$. To define $f(x)$ on the points of CE we write
$$f(x) = f(a) + \varphi(x) + \psi(x).$$

The function $f(x)$, so extended, is of limited variation on (a, b); moreover $f(x+0)$ and $f(x-0)$ have the values which are defined merely by the points of E, and also $f(x-0) = f(x)$ on the points of CE. The functions $\varphi(x)$ and $\psi(x)$ are evidently the positive and negative variation functions respectively for $f(x)$, and the function $t(x) = \varphi(x) - \psi(x)$ is the total variation function.

If $g(x)$, of limited variation, is any other extension of $f(x)$ which agrees with the common value of $f(x+0)$ and $f(x-0)$, whenever these have a common value as determined by points of E, then $g(x) = f(x)$ throughout, except at the points of discontinuity of $f(x)$, which of course, constitute at most a denumerable infinity of points in E and CE; for $f(x+0)$ and $f(x-0)$ defined by means of E are the same as when defined by means of all the points of (a, b), and therefore if x is a point of continuity or an unnecessary discontinuity of $f(x)$, the function $g(x)$ will be $f(x-0)$.

We have also immediately the following corollaries, of which the proofs are obvious.

COROLLARY 1. Any function of limited variation on (a, b) which is an extension of $f(x)$ on the points of E agrees with the extension above defined except at a denumerable infinity of points; if the discontinuities are made regular on (a, b) the extension is uniquely determined (except at a and b, if a, b are members of CE).*

COROLLARY 2. The proofs of these theorems may be applied at once to the case where the set of points E is dense on the circumference of a circle, and f is given, of limited variation, on E.† It is evident then that f may be so defined as to be of limited variation on the whole circumference, that any two such extensions differ only on a denumerable set, and if the discontinuities are made regular the function is uniquely determined on the circumference.

COROLLARY 3. If $f(x)$ is the difference of two non-decreasing *bounded* functions on E it is of limited variation on E.

EXERCISE. The figure illustrates a process of arriving at a function of limited variation (non-decreasing) which is continuous, whose derivative is zero for points in intervals whose lengths add to-

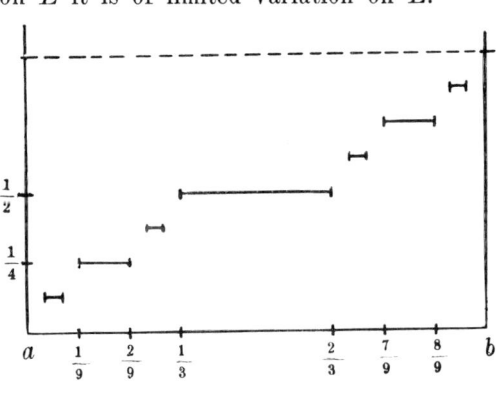

gether into the total length of (a, b), and yet which is not constant in (a, b). The reader may demonstrate these facts. This type of example is due to Cantor, Vitali and Borel.

3. Integrals with respect to a function of limited

---

* The function $f(x)$ has a regular discontinuity at x_0 if $f(x_0) = \frac{1}{2}\{f(x_0+0)+f(x_0-0)\}$.

† E is such that every circular arc contains at least one point of E; f is a function of the central angle θ or of some variable which is a continuous monotonic function of θ.

variation. We turn now to the definition of our integral. Let $u(x)$ be a bounded function and $f(x)$ a monotonic function, say non-decreasing, in the closed interval (a, b). Let M_i be the upper, and m_i the lower, bound of $u(x)$ in the subinterval (x_i, x_{i+1}), and form the sums

$$(1) \quad \begin{aligned} S_n &= \sum_{0}^{n-1} M_i \{f(x_{i+1}) - f(x_i)\}, \\ s_n &= \sum_{0}^{n-1} m_i \{f(x_{i+1}) - f(x_i)\}. \end{aligned}$$

Evidently S_n cannot be increased by adjoining new points of subdivision, nor can s_n be decreased; further $S_n \geq s_n$. The lower bound S of S_n is called the superior integral, the upper bound s of s_n is called the inferior integral, and if the two bounds are the same the common value, $S = s$, is called the Stieltjes, (or sometimes the Riemann-Stieltjes) integral. It is denoted by the symbol

$$\int_a^b u(x)\, df(x).$$

If $u(x)$ is continuous, *the Stieltjes integral exists*. In fact, since $f(x)$ is non-decreasing, the classical method of Jordan for the Riemann integral applies precisely to this case. Moreover, the law of the mean is justified, that

$$(2) \quad \int_a^b u(x)\, df(x) = u(\xi)\,[f(b) - f(a)], \quad a \leq \xi \leq b.$$

EXERCISE. In the inequality $a \leq \xi \leq b$, the equal signs may not be omitted. The reader may find an example to verify this fact.

EXERCISE. Show that

$$\left| \int_a^b u(x)\, df(x) \right| \leq T \max |u(x)|.$$

If $u(x)$ is continuous and $f(x)$ of bounded variation, not necessarily monotonic, we *define*

$$(3) \quad \int_a^b u(x)\, df(x) = \int_a^b u(x)\, d\varphi(x) + \int_a^b u(x)\, d\psi(x)$$

where φ and ψ are the positive and negative variations of $f(x)$. If we insert the points of subdivision $a = x_0$,

$x_1, \cdots, x_n = b$, with $x_{i+1} - x_i < \delta$, and let x'_i be a point in the interval $x_i \leq x'_i \leq x_{i+1}$, we have

(4) $\quad \left| \int_a^b u(x)\, df(x) - \sum_0^{n-1} u(x'_i) \{f(x_{i+1}) - f(x_i)\} \right| \leq T\omega_\delta$

where ω_δ is the oscillation of $u(x)$ in an interval of length δ in (a, b). Hence, in particular,

(5) $\quad \lim_{\delta = 0} \sum_{i=0}^{n-1} u(x'_i) \{f(x_{i+1}) - f(x_i)\} = \int_a^b u(x)\, df(x),$

a result which conforms to a definition of the Stieltjes integral according to the method of Cauchy in the ordinary case.

We may take equation (5) as an alternative definition of the Stieltjes integral, entirely suitable for our purposes.

A limit of this sort may exist even if $f(x)$ is not of limited variation and $u(x)$ continuous; in fact it always does exist if the roles of the two functions are reversed, $u(x)$ being of limited variation and $f(x)$ continuous. This is demonstrated by means of Stieltjes's celebrated integration by parts, which we now proceed to carry through.

We have, assuming $f(x)$ to be of limited variation, and $u(x)$ continuous, taking x'_i in the interval $x_i \leq x'_i \leq x_{i+1}$, the equality

$$\sum_{k=0}^{n-1} f(x'_k) \{u(x_{k+1}) - u(x_k)\} - f(x_n) u(x_n) + f(x_0) u(x_0)$$
$$= u(x_0) \{f(x_0) - f(x'_0)\} + \sum_{k=1}^{n-1} u(x_k) \{f(x'_{k-1}) - f(x'_k)\}$$
$$+ u(x_n) \{f(x'_{n-1}) - f(x_n)\}$$
$$= -\sum_{k=0}^{n} u(x_k) \{f(x'_k) - f(x'_{k-1})\},$$

where for convenience we write $x'_{-1} = x_0 = a$, $x'_n = x_n = b$. Since as $\delta \doteq 0$ the limit of the summation in the last member exists, and is a Stieltjes integral, the same applies to the first summation, whose limit gives a Cauchy definition of the other Stieltjes integral; hence, if we take the limit,

(6) $\quad \int_a^b f(x)\, du(x) = f(x) u(x) \Big]_a^b - \int_a^b u(x)\, df(x).$

If $f(x)$ and $u(x)$ are periodic, with period $b-a$, or if they are defined on the circumference of a circle, x being a continuous monotonic function of θ, we have $\int f(x)\,du(x) = -\int u(x)\,df(x)$.

As a concluding paragraph in this section, we define the Stieltjes integral for an open interval (a, b) and thus interpret the symbol $\int_{a+0}^{b-0} u(x)\,df(x)$. We take $f(x)$ of limited variation in the open interval (a, b). In the sums S_n and s_n we replace $f(a) = f(x_0)$ by $f(a+0) = f(x_0+0)$ and $f(b) = f(x_n)$ by $f(b-0) = f(x_n-0)$, and say that the Stieltjes integral on the open interval exists if $S = s$, and we take that common value as its value. The integral exists if $u(x)$ is *uniformly* continuous in the open interval, and in this case all the formulae previously given may be carried over merely changing $f(x_0)$ to $f(a+0)$ and $f(x_n)$ to $f(b-0)$ throughout. *In the law of the mean, ξ lies actually between a and b.* In particular, we have the integration by parts

(7) $\quad \int_{a+0}^{b-0} f(x)\,du(x) = f(x)u(x)\Big]_{a+0}^{b-0} - \int_{a+0}^{b-0} u(x)\,df(x).$

There is however an essential difference in the characters of the two integrals appearing in (7). We may verify directly that

(7') $\quad \begin{aligned} \int_{a+0}^{b-0} f(x)\,du(x) &= \lim_{\varepsilon \doteq 0} \int_{a+\varepsilon}^{b-\varepsilon} f(x)\,du(x), \\ \int_{a+0}^{b-0} u(x)\,df(x) &= \lim_{\varepsilon \doteq 0} \int_{a+\varepsilon}^{b-\varepsilon} u(x)\,df(x), \end{aligned}$

but if we complete the definitions of $u(x)$ and $f(x)$ in the closed interval (a, b) by defining $u(a)$ as $u(a+0)$ and $u(b)$ as $u(b-0)$ (in order to keep $u(x)$ continuous), and by defining $f(a)$ and $f(b)$ arbitrarily, we shall have

(8) $\quad \int_a^b f(x)\,du(x) = \int_{a+0}^{b-0} f(x)\,du(x)$

although in general

$$\int_a^b u(x)\,df(x) \neq \int_{a+0}^{b-0} u(x)\,df(x).$$

PRELIMINARY CONCEPTS

4. Note on the second law of the mean. *If $f(x)$ is monotonic in the closed interval (a, b) and $u(x)$ continuous, there is a value ξ, $a \leq \xi \leq b$, and a value ξ', $a < \xi' < b$, such that*

$$\text{(9)} \quad \int_a^b f(x)\,u(x)\,dx = f(a)\int_a^\xi u(x)\,dx + f(b)\int_\xi^b u(x)\,dx$$
$$= f(a+0)\int_a^{\xi'} u(x)\,dx + f(b-0)\int_{\xi'}^b u(x)\,dx.$$

Let $U(x)$ be an indefinite integral of $u(x)$. Then from (6) and (2) we have

$$\int_a^b f(x)\,dU(x) = f(x)\,U(x)\Big]_a^b - \int_a^b U(x)\,df(x)$$
$$= f(x)\,U(x)\Big]_a^b - U(\xi)\,[f(b) - f(a)]$$
$$= f(a)\,[U(\xi) - U(a)] + f(b)\,[U(b) - U(\xi)].$$

But as is seen directly from the Cauchy definition,

$$\int_a^b f(x)\,dU(x) = \int_a^b f(x)\,u(x)\,dx.$$

Hence

$$\int_a^b f(x)\,u(x)\,dx = f(a)\int_a^\xi u(x)\,dx + f(b)\int_\xi^b u(x)\,dx.$$

In a similar fashion, we obtain the other form, given above, by utilizing the open interval (a, b). That is,

$$\int_a^b f(x)\,dU(x) = f(x)\,U(x)\Big]_{a+0}^{b-0} - \int_{a+0}^{b-0} U(x)\,df(x)$$
$$= f(x)\,U(x)\Big]_{a+0}^{b-0} - U(\xi')\,[f(b-0) - f(a+0)], \quad (a < \xi' < b),$$
$$= f(a+0)\,[U(\xi) - U(a+0)] + f(b-0)\,[U(b-0) - U(\xi')],$$

from which the desired conclusion follows, since $U(a+0) = U(a)$, $U(b-0) = U(b)$.

5. Classical theorems on integrals and limits of integrals. A sequence of functions $f_n(x)$ is said to be bounded in their set, if there is a constant N, independent of n, such that $|f_n(x)| \leq N$, in the interval in question. If we have a sequence of functions $f_n(x)$ integrable over an interval (a, b) and bounded in their set, such that $\lim_{n=\infty} f_n(x) = f(x)$, we know that

$$\lim_{n=\infty} \int_a^b f_n(x)\, dx = \int_a^b f(x)\, dx,$$

provided that $f(x)$ is also integrable. This is Osgood's theorem. With integration in the Riemann sense we do not know that the limit function $f(x)$ is integrable. But the definition of integrability has been so extended by Borel and Lebesgue that if $f_n(x)$ is set-bounded and integrable, and $\lim_{n=\infty} f_n(x) = f(x)$, then $f(x)$, which of course is bounded, is also integrable. This theory has become classical.

If the $f_n(x)$ are not bounded in their set the corresponding theorems are more difficult to state. But the Lebesgue theory in the first place extends the definition of integration to functions which are not bounded, by defining the integral of such a function, if it is nowhere negative, as itself a limit; namely, the limit of the integral of the function

$$f_n(x) = f(x) \text{ when } f(x) < n$$
$$= n \quad \text{ when } f(x) \geq n$$

as n becomes infinite, if this limit exists. The function $f(x)$ is then said to be *summable* in (a, b). If $f(x)$ is not ≥ 0, it is said to be summable if the two functions $\frac{1}{2}\{|f(x)|+f(x)\}$ and $\frac{1}{2}\{|f(x)|-f(x)\}$ are summable, and the integral is defined as the difference of these two integrals. Sequences of summable functions are then dealt with by means of sufficient conditions; thus if $\lim_{n=\infty} f_n(x) = f(x)$, where the $f_n(x)$ are summable, a sufficient condition that $f(x)$ be summable and that its integral be the limit of the integral of $f_n(x)$ is that $|f_n(x)| \leq \psi(x)$ were $\psi(x)$ is some summable function.

It is useful, however to have some form of necessary and sufficient condition. This may be obtained in terms of the concept of *absolute continuity*; a function $F(x)$ is *absolutely continuous* if when we take a finite number of non-overlapping intervals (x'_i, x''_i), $a \leqq x'_i < x''_i \leqq b$, of total length $\leqq \delta$, we have

$$\sum_i |F(x''_i) - F(x'_i)| \leqq m(\delta),$$

where $m(\delta)$ is a function of δ alone, which approaches zero with δ.

The function $m(\delta)$ is obviously unchanged if we take a denumerable infinity, instead of a finite number, of intervals of total length $\leqq \delta$. For if there were such a collection of intervals for which the given sum were $> m(\delta)$, there would be a finite number of them for which that sum would be greater than $m(\delta)$.

The absolute continuity of a set of such functions $F_n(x)$ is *uniform* if the function $m(\delta)$ is independent of n.

These concepts enable us to state what we may call De la Vallée Poussin's theorem, of which those already given are special cases.

THEOREM OF DE LA VALLÉE POUSSIN.* *Let $f_n(x)$ constitute a denumerable sequence of summable functions in an interval (a, b), not negative, with $\lim_{n=\infty} f_n(x) = f(x)$. A necessary and sufficient condition that $f(x)$ shall be summable and that we shall have*

$$\lim_{n=\infty} \int_a^b f_n(x)\, dx = \int_a^b f(x)\, dx$$

is that the absolute continuity of $F_n(x) = \int f_n(x)\, dx$ shall be uniform. If the $f_n(x)$ are of variable sign in (a, b), or if the sequence is non-denumerable, the condition is still sufficient.

A set of points on a line (or on a measurable arc) is said to be *of zero measure* if it can be contained in a denumerable set of intervals (or arcs) whose total length is arbitrarily small. A property is said to hold *almost everywhere* if it holds for all the points of a given segment except at most those which constitute some set of zero measure. A function is said to be *bounded almost everywhere*, or is still said to be *bounded*, if it may be defined or re-defined on a set of

* Or, Vitali's theorem.

zero measure so as to become bounded in the strict sense. To change the value of a function on a set of zero measure does not change the value of its Lebesgue integral, if it has one. A function is still said to be *summable* if it may be defined or re-defined on a set of zero measure so as to become summable.

It may be shown that a function of limited variation has a derivative almost everywhere, in spite of its possible discontinuities, and that this derivative is summable. An absolutely continuous function, which we see from the definition to be merely a special kind of continuous function of limited variation, is an indefinite integral of its derivative; conversely, the indefinite integral of a summable function is absolutely continuous. The total variation of an absolutely continuous function is the integral of the absolute value of its derivative.

The theorem of De la Vallée Poussin remains true if $\lim_{n=\infty} f_n(x) = f(x)$ almost everywhere, instead of everywhere.

6. The limits of Stieltjes integrals. THEOREM 2. *If $u_k(x)$ approaches $u(x)$ uniformly, both functions being continuous in (a, b), and $f(x)$ is of limited variation, then*

$$\lim_{k=\infty} \int_a^b u_k(x) df(x) = \int_a^b u(x) df(x).$$

In fact the absolute value of the difference between the integral for $u(x)$ and that for $u_k(x)$ is not greater than ηT, where $\eta \geqq |u - u_k|$ and approaches 0 as k becomes infinite. Similarly

$$\lim_{k=\infty} \int_{a+0}^{b-0} u_k(x) df(x) = \int_{a+0}^{b-0} u(x) df(x).$$

We say that $f_m(x)$ is of uniformly limited variation in (a, b), closed or open, if the total variation T_k is bounded for all k. This concept enables us to give a theorem proved independently by Helly and Bray, which we shall call the Helly-Bray theorem.*

* This theorem is discussed by Hildebrandt [Bulletin of the American Mathematical Society, vol. 28 (1922), p. 54] who gives a condition,

THEOREM 3. *Let (a, b) be a closed interval, $u(x)$ a continuous function in the interval, and E a set of points dense in the interval and including a and b. In the same interval let $f(x)$ be of limited variation and $f_m(x)$ a sequence of functions of uniformly limited variation, such that $\lim_{m=\infty} f_m(x) = f(x)$ if x is in E. Then*

$$\lim_{m=\infty} \int_a^b u(x)\, df_m(x) = \int_a^b u(x)\, df(x).$$

To prove this theorem, we notice that by taking δ small enough, and successive points of subdivision $x_1, x_2 \cdots, x_{n-1}$ close enough, in E, so that $x_{i+1} - x_i < \delta$, we can make

$$\left| \int_a^b u(x)\, df_m(x) - \sum_{i=0}^{n-1} u(x_i')\{f_m(x_{i+1}) - f_m(x_i)\} \right| < \varepsilon/4,$$

$$\left| \int_a^b u(x)\, df(x) - \sum_{i=0}^{n-1} u(x_i')\{f(x_{i+1}) - f(x_i)\} \right| < \varepsilon/4,$$

where ε is given arbitrarily, on account of the inequality (4). But also, $x, \cdots, x_{n-1}$ now being fixed and finite in number, by taking $m > m_0$, large enough, we can make

$$\left| \sum u(x_i')\{(f_m(x_{i+1}) - f_m(x_i)) - (f(x_{i+1}) - f(x_i))\} \right| < \varepsilon/2,$$

from our hypothesis about $\lim_{m=\infty} f_m(x)$ on E. From this the conclusion follows.

COROLLARY 1. *If in the above hypotheses we substitute*

$$\lim_{m=\infty} f_m(a+0) = f(a+0), \quad \lim_{m=\infty} f_m(b-0) = f(b-0),$$

then also

$$\lim_{m=\infty} \int_{a+0}^{b-0} u(x)\, df_m(x) = \int_{a+0}^{b-0} u(x)\, df(x).$$

---

necessary as well as sufficient, for the limit of the integral to be the integral of the limit. Helly's formulation of the theorem [Wiener Sitzungsberichte, vol. 121 (IIa) (1912), p. 288] assumes $\lim_{m=\infty} f_m(x) = f(x)$ for all x. Bray's theorem [Annals of Mathematics, vol. 20 (1919), p. 180] is more easily applicable.

Let us note however that we can not deduce the hypothesis of this corollary merely from the properties of the set E.

COROLLARY 2. *If the conditions of both theorems on limits are assumed, we have*

$$\lim_{\substack{k=\infty \\ m=\infty}} \int_a^b u_k(x)\, df_m(x) = \int_a^b u(x)\, df(x),$$

without regard to the order in which k and m become infinite.

The theorems on the limit of a Stieltjes integral have a special interpretation if the interval (a, b) comprises the whole of a closed curve. For simplicity, let the curve be a circle, although we may use any simple closed rectifiable curve as well.

COROLLARY 3. *Let E denote a set of points dense on the circumference, and x the central angle or a continuous monotonic function of the central angle. Let $f(x)$, $f_m(x)$ denote functions of uniformly limited variation such that $\lim_{m=\infty} f_m(x) = f(x)$ if x is in E, and $u(x)$, $u_k(x)$ continuous functions on the circumference, such that $\lim_{k=\infty} u_k(x) = u(x)$ uniformly. Then for integration over the whole circumference, we have*

$$\int u(x)\, df(x) = \lim_{\substack{k=\infty \\ m=\infty}} \int u_k(x)\, df_m(x),$$

no matter how k and m become infinite.

7. Note on Lebesgue Integrals. Let now $u(x)$ be continuous, and $f(x)$ absolutely continuous, so that $f(x) = \int_a^x f'(x)\, dx + f(a)$. Then

$$\int_a^b u(x) f'(x)\, dx = \lim_{\delta=0} \sum_{i=0}^{n-1} u(x_i') \int_{x_i}^{x_{i+1}} f'(x)\, dx,$$

from the limit theorem on the Lebesgue integral; in fact, $u(x)$ is bounded. But also, from (5)

$$\lim_{\delta=0} \sum_{i=0}^{n-1} u(x_i') \int_{x_i}^{x_{i+1}} f'(x)\, dx = \int_a^b u(x)\, df(x).$$

PRELIMINARY CONCEPTS   17

Hence
$$\text{(10)} \qquad \int_a^b u(x)\,df(x) = \int_a^b u(x)f'(x)\,dx.$$

Also if $f(x)$ is of limited variation and $U(x)$ is absolutely continuous so that $U(x) = \int_a^x u(x)\,dx + U(a)$, where $u(x) = U'(x)$ almost everywhere and is summable, we have

$$\text{(11)} \qquad \int_a^b f(x)\,u(x)\,dx = \int_a^b f(x)\,dU(x).$$

In fact, it is sufficient to consider a non-decreasing function $U(x)$ and write, by the law of the mean,

$$\int_a^b f(x)\,dU(x) = \lim_{\varepsilon=0} \sum_{i=0}^{n-1} f(x_i') \int_{x_i}^{x_{i+1}} u(x)\,dx.$$

But the function $f_i(x)$ defined as follows

$$f_i(x) = f(x_i'),$$

for $x_i \leq x < x_{i+1}$, when $i+1 < n$, and for $x_{n-1} \leq x \leq x_n = b$ when $i+1 = n$, remains bounded, and has as a limit the function $f(x)$ on a set of points which includes those where $f(x)$ is continuous, that is, almost everywhere in (a, b). Hence

$$\sum_{i=0}^{n-1} f(x_i') \int_{x_i}^{x_{i+1}} u(x)\,dx = \int_a^b f_i(x)\,u(x)\,dx$$

and

$$\lim_{\varepsilon=0} \int_a^b f_i(x)\,u(x)\,dx = \int_a^b f(x)\,u(x)\,dx,$$

whence the conclusion follows.

By applying the idendity (11) to a monotonic function $f(x)$ it may be used in the proof of the second law of the mean. Hence we have the

COROLLARY TO THE SECOND LAW OF THE MEAN. *The second law of the mean holds if $f(x)$ is monotonic and $u(x)$ is summable in the Lebesgue sense.*

8. Convergence of Fourier series. We turn now to the subject of Fourier series, which is the second main subject of this chapter. We write

(12) $$S_m(x) = \frac{a_0}{2} + \sum_{1}^{m}{}_k (a_k \cos kx + b_k \sin kx)$$

with a_k and b_k the Fourier coefficients,

$$a_k = \frac{1}{\pi}\int_{-\pi}^{\pi} f(\alpha) \cos k\alpha \, d\alpha, \qquad b_k = \frac{1}{\pi}\int_{-\pi}^{\pi} f(\alpha) \sin k\alpha \, d\alpha,$$

$$a_0 = \int_{-\pi}^{\pi} f(\alpha) \, d\alpha.$$

Thus

$$S_m(x) = \int_{-\pi}^{\pi} f(\alpha) \left\{ \frac{1}{2} + \sum_{k=1}^{m} \cos kx \cdot \cos k\alpha + \sin kx \cdot \sin d\alpha \right\} d\alpha,$$

where the bracket { } reduces to $\dfrac{1}{2} + \sum_{k=1}^{m} \cos k(\alpha - x)$ and

(12.1) $$S_m(x) = \frac{1}{\pi}\int_{-\pi}^{\pi} f(x + \alpha) \left\{ \frac{1}{2} + \sum_{k=1}^{m} \cos k\alpha \right\} d\alpha$$

the α having been replaced by $x + \alpha$. We turn first to

THEOREM 4.*

(13) $$S_m(x) = \frac{1}{\pi}\int_{-\pi}^{\pi} f(x + \alpha) \frac{\sin\left(m + \dfrac{1}{2}\alpha\right)}{2\sin\dfrac{1}{2}\alpha} d\alpha,$$

assuming for the present that $f(\alpha)$ is bounded and continuous.

Write $x = 2(\cos \alpha + \cos 2\alpha + \cdots + \cos m\alpha)$. By means of the

*This theorem and its proof is adapted from Ruel V. Churchill, *Fourier Series and Boundary Value Problems,* 2nd ed., New York, 1963, pp. 64–65. I am indebted for this reference to Professor George W. Evans, II.

formula*

$$\cos x = \frac{1}{2}(e^{xi} + e^{-xi})$$

we obtain an expression X:

$$\begin{aligned}
X &= \sum_{k=1}^{m} e^{ik\alpha} + \sum_{k=1}^{m} e^{-ik\alpha} \\
&= \sum_{k=1}^{m} \frac{e^{ik\alpha}(1 - e^{i\alpha})}{1 - e^{i\alpha}} + \sum_{k=1}^{m} \frac{e^{-ik\alpha}(1 - e^{-i\alpha})}{1 - e^{-i\alpha}}.
\end{aligned}$$

For example,

$$\begin{aligned}
\sum_{k=1}^{m} \frac{e^{ik\alpha}(1 - e^{i\alpha})}{1 - e^{i\alpha}} &= \frac{1}{1 - e^{i\alpha}} \\
&\cdot \{e^{i\alpha} - e^{2i\alpha} + e^{2i\alpha} - e^{3i\alpha} + \cdots + e^{mi\alpha} - e^{(m+1)i\alpha}\} \\
&= \frac{e^{i\alpha}}{1 - e^{i\alpha}} \{1 - e^{mi\alpha}\}.
\end{aligned}$$

Thus

$$\begin{aligned}
X &= \frac{e^{i\alpha}(1 - e^{mi\alpha})}{1 - e^{i\alpha}} - \frac{e^{i\alpha}(1 - e^{mi\alpha})}{1 - e^{-mi\alpha}} \\
&= \frac{1}{e^{(1/2)i\alpha} - e^{-(1/2)i\alpha}} \{-e^{(1/2)i\alpha} + e^{(m+1/2)i\alpha} + e^{-(1/2)i\alpha} - e^{-(m+1/2)i\alpha}\} \\
&= -1 + \frac{1}{e^{(1/2)i\alpha} - e^{-(1/2)i\alpha}} \{-e^{(m+1/2)i\alpha} - e^{-(m+1/2)i\alpha}\},
\end{aligned}$$

where†

$$e^{(m+1/2)i\alpha} - e^{-(m+1/2)i\alpha} = 2i \sin\left(m + \frac{1}{2}\right)\alpha.$$

*B. O. Peirce, *A Short Table of Integrals*, 2nd ed., formula 613, page 76.
†B. O. Peirce, *loc. cit*, formula 612.

Hence, finally,

$$2\{\cos \alpha + \cos 2\alpha + \cdots + \cos m\alpha\} = -1 + \frac{\sin\left(m + \frac{1}{2}\right)\alpha}{\sin \frac{1}{2}\alpha}$$

for sin $1/2\ \alpha \neq 0$, — a relation known as 'Lagrange's Trigonometric Identity.'

To continue with Churchill:

$$S_m(x) = \frac{1}{\pi}\int_{-\pi}^{\pi} f(\alpha)\left[\frac{1}{2} + \sum_{k=1}^{m} \cos k(\alpha - x)\right] d\alpha$$
$$= \frac{1}{\pi}\int_{-\pi}^{\pi} f(x + \alpha)\left[\frac{1}{2} + \sum_{k=1}^{m} \cos k\alpha\right] d\alpha,$$

interchanging α and $x + \alpha$, x being fixed. Therefore, finally, $f(\alpha)$ being continuous,

$$S_m(x) = \frac{1}{\pi}\int_{-\pi}^{\pi} f(x + \alpha)\frac{\sin\left(m + \frac{1}{2}\right)\alpha}{2\sin\frac{1}{2}\alpha} d\alpha.$$

By taking limits we can arrive at the generality of our formula (13), page 18, and return the time to 1927.

THEOREM 4, COROLLARY.

(13.1) $$1 = \frac{1}{\pi}\int_{-\pi}^{\pi} \frac{\sin\left(m + \frac{1}{2}\right)\alpha}{2\sin\frac{1}{2}\alpha} d\alpha.$$

As in the previous theorem, we make use of Lagrange's Trigonometric Identity

$$2(\cos \alpha + \cos 2\alpha + \cdots + \cos m\alpha) = -1 + \frac{\sin\left(m + \frac{1}{2}\right)\alpha}{\sin\frac{1}{2}\alpha}$$

for $\sin 1/2\, \alpha \neq 0$.

$$S_m = \frac{1}{\pi} \int_{-\pi}^{\pi} \left[\frac{1}{2} + \sum_{k=1}^{m} \cos k\alpha \right] d\alpha = 1$$

$$S_m = \frac{1}{\pi} \int_{-\pi}^{\pi} \frac{\sin\left(m + \frac{1}{2}\alpha\right)}{2 \sin\frac{1}{2}\alpha} d\alpha,$$

Therefore

$$1 = \frac{1}{\pi} \int_{-\pi}^{\pi} \frac{\sin\left(m + \frac{1}{2}\alpha\right)}{2 \sin\frac{1}{2}\alpha} d\alpha. \qquad \text{Q.E.D.}$$

We shall prove the convergence of Fourier series, that is, the finiteness of

$$\lim_{m \to \infty} S_m(x)$$

In the form

$$\lim_{m \to \infty} S_m(x) = \frac{1}{\pi} \int_{-\pi}^{\pi} f(x + \alpha) \cdot \left\{ \frac{1}{2} + \cos \alpha + \cos 2\alpha + \cdots \cos m\alpha + \cdots \right\} d\alpha$$

the summation would be difficult, if at all possible. Therefore we turn to the form

(13) $$S_m(x) = \frac{1}{\pi} \int_{-\pi}^{\pi} f(x + \alpha) \frac{\sin\left(m + \frac{1}{2}\right)\alpha}{2 \sin \frac{1}{2}\alpha} d\alpha$$

$$= \frac{1}{\pi} \int_0^{\pi} \{f(x + \alpha) + f(x - \alpha)\} \frac{\sin\left(m + \frac{1}{2}\alpha\right)}{2 \sin \frac{1}{2}\alpha} d\alpha$$

and introduce the terms

$$\varphi(\alpha) = f(x + \alpha) + f(x - \alpha) - s(x)$$
$$s(x) = f(x + 0) + f(x - 0).$$

With regard to $s(x)$ we have for its integral

$$\frac{1}{\pi} \int_0^{\pi} \{f(x + 0) + f(x - 0)\} \frac{\sin\left(m + \frac{1}{2}\right)\alpha}{2 \sin \frac{1}{2}\alpha} d\alpha,$$

where the $\{f(x + 0) + f(x - 0)\}$ may be taken outside the integral, and $\int_0^{\pi} \{\sin(m + 1/2)\alpha/2 \sin 1/2\, \alpha\} d\alpha$ is a constant, independent of m.

Hence for $S_m(x)$ may be substituted the integral

$$\begin{cases} \overline{S}_m(x) = \frac{1}{\pi} \int_0^{\pi} \varphi(\alpha) \dfrac{\sin\left(m + \frac{1}{2}\right)\alpha}{2 \sin \frac{1}{2}\alpha} d\alpha \\ \varphi(\alpha) = \{f(x + \alpha) - f(x + 0)\} + \{f(x - \alpha) - f(x - 0)\}. \end{cases}$$

The function $\varphi(\alpha)$ is a function of limited variation of α, and vanishes continuously at $\alpha = 0$. Its positive and negative variation functions therefore also vanish continuously at $\alpha = 0$.

For arbitrary x, as m changes in brief stages, we may divide up the interval of integration into small elements:

$$\frac{1}{\pi}\int_0^\epsilon \varphi(\alpha)\frac{\sin\left(m+\frac{1}{2}\right)\alpha}{\alpha}\,d\alpha,$$

replacing $2\sin 1/2\,\alpha$ by α. Finally, as an application of the second law of the mean, we may rewrite the element in the form:

$$(14)\quad K_m(x) = \frac{\varphi(\epsilon)}{\pi}\int_{\epsilon'}^\epsilon \frac{\sin\left(m+\frac{1}{2}\right)\alpha}{\alpha}\,d\alpha,\quad 0 < \epsilon' < \epsilon.$$

And by reference to the accompanying figure it is seen that the greatest value that

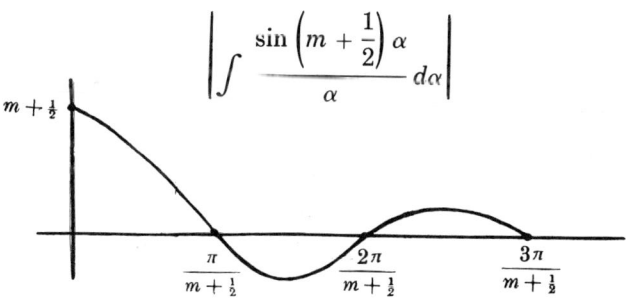

can take on for any finite interval of integration is

$$\int_0^{\pi/(m+1/2)} \frac{\sin(m+\tfrac{1}{2})\alpha\,d\alpha}{\alpha} = \int_0^\pi \frac{\sin\alpha}{\alpha}\,d\alpha = C < \pi$$

Hence the integral above considered is $\leq C\varphi(\varepsilon)/\pi$, which can be made as small as we please by taking ε small enough. In other words $S_m(x) - S(x)$ can be made as small as we please, by taking $m \geq m_0$, large enough. The same conclusion holds if $\varphi(a)$ is not monotonic, since the contributions to $S_m(x) - S(x)$ by the positive and negative variations of $\varphi(a)$ taken separately may be made as small as we please. Hence, in general

$$\lim_{m=\infty} S_m(x) - S(x) = 0.$$

We have thus proved the following theorem.

THEOREM 4.1. *If $f(x)$ is a function of limited variation, periodic, with period 2π, its Fourier series converges everywhere and to the value*

$$\frac{1}{2} \{f(x+0) + f(x-0)\}.$$

The second theorem of the mean gives also a measure of the law of decrease of the Fourier coefficients for a function of limited variation. In fact, if $f(x)$ is monotonic, $-\pi < x < \pi$, non-decreasing, we have

$$\frac{1}{\pi} \int_{-\pi}^{\pi} f(x) \cos kx \, dx$$
$$= \frac{1}{\pi} f(-\pi+0) \int_{-\pi}^{\xi'} \cos kx \, dx + \frac{1}{\pi} f(+\pi-0) \int_{\xi'}^{\pi} \cos kx \, dx$$
$$= (1/\pi) \{f(\pi-0) - f(-\pi+0)\} \frac{\sin k\xi'}{k}, \qquad -\pi < \xi' < \pi,$$

and

$$|a_k| \leq \frac{1}{k\pi} \{f(\pi-0) - f(-\pi+0)\}.$$

Similarly

$$b_k = \frac{1}{\pi} f(-\pi+0) \left[-\frac{\cos kx}{k}\right]_{-\pi}^{\xi''}$$
$$+ \frac{1}{\pi} f(\pi-0) \left[-\frac{\cos kx}{k}\right]_{\xi''}^{\pi}, \qquad -\pi < \xi'' < \pi,$$

whence

$$|b_k| \leq \frac{2}{k\pi} \{f(\pi-0) - f(-\pi+0)\}.$$

If $f(x)$ is not necessarily monotonic, the same relations hold separately for the Fourier coefficients a'_k, b'_k, a''_k, b''_k of the positive and negative variations of $f(x)$; and since a_k is $a'_k + a''_k$ and b_k is $b'_k + b''_k$, if we denote by T' the total variation of $f(x)$ in the open interval $-\pi < x < \pi$ we have

$$|a_k| \leq \frac{T'}{k\pi}, \qquad |b_k| \leq \frac{2T'}{k\pi}.$$

If T is the total variation in the corresponding closed interval, it may be shown that also $b_k \leq T/k\pi$.

In fact
$$b_k = -\frac{1}{\pi k} \int_{-\pi}^{\pi} f(x) \, d\cos kx$$
$$= -\frac{1}{\pi k} [f(x) \cos kx]_{-\pi}^{\pi} + \frac{1}{\pi k} \int_{-\pi}^{\pi} \cos kx \, df(x),$$

by an integration by parts of the Stieltjes integral. But the first term of this expression is zero, $f(x)$ being periodic in the closed interval. Moreover

$$\left| \int_{-\pi}^{\pi} \cos kx \, df(x) \right| \leq \int_{-\pi}^{\pi} |\cos kx| \, dt(x) \leq T,$$

which proves the desired inequality.*

We have thus established the following corollary for the functions concerned in Theorem 4.1.

COROLLARY. Let $f(x)$ be a function of limited variation, periodic with period 2π, and let T and T' be the total variations of $f(x)$ in any closed and any open interval, respectively, of length 2π. Then the following inequalities hold for the Fourier coefficients a_k, b_k

(15) $\quad |a_k| \leq \dfrac{T'}{k\pi}, \qquad |b_k| \leq \dfrac{T}{k\pi}, \qquad |b_k| \leq \dfrac{2T'}{k\pi}.$

EXERCISE. The reader may show by examples that these are the best inequalities obtainable in terms of the total variation.

* This short proof was given by Professor D. A. F. Robinson in my course at the University of Chicago, 1925.

A kind of converse problem to that of Fourier series is that of deciding when a given trigonometric series is convergent, and is a Fourier series for some function. Lebesgue's idea of measure has been a fundamental concept here also; and the problem has been extensively treated in a series of memoirs from Fatou (Acta Mathematica 1907) to Rademacher (Mathematische Annalen, 1922). For *normal functions* in general, the following theorem has been established.

THEOREM 5. *Given the series in terms of normal functions*

$$S = \sum_{1}^{\infty} c_i \, \varphi_i(x),$$

let $\sum c_i^2 (\log i)^2$ converge. Then S converges to a function $f(x)$ which is summable with its square; and the c_i are the generalized Fourier coefficients for that function.

9. Summability of series. We wish to discuss here merely one type of method of attributing a value to a divergent series, which is a natural extension of the concept of sum for a convergent series. This is a method of Borel.

If we have a series

(16) $$u_0 + u_1 + u_2 + \cdots$$

we denote by S_n the sum

(17) $$S_n = u_0 + u_1 + u_2 + \cdots + u_n$$

and if the series is convergent we define its value by the formula

$$S = \lim_{n=\infty} S_n.$$

In general, consider a sequence of values of r, —$r_1, r_2, \cdots$ with limit r', finite or infinite,—and let $a_k(r)$ be a sequence of not negative functions of r, non-increasing with k (i. e., such that $a_{k+1}(r) \leqq a_k(r)$ for all r_i) and such that

$$\lim_{r_i=r'} a_k(r_i) = 1.$$

If the series
$$\sigma(r) = \sum_0^\infty a_k(r) u_k \tag{18}$$

converges for all r_i and if $\lim_{i=\infty} \sigma(r_i)$ exists, we say that this limit is the sum, or sum in the extended sense, of the series (17). The following is the central theorem.

THEOREM 6. *If S exists, then $\sigma(r_i)$ is uniformly convergent for all r_i, and $S = \lim_{i=\infty} \sigma(r_i)$.*

Let $R_n = u_n + u_{n+1} + \cdots$

$$\varrho_{n,m}(r) = a_n(r) u_n + a_{n+1}(r) u_{n+1} + \cdots + a_{n+m}(r) u_{n+m}.$$

Given ε we can choose N so that $|R_n| < \varepsilon$ for $n > N$, $m \geq 0$. We have

$$\begin{aligned}\varrho_{n,m}(r) &= a_n(r)(R_n - R_{n+1}) + a_{n+1}(r)(R_{n+1} - R_{n+2}) + \cdots \\ &\quad \cdots + a_{n+m}(r)(R_{n+m} - R_{n+m+1}) \\ &= a_n R_n + (a_{n+1} - a_n) R_{n+1} + (a_{n+2} - a_{n+1}) R_{n+2} + \cdots \\ &\quad \cdots + (a_{n+m} - a_{n+m-1}) R_{n+m} - a_{n+m} R_{n+m+1}\end{aligned}$$

and

$$\begin{aligned}|\varrho_{n,m}(r)| &\leq \varepsilon[a_n + (a_n - a_{n+1}) + (a_{n+1} - a_{n+2}) + \cdots \\ &\quad \cdots + (a_{n+m-1} - a_{n+m}) + a_{n+m}] \\ &\leq \varepsilon(2 a_n + a_{n+m}) \\ &\leq 3 \varepsilon a_n(r).\end{aligned}$$

But $a_1(r_i)$ has a finite upper bound M, since $\lim_{i=\infty} a_1(r_i) = 1$; and $a_n(r_i) \leq a_1(r_i)$. Hence $a_n(r_i) \leq M$ and

$$|\varrho_{n,m}(r)| < 3 M \varepsilon, \qquad \left.\begin{array}{c} n \\ n+m \end{array}\right\} > N,$$

uniformly in r. In this range of uniformity is included the limit value $r = r'$. The series is therefore uniformly convergent for all values of r_i, including r', and the limit of the series as r_i tends to r' is S.

We note in passing that a similar theorem holds if $a_k(r_i)$ is non-decreasing, instead of non-increasing with k, and bounded, but the theorem is not so useful.

We shall meet in the next chapter a type of this summation specially adapted to harmonic functions. One kind which has a peculiar interest in the case of trigonometric series is that of the arithmetic mean, applied by Fejér. In this case we write

$$(19) \quad \sigma_n = \frac{S_0 + S_1 + \cdots S_{n-1}}{n} = \sum_{k=0}^{n-1} (1 - k/n) u_k$$

identical with the $\sigma(r_n)$ previously discussed if we let $r_n = n$, and

$$a_k(r_n) = (1 - k/n) \quad k = 0, 1, \cdots, n-1,$$
$$= 0 \quad k \geq n.$$

Hence whenever the Fourier series converges, $\lim_{n=\infty} \sigma_n$ converges and to the same value.

It may be readily calculated that

$$(19') \quad \sigma_m = \frac{1}{m\pi} \int_0^{\pi/2} \{f(x + 2\alpha) + f(x - 2\alpha)\} \left(\frac{\sin m\alpha}{\sin \alpha}\right)^2 d\alpha$$

from which it follows immediately that *if l and L are respectively the lower and upper bounds of the bounded function $f(x)$, then*

$$l \leq \sigma_m \leq L.$$

In fact (by putting $f(x) \equiv 1$) we have the identity

$$1 = \frac{2}{m\pi} \int_0^{\pi/2} \left(\frac{\sin m\alpha}{\sin \alpha}\right)^2 d\alpha.$$

From this follows Fejér's theorem:

THEOREM OF FEJÉR. *If $f(x)$ is bounded, by limits l and L, and if $|m a_m| \leq A$ and $|m b_m| \leq B$ for all m, a_m and b_m being the Fourier coefficients for $f(x)$, then any sum S_m of the Fourier series satisfies the inequality*

$$(20) \quad l - (A + B) \leq S_m \leq L + A + B.$$

In fact, from (19),

$$\sigma_n = S_{n-1} - \frac{1}{n}\sum_0^{n-1} k u_k,$$

$$S_{n-1} = \sigma_n + \frac{1}{n}\sum_0^{n-1} k u_k,$$

and the desired inequality follows. It applies in particular to all functions of limited variation.

It is worth while to notice another simple property of this method of summation.

THEOREM 7. *If $f(x)$, periodic, of period 2π, is of limited variation, then $\sigma_n(x)$ is of uniformly limited variation for all n.*

In fact, from

$$\sigma_m = \frac{1}{2m\pi}\int_{-\pi}^{\pi} f(x+\alpha)\left(\frac{\sin m\alpha/2}{\sin \alpha/2}\right)^2 d\alpha$$

which is equivalent to (19'), we have

$$\sum_1^n |\sigma_m(x_i') - \sigma_m(x_i'')|$$
$$\leq \frac{1}{2m\pi}\int_{-\pi}^{\pi}\sum_1^n |f(x_i'+\alpha) - f(x_i''+\alpha)|\left(\frac{\sin m\alpha/2}{\sin \alpha/2}\right)^2 d\alpha.$$
$$\leq T.$$

CHAPTER II

FUNCTIONS HARMONIC WITHIN A CIRCLE

10. Preliminary theorems. We say that a function $u(M)$ is harmonic at a point M if it is continuous there with its first partial derivatives and has finite second derivatives with respect to x and y which satisfy Laplace's equation,*

$$(1) \qquad \frac{\partial^2 u}{\partial x^2} + \frac{\partial^2 u}{\partial y^2} = 0.$$

If a function is harmonic at every point of a region, it is said to be harmonic in the region. There are some elementary facts about such functions which we may take for granted.

If a function is harmonic in a region Σ which contains the whole of a circle and its circumference, the value of the function at the center of the circle is the mean of its values on the circumference. Hence such a function cannot have a maximum or a minimum at an interior point of the region; for such a point could be made the center of a circle small enough to be entirely contained, with its circumference, in the region. It follows therefore that if we have a simple closed curve l, contained, with its interior, in the given region, there cannot be two functions, harmonic in the region, which are different at some point interior to l and have identical values on all the points of l; for their difference, which would be harmonic, would have a maximum or a minimum at an interior point of l. In particular l may be a circle.

If we make a transformation of the points of the plane which is conformal, directly or reversely, so that the region Σ is mapped univocally and continuously on another region Σ', a function harmonic in Σ is carried into a function harmonic

---

*In polar coördinates Laplace's equation is

$$\frac{\partial^2 u}{\partial r^2} + \frac{1}{r}\frac{\partial u}{\partial r} + \frac{1}{r^2}\frac{\partial^2 u}{\partial \theta^2} = 0.$$

FUNCTIONS HARMONIC WITHIN A CIRCLE 27

in Σ'. In general, an analytic function of a complex variable effects such a transformation. An example of a transformation which reverses the angles is furnished by inversion in a circle.

11. A preliminary result. Let us assume that our function $u(M)$ is harmonic inside a unit circle. The function $v(M)$ conjugate to $u(M)$ is then harmonic at every point inside the same circle; in particular it is single valued; and the function of a complex variable

$$\gamma(z) = u + iv,$$
$$z = x + iy$$

is analytic inside the circle, and may be developed in a power series about the origin,

$$\gamma(z) = C_0 + C_1 z + C_2 z^2 + \cdots$$

convergent for $|z| = \sqrt{x^2 + y^2} < 1$. If we write $C_k = A_k + i B_k$, and form the corresponding developments of the real and pure imaginary parts of $\gamma(z)$ they will also necessarily be convergent for $r = \sqrt{x^2 + y^2} < 1$.

We write then

$$C_k z^k = (A_k + i B_k) r^k (\cos k\theta + i \sin k\theta)$$
$$= r^k \{A_k \cos k\theta - B_k \sin k\theta\} + i r^k \{B_k \cos k\theta + A_k \sin k\theta\}.$$

Hence, for all points inside the unit circle,

(2) $\quad u(M) = u(r, \theta) = \quad a_0/2 + \sum_1^\infty r^k(\quad a_k \cos k\theta + b_k \sin k\theta),$

(2') $\quad\quad\quad v(r, \theta) = -b_0/2 + \sum_1^\infty r^k(-b_k \cos k\theta + a_k \sin k\theta),$

where we have written $a_0/2 = A_0$, $b_0/2 = -B_0$; $a_k = A_k$, $b_k = -B_k$, for $k \geq 1$. We note that

$$a_k\, r^k, \quad b_k\, r^k$$

both tend to 0 as k becomes infinite if $r < 1$, whereas one of these expressions fails to remain finite if $r > 1$, if the unit circle is the largest circle throughout the interior of which $u(M)$ is harmonic. That is the same as saying that the unit circle is the circle of convergence of $\gamma(z)$.

These inequalities show that the series for $u(M)$ is uniformly convergent for the points of any circle, of radius $r<1$. Hence we may determine the coefficients a_k, b_k by multiplying through by $\cos kx$ and $\sin kx$ respectively and integrating from 0 to 2π. We find then that $r^k a_k$ and $r^k b_k$ are the Fourier coefficients of $u(M)$ considered as a function of θ on the circle of radius $r<1$.

A second traditional method for the representation of a harmonic function is by means of Poisson's integral. In fact, if R is some fixed value, <1, and $r<R$, we have

$$(3) \quad u(r,\theta) = \frac{1}{2\pi}\int_0^{2\pi} \frac{R^2-r^2}{R^2+r^2-2Rr\cos(\varphi-\theta)} \cdot u(R,\varphi)\,d\varphi.$$

In order to prove this identity, we prove first the special cases of it where $u(r,\theta)$ is taken successively as 1, $r^n\cos n\theta$, and $r^n\sin n\theta$, functions which we know to be harmonic by direct differentiation or because they are the real parts of simple functions of a complex variable.

If z is the complex number

$$z = r\cos\theta + ir\sin\theta = re^{\theta i}$$

and, correspondingly, $t = Re^{\varphi i}$, and we let $\Re f(z)$ denote the real part of $f(z)$, we see by direct calculation that

$$\Re\left(\frac{t+z}{t-z}\right) = \frac{R^2-r^2}{R^2+r^2-2Rr\cos(\varphi-\theta)}$$

so that, if we denote this fraction by $Q(\varphi)$, we have

$$Q(\varphi) = \Re\left(\frac{1+z/t}{1-z/t}\right) = \Re\left[1+\frac{2z}{t}\left(\frac{1}{1-z/t}\right)\right]$$
$$= 1 + \Re\left(\frac{2z}{t} + \cdots + \frac{2z^{p+1}}{t^{p+1}} + \cdots\right)$$

since $|z| = r < |t| = R$. Hence

$$Q(\varphi) = 1 + 2\sum_{p=0}^{\infty}\left(\frac{r}{R}\right)^{p+1}\Re\{e^{(p+1)(\theta-\varphi)i}\}$$
$$= 1 + 2\sum_{p=0}^{\infty}\left(\frac{r}{R}\right)^{p+1}\{\cos(p+1)\varphi\,\cos(p+1)\theta + \sin(p+1)\varphi\,\sin(p+1)\theta\},$$

a series which for a given value of $r < R$ is uniformly convergent in θ and φ. Hence we may integrate term by term.

We have then the following identities

(4)
$$\int_0^{2\pi} Q(\varphi)\,d\varphi = 2\pi,$$
$$\int_0^{2\pi} Q(\varphi)\cos n\varphi\,d\varphi = 2\pi\left(\frac{r}{R}\right)^n \cos n\theta,$$
$$\int_0^{2\pi} Q(\varphi)\sin n\varphi\,d\varphi = 2\pi\left(\frac{r}{R}\right)^n \sin n\theta,$$

remembering that

(4')
$$\int_0^{2\pi}\sin n\varphi\,\cos p\varphi\,d\varphi = 0,$$
$$\int_0^{2\pi}\sin n\varphi\,\sin p\varphi\,d\varphi = \int_0^{2\pi}\cos n\varphi\,\cos p\varphi\,d\varphi$$
$$= \begin{cases} 0, & n \neq p, \\ \pi, & n = p. \end{cases}$$

But equations (4) are merely special cases of the identity (3) which we wish to prove.

By means of (4) we have, for a finite sum,

(5)
$$\frac{a_0}{2} + \sum_1^n r^k(a_k\cos k\theta + b_k\sin k\theta)$$
$$= \frac{1}{2\pi}\int_0^{2\pi}\frac{R^2-r^2}{R^2+r^2-2Rr\cos(\varphi-\theta)}$$
$$\times\left\{\frac{a_0}{2} + \sum_1^n R^k(a_k\cos k\varphi + b_k\sin k\varphi)\right\}d\varphi.$$

From this we proceed to deduce first the following theorem.

THEOREM 1. *Let $f(\varphi)$ be a function of limited variation with regular discontinuities, and periodic with period 2π. Then the function*

$$(6) \quad u(r, \theta) = \frac{1}{2\pi} \int_0^{2\pi} \frac{1-r^2}{1+r^2-2r\cos(\varphi-\theta)} f(\varphi)\, d\varphi$$

is harmonic inside the circle of radius 1 and satisfies the relation

$$\lim_{r=1} u(r, \theta) = f(\theta)$$

for every value of θ. The same is true of the function $u(r, \theta)$ given by (2), when a_k and b_k are the Fourier coefficients for $f(\theta)$; and these two functions are the same.

That these functions are harmonic follows by direct differentiation, since $r < 1$. Now the Fourier series for $f(\theta)$ converges everywhere, and to $f(\theta)$, by Theorem 4.1, Chap. I. But the representation given in (2) is merely a summation process, as described in Chap. I; hence it converges for $r < 1$, and the limit as r tends to 1 is precisely $f(\theta)$. In fact, this is true for any sequence of values of r tending to 1.

Let us now apply the identity (5), taking $R = 1$, to a finite number of terms of the expansion (2). We have then for the function $u(r, \theta)$ given by (2) the expression

$$u(r, \theta) = \lim_{n=\infty} \frac{1}{2\pi} \int_0^{2\pi} \frac{1-r^2}{1+r^2-2r\cos(\varphi-\theta)} \\ \times \left\{ \frac{a_0}{2} + \sum_1^n (a_k \cos k\varphi + b_k \sin k\varphi) \right\} d\varphi,$$

where the quantity in brackets is the sum S_n of the first $2n+1$ terms of the Fourier series for $f(\varphi)$. But $f(\varphi)$ is of limited variation, and we may therefore apply Fejér's theorem, and write

$$|S_n| \leq |l| + |L| + \frac{2T}{\pi},$$

where the symbols have the significance already given them. Hence the sums S_n remain bounded in their set, and since they approach an integrable limit function, Osgood's theorem tells us that the limit of the integral is the integral of the limit. But this last is precisely (6). In other words the function given by (2) is also given by (6), and the theorem is proved.

The function represented by (6) would be harmonic if $f(\varphi)$ were merely summable in the Lebesgue sense. But we do not yet know what values would be approached as r approached 1. And even in the case of Theorem 1 we have not discussed the question of the uniqueness of a harmonic function as determined by the boundary values $f(\varphi)$. But the preliminary result obtained in Theorem 1 will enable us to formulate necessary and sufficient conditions for such determinateness.

Let us return now to the proof of (3). According to (2), $r^k a_k$ and $r^k b_k$, and $R^k a_k$ and $R^k b_k$ are the Fourier coefficients of $u(r, \theta)$ and $u(R, \varphi)$ respectively. If we insert these values in (5), and take the limit as n becomes infinite, the identity (3) will be established, since $u(R, \varphi)$ is a function of φ of limited variation.

EXERCISE. Show from (3) that if $|u(R, \varphi)| < N$,

$$\left|\frac{\partial u}{\partial \theta}\right| < 2N \frac{r}{R} \frac{1+r/R}{(1-r/R)^3}, \qquad r < R.$$

Hence show that if $u(r, \theta)$ is harmonic in the entire plane and bounded it must reduce to a mere constant.*

12. Note on integral identities. A particular case of the equation so far considered is obtained by putting $f(\theta) = 1$ for all values of θ. In this case $u(r, \theta)$, as given by (2), where a_k and b_k are the Fourier constants for $f(\theta)$, reduces merely to unity, and the theorem just proved yields the identity

$$(7) \qquad 1 = \frac{1}{2\pi} \int_0^{2\pi} \frac{1-r^2}{1+r^2-2r\cos(\varphi-\theta)} d\varphi.$$

Let now (θ_1, θ_2), $\theta_2 > \theta_1$, be an arc of the circumference of length less than 2π, and apply the theorem to the function $f(\theta)$ which is defined as 1 within the arc and 0 outside the

---

* A closer inequality from the same source yields the quantity

$$(r/R)(4N/\pi)(1-r/R)^{-1}.$$

arc, the definition at the end points being arbitrary without affecting the value of the integral. We have then the identity

(8)
$$\lim_{r=1} \frac{1}{2\pi} \int_{\theta_1}^{\theta_2} \frac{1-r^2}{1+r^2-2r\cos(\varphi-\theta)} \, d\theta$$
$$= \begin{cases} 1, & \theta_1 < \theta < \theta_2, \\ \frac{1}{2}, & \theta = \theta_1 \text{ or } \theta = \theta_2, \\ 0, & \theta \text{ outside } (\theta_1, \theta_2). \end{cases}$$

13. Digression. Functions of points and of point sets.

If we imagine a distribution of positive and negative matter on the circumference of a circle, that distribution will be an additive function of point sets, and therefore the difference of two not negative additive functions of point sets. The essential character of the Poisson integral is that of being a sort of potential of a mass of this kind; in fact, the quantity

$$\frac{(1-r^2)m}{1+r^2-2r\cos(\varphi-\theta)}$$

represents the potential of what might be called a "circular doublet" of mass m; i. e., except for a constant it is the limit of the potential of two point masses made to increase in charge as they draw together, in such a way as to keep the potential at zero around the circle of radius 1. Hence the Poisson integral (3) represents the potential of circular doublets of strength density $f(\varphi)$. *But it does not represent the potential of the most general distribution of such doublets.*

EXERCISE. 1. Let B and B' be two points inverse with respect to the unit circle, and write $b = OB < 1$, $b' = OB' > 1$. At B place a mass of amount $-1/\log b = 2/(b'-b)$ approx., and at B' an equal mass of opposite sign. Show, by means of similar triangles, that the potential of these two masses becomes 1 at a point on the circumference of the circle, taking the potential of mass 1 as $\log 1/\varrho$ at a distance ϱ from the mass. Find the limit of the potential due to these two masses as b approaches 1, at an arbitrary point (r, θ)

interior to the circle. What is the value of this limit on the circumference?

2. At an arbitrary distance x on the axis OBB' place a mass of amount $\frac{1}{\log x}$. What is the limit of the potential due to this mass within the circle and on the circumference, as x is made to become infinite? The resulting potential may be called a level of strength -1.

3. Add the potential of problem 1 to a level of strength -1, and describe the result within and on the circumference of the circle.

Let then $\Phi(e)$ be an arbitrary additive function of point sets e on the circumference of radius 1; we can picture such a function $\Phi(e)$ as a total mass on the set e. The corresponding integral will be the Stieltjes integral

$$(9) \qquad u(r, \theta) = \frac{1}{2\pi} \int_C \frac{(1-r^2)\, d\Phi(e)}{1+r^2-2r\cos(\varphi-\theta)},$$

formed for the complete boundary of the circle, and defined as the limit of a sum in the same sort of way as the Stieltjes integral treated in Chap. I.

There is obviously only a denumerable infinity of points P_i, where the function of point sets will not vanish when the set is taken as a single point, i. e. $\Phi(P_i) \neq 0$. If we take the point a as distinct from such a point, we may define a function of θ, $F(\theta)$, almost everywhere by the equation

$$(9') \qquad F(\theta) = \Phi(a, \theta) + \text{const.}$$

where (a, θ) stands for the interval from a to θ. In fact this equation will yield a unique definition except at the points P_i; for these last points also we should get a unique definition if we specified the interval as open or closed. In this way, the value of $F(\theta)$ would be determined at every point of discontinuity as the value $F(\theta-0)$, or for every point of discontinuity as the value $F(\theta+0)$, respectively, according as we took the interval as open or closed. But it is better to leave the choice more arbitrary. We shall limit ourselves, however, to such choices of the value of the function at a point of discontinuity as will render the sum total of all the numerical values of the jumps of the function bounded in magnitude,

$$\sum_i (\text{Oscill. at } P_i \text{ of } F(\theta)) < N.$$

In this way the function $F(\theta)$ will satisfy the definition of function of limited variation, given in the introduction. We may thus introduce extra points of discontinuity (which may be called unnecessary ones) if we desire.

We shall find that the most convenient choice for many purposes is to make the discontinuities regular, so that the equation

$$F(\theta) = \frac{1}{2} \{F(\theta+0) + F(\theta-0)\},$$

will be satisfied throughout. It will also, naturally, be satisfied for the positive, negative, and total variation functions of $F(\theta)$.

If we turn to the reciprocal problem, we find as a familiar theorem in the theory of functions of a real variable, that given a function $F(\theta)$ of limited variation, there is one and only one additive function of point sets $\Phi(e)$ which accords with it by the equation

(10) $$F(\theta_1) - F(\theta_2) = \Phi(\theta_1, \theta_2),$$

θ_1 and θ_2 being points of continuity for $F(\theta)$.*

There is a further point that is worthy of remark. Since our fundamental interval is the circumference of a circle, we may have to consider values of θ less than 0 and greater than 2π. For this reason we must assign the values of $F(\theta)$ outside this interval in accordance with the values which it takes on within it. Accordingly we write

---

* In fact, an additive function of point sets is completely determined by the values which it takes on the meshes of some net [De la Vallée Poussin, *Intégrales de Lebesgue*, Paris (1916), Chap. VI].

(11) $$F(2\pi + \theta) = F(\theta) + \Phi(c)$$

where $\Phi(c)$ is the value of $\Phi(e)$ for the whole circumference, and by (9') the value of $F(2\pi) - F(0)$.

Let us form now the integral

(12) $$u(M) = \frac{1}{2\pi} \int_0^{2\pi} \frac{(1-r^2)\,dF(\varphi)}{1+r^2-2r\cos(\varphi-\theta)},$$

where $F(\varphi)$ corresponds to $\Phi(e)$ by (10) and (11). We see that (12) and (9) are identical. In fact, both may be compared with the same Riemann sum, formed on points φ_i which are points of continuity for $F(\varphi)$. We are about to get necessary and sufficient conditions for the integral (12), and thus for (9), and incidentally for (6).

14. Properties of the Poisson-Stieltjes integral. *We consider the fundamental formula, in terms of a Poisson-Stieltjes integral,*

(A) $$u(r, \theta) = \frac{1}{2\pi} \int_0^{2\pi} \frac{(1-r^2)\,dF(\varphi)}{1+r^2-2r\cos(\varphi-\theta)},$$

in which $F(\varphi)$ is of limited variation, such that $F(2\pi + \theta) = F(\theta) + F(2\pi) - F(0)$, and $r < 1$. We proceed to deduce from (A) *a number of significant properties of the function $u(r, \theta)$.*

We have immediately

THEOREM 2. *The function given by* (A) *is harmonic inside the circle of radius* 1.

As a matter of fact we would be justified in proving this by differentiating the Stieltjes integral under the integral sign, since $r < 1$. In order to avoid the unfamiliar process, however, let us integrate (A) by parts. We have

$$u(r, \theta) = \frac{1}{2\pi} \frac{F(\varphi)(1-r^2)}{1+r^2-2r\cos(\varphi-\theta)}\bigg]_0^{2\pi}$$
$$- \frac{1}{2\pi} \int_0^{2\pi} F(\varphi)\, d\,\frac{1-r^2}{1+r^2-2r\cos(\varphi-\theta)},$$

or

(13)
$$u(r, \theta) = \frac{F(2\pi) - F(0)}{2\pi} \frac{1-r^2}{1+r^2-2r\cos\theta}$$
$$+ \frac{1}{2\pi} \int_0^{2\pi} F(\varphi) \frac{2r(1-r^2)\sin(\varphi-\theta)}{(1+r^2-2r\cos(\varphi-\theta))^2} d\varphi.$$

But (13) gives $u(r, \theta)$ in terms of the usual type of integral. We may then differentiate under the integral sign and substitute in Laplace's equation. Thus the fact will be verified.

In the expression just written, r being <1, we may also integrate under the integral sign with respect to θ. This gives us

$$\int_{\theta_1}^{\theta_2} u(r, \theta) d\theta = \frac{F(2\pi) - F(0)}{2\pi} \int_{\theta_1}^{\theta_2} \frac{1-r^2}{1+r^2-2r\cos\theta} \cdot d\theta$$
$$+ \frac{1}{2\pi} \int_0^{2\pi} F(\varphi) \frac{1-r^2}{1+r^2-2r\cos(\varphi-\theta_2)} \cdot d\varphi$$
$$- \frac{1}{2\pi} \int_0^{2\pi} F(\varphi) \frac{1-r^2}{1+r^2-2r\cos(\varphi-\theta_1)} \cdot d\varphi,$$

and if we denote by $F(r, \theta)$ the quantity

$$F(r, \theta) = \int_0^\theta u(r, \theta) d\theta,$$

we may let r approach 1 in the above formula, and with reference to (8) of Art. 12, obtain for $\theta \neq 0$, $\theta \neq 2\pi$, the result

$$\lim_{r=1} F(r, \theta) = \frac{F(2\pi) - F(0)}{2} + \frac{1}{2} \{F(\theta+0) + F(\theta-0)\}$$
$$- \frac{1}{2} \{F(2\pi-0) + F(0+)\}$$
$$= \frac{F(2\pi) - F(0)}{2} + \frac{1}{2} \{F(\theta+0) + F(\theta-0)\}$$
$$- \frac{1}{2} \{F(2\pi) - F(0) + F(0-) + F(0+)\}$$

or

(14)
$$\lim_{r=1} F(r, \theta) = \frac{1}{2}\{F(\theta+0)+F(\theta-0)\} \\ -\frac{1}{2}\{F(0+)+F(0-)\},$$

a formula which may also be verified directly for $\theta = 0$ and $\theta = 2\pi$, reducing to 0 in the former case and to $F(2\pi) - F(0)$ in the latter. Hence we have the theorem:

THEOREM 3. *If $u(r, \theta)$ is given by* (A), *the function* $F(r, \theta) = \int_0^\theta u(r, \theta)\,d\theta$ *has a limit, as r approaches* 1, *for every value of θ. This limit is given by* (14).

It may be noticed that if the discontinuities of $F(\theta)$ are regular in an open interval which includes the closed interval $(0, 2\pi)$, (14) yields the formula

(15) $\quad \lim\limits_{r=1} F(r, \theta) = F(\theta) - F(0), \quad 0 \leq \theta \leq 2\pi.$

We can also obtain important results about $u(r, \theta)$ itself. Since

$$u(r, \theta) = \frac{1}{2\pi}\int_0^{2\pi} \frac{(1-r^2)\,dF_1(\varphi)}{1+r^2-2r\cos(\varphi-\theta)} \\ - \frac{1}{2\pi}\int_0^{2\pi} \frac{(1-r^2)\,dF_2(\varphi)}{1+r^2-2r\cos(\varphi-\theta)}.$$

where $F_1(\varphi)$ is the positive, and $-F_2(\varphi)$ the negative variation of $F(\varphi)$, we see that $u(r, \theta)$ is, within the circle, the difference of two not negative harmonic functions. For each of the integrals (as we see by the corresponding Riemann sum) yields an essentially not negative quantity. Hence:

THEOREM 4. *If $u(r, \theta)$ is given by* (A) *it is the difference of two not negative functions harmonic within the unit circle.*

If we denote by $T(\theta)$ the total variation function of $F(\theta)$, and by $U(r, \theta)$ the harmonic function

$$U(r, \theta) = \frac{1}{2\pi}\int_0^{2\pi} \frac{1-r^2}{1+r^2-2r\cos(\varphi-\theta)} \cdot dT(\varphi),$$

then obviously
$$|u(r, \theta)| \leq U(r, \theta),$$
and
$$\int_0^{2\pi} |u(r, \theta)|\, d\theta \leq \int_0^{2\pi} U(r, \theta)\, d\theta.$$

But we know from the mean value property of harmonic functions given in Art. 10, that this latter quantity is a constant, independent of r, equal in fact to 2π times the value of U at the center of the circle. By (14), however, this value is also $T(2\pi) - T(0)$. We have then the following result.

THEOREM 5. *The quantity* $\int_0^{2\pi} |u(r, \theta)|\, d\theta$ *is bounded for* $r < 1$, $\leq T(2\pi) - T(0)$. *In other words the function* $F(r, \theta)$ *is of uniformly limited variation in* θ *for all* r.

In fact, the total variation of $F(r, \theta)$ in the closed interval 0 to 2π is equal precisely to $\int_0^{2\pi} |u(r, \theta)|\, d\theta$.

We can, as a further property, evaluate directly the limit as r approaches 1 of the integral $\int_{\theta_1}^{\theta_2} |u(r, \theta)|\, d\theta$. For this purpose let us write $\overline{F}(\theta) = (1/2)\,[F(\theta+0) + F(\theta-0)]$, and take $\overline{T}(\theta)$ as the total variation function of $F(\theta)$. By introducing again the function $\overline{U}(r, \theta)$ corresponding to the $U(r, \theta)$ above, we have by (14) immediately

$$\lim_{r=1} \int_{\theta_1}^{\theta_2} |u(r, \theta)|\, d\theta \leq \overline{T}(\theta_2) - \overline{T}(\theta_1), \quad \theta_2 \geq \theta_1 \geq 0.$$

But also, we can show that

$$\lim_{r=1} \int_{\theta_1}^{\theta_2} |u(r, \theta)|\, d\theta > \overline{T}(\theta_2) - \overline{T}(\theta_1) - \varepsilon$$

no matter what $\varepsilon > 0$ is given, by taking r near enough to 1, i. e. $r > r_\varepsilon$. From this the theorem will follow.

In fact, we can choose a finite number of subintervals in (θ_1, θ_2), namely (θ_j', θ_j''), $j = 1, 2, \cdots, n$, so that

$$\overline{T}(\theta_2) - \overline{T}(\theta_1) - \sum_1^n {}_j\, |[\overline{F}(\theta_j'') - \overline{F}(\theta_j')]| < \varepsilon/2.$$

Now $\lim_{r=1} F(r, \theta) = \bar{F}(\theta) - \bar{F}(0)$, hence we can take r near enough to 1 so that for every one of this finite number $2n$ of values of θ'_j, θ''_j, $|[\bar{F}(\theta) - \bar{F}(0)] - F(r, \theta)| < \dfrac{\varepsilon}{4n}$. Therefore

$$\big||F(\theta''_j) - F(\theta'_j)| - |F(r, \theta''_j) - F(r, \theta'_j)|\big| < \varepsilon/2n,$$

and finally, with the help of the previous inequality,

$$\int_{\theta_1}^{\theta_2} |u(r, \theta)| \, d\theta > \sum_{1}^{n}{}_j |F(r, \theta''_j) - F(r, \theta'_j)| > \bar{T}(\theta_2) - \bar{T}(\theta_1) - \varepsilon.$$

Hence $\lim_{r=1} \int_{\theta_1}^{\theta_2} |u(r, \theta)| \, d\theta > \bar{T}(\theta_2) - \bar{T}(\theta_1) - \varepsilon$, whatever ε.
Hence we have the theorem:

THEOREM 6. *Let $\bar{T}(\theta)$ be the total variation function of $F(\theta)$ after its discontinuities have been made regular. Then*

$$\lim_{r=1} \int_{\theta_1}^{\theta_2} |u(r, \theta)| \, d\theta = \bar{T}(\theta_2) - \bar{T}(\theta_1).$$

15. Continuation of the preceding. Behaviour of $u(r, \theta)$ in the neighborhood of the boundary.

We assume that $u(r, \theta)$ is still given by (A) with $F(2\pi + \theta) = F(\theta) + F(2\pi) - F(0)$ and proceed to establish the following result.

THEOREM 7. *Let P be a point on the circumference where $F(\varphi)$ is continuous and has a unique derivative $F'(\varphi) = f$. We draw any two chords of the circle at P. If M approaches P in such a way as to remain between these two chords, then $\lim_{M \to P} u(M) = f$.*

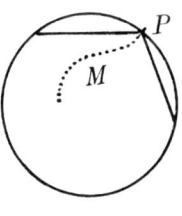

We shall say that M approaches P *in the wide sense*, if it approaches P as in Theorem 7, and that $u(M)$ takes on the values $f(P)$ *in the wide sense*.

Since $F(\varphi)$ is continuous and possesses a unique derivative *almost everywhere*, this result applies almost everywhere on the circumference. Hence the proposition:

COROLLARY. *The harmonic function given by* (A) *takes on the boundary values given by* $F'(\varphi) = f(\varphi)$ *in the wide sense at almost all the points of the boundary.*

In order to prove the theorem we shall take the point P at the position $(1,0)$ and by making use of the relation $F(2\pi + \theta) = F(\theta) + \text{const.}$, write the integral in the form (16)

$$(16) \quad u(r, \theta) = \frac{1}{2\pi} \int_{-\pi}^{\pi} \frac{(1-r^2)\, dF(\varphi)}{1 + r^2 - 2r\cos(\varphi - \theta)}.$$

Consider a sequence of points $(r, \theta_r) = P_r$ such that $\lim_{r=1} \theta_r = 0$. The hypothesis of the theorem, relative to the manner of approach, is equivalent to saying that there is some finite N such that $|\theta_r|/(1-r) < N$.

We may split the integral into three, corresponding to the equation

$$F(\varphi) = F(0) + \varphi f(0) + \varphi \eta(\varphi),$$

where $f(0) = F'(0)$ and $\varphi \eta(\varphi)$ is a function of limited variation such that $\lim_{\varphi=0} \eta(\varphi) = 0$. In fact

$$\eta(\varphi) = \frac{F(\varphi) - F(0)}{\varphi} - f(0).$$

We have therefore the relation

$$(17) \quad \begin{aligned} u(P_r) = &\frac{f(0)}{2\pi} \int_{-\pi}^{\pi} \frac{1-r^2}{1 + r^2 - 2r\cos(\varphi - \theta_r)} \cdot d\varphi \\ &+ \frac{1}{2\pi} \int_{-\pi}^{\pi} \frac{(1-r^2)\, d(\varphi \eta(\varphi))}{1 + r^2 - 2r\cos(\varphi - \theta_r)}, \end{aligned}$$

in which the first term of the right hand member is identically equal to $f(0)$ for all values of $r < 1$.

We wish to show that the second term of (17) has a limiting value ξ as r tends to 1 such that $|\xi| \leq \varepsilon$, no matter what ε has been assigned. For this purpose, define η_α as the upper bound of $\eta(\varphi)$ in the interval $|\varphi| \leq \alpha$, and fix α as a quantity small enough so that

$$\eta_\alpha \leq \frac{\varepsilon}{\frac{2N}{\pi}+1}$$

which is of course possible in accordance with our hypothesis. We have, if the limit exists,

$$\xi = \lim_{r=1} \frac{1}{2\pi} \int_{-\alpha}^{\alpha} \frac{(1-r^2)\, d(\varphi \eta(\varphi))}{1+r^2-2r\cos(\varphi-\theta_r)},$$

since ultimately $|\theta_r| < \alpha$, say $\leq \alpha/2$.

If we denote the integral in the above expression by I_r, and perform an integration by parts, we have

$$I_r = \left[\frac{(1-r^2)\varphi\eta(\varphi)}{1+r^2-2r\cos(\varphi-\theta_r)}\right]_{-\alpha}^{\alpha} + \int_{-\alpha}^{\alpha} \frac{2r(1-r^2)\sin(\varphi-\theta_r)}{[1+r^2-2r\cos(\varphi-\theta_r)]^2}\, \varphi\eta(\varphi)\, d\varphi,$$

of which the first term has the limit 0 as r approaches 1. Consider now the integral which is left, and denote it by J_r. We have

$$J_r = \theta_r \int_{-\alpha}^{\alpha} \frac{2r(1-r^2)\sin(\varphi-\theta_r)\eta(\varphi)}{[1+r^2-2r\cos(\varphi-\theta_r)]^2}\, d\varphi + \int_{-\alpha}^{\alpha} \frac{2r(1-r^2)\sin(\varphi-\theta_r)(\varphi-\theta_r)\eta(\varphi)}{[1+r^2-2r\cos(\varphi-\theta_r)]^2}\cdot d\varphi.$$

The second integral, which we may call I_r'', has an integrand which is constantly of one sign except for the factor $\eta(\varphi)$. Hence, by the law of the mean,

$$|I_r''| \leq \eta_\alpha \int_{-\alpha}^{\alpha} \frac{2r(1-r^2)(\varphi-\theta_r)\sin(\varphi-\theta_r)\, d\varphi}{[1+r^2-2r\cos(\varphi-\theta_r)]^2}$$
$$\leq \eta_\alpha \left[(\varphi-\theta_r)\frac{1-r^2}{1+r^2-2r\cos(\varphi-\theta_r)}\right]_{-\alpha}^{\alpha} + \eta_\alpha \int_{-\alpha}^{\alpha} \frac{(1-r^2)\, d\varphi}{1+r^2-2r\cos(\varphi-\theta_r)}.$$

The first term of this expression has the limit 0, and the second term is less than the value it would have if the integration were from $-\pi$ to π, that is, $< 2\pi\eta_\alpha$. Hence, if a limit exists,
$$\lim_{r=1} |I_r''| \leq 2\pi\eta_\alpha.$$

With regard to the first term in the expression for J_r, which we shall call I_r', we have
$$|I_r'| \leq \eta_\alpha |\theta_r| \left\{ \int_{-\alpha}^{\theta_r} -\frac{2r(1-r^2)\sin(\varphi-\theta_r)\,d\varphi}{[1+r^2-2r\cos(\varphi-\theta_r)]^2} \right.$$
$$\left. + \int_{\theta_r}^{\alpha} \frac{2r(1-r^2)\sin(\varphi-\theta_r)\,d\varphi}{[1+r^2-2r\cos(\varphi-\theta_r)]^2} \right\}.$$

The first term of this is merely
$$\eta_\alpha |\theta_r| \left\{ \frac{1+r}{1-r} - \frac{1-r^2}{1+r^2-2r\cos(\alpha-\theta_r)} \right\}.$$

of which the second part has the limit 0 and the first part is in numerical value $< 2N\eta_\alpha$, by hypothesis. The second term in the preceding inequality can be handled in the same way, and therefore, if a limit exists,
$$\lim_{r=1} |I_r'| \leq 4\eta_\alpha N.$$

Consequently, if a limit exists,
$$|\xi| \leq \frac{2\pi\eta_\alpha + 4N\eta_\alpha}{2\pi} = \varepsilon$$

and therefore all these limits do exist, and
$$\xi = 0.$$

The inequalities given above make it possible to investigate the approach when the conditions of the hypothesis are not satisfied as to the manner of approach, and also when $F(\varphi)$ is discontinuous in certain ways at $\varphi = 0$. The result just given is a special case of Fatou's well known theorem. Fatou shows that if

$$u(r,\theta) = \frac{1}{2\pi}\int_0^{2\pi} \frac{(1-r^2)\,2\,r\sin(\varphi-\theta)}{[1+r^2-2r\cos(\varphi-\theta)]^2}\,F(\varphi)\,d\varphi,$$

with $F(\varphi)$ summable in the Lebesgue sense, $\lim u(M) = F'(P)$ in the wide sense, wherever $F'(P)$ exists.* Theorem 7 may be obtained from this by means of an integration by parts.

EXERCISE. Extend the above theorem to the case where $F(\varphi)$ is bounded, continuous in the neighborhood of $(1,0)$, with a unique derivative at $(1,0)$, but not necessarily of bounded variation. For convenience take $F(\varphi)$ as continuous on the circle except for a finite number of discontinuities, and extend siutably the definition of the Stieltjes integral.

16. The Poisson integral: $F(\varphi)$ absolutely continuous.

If $F(\varphi)$ is absolutely continuous, equal to a constant plus $\int_0^\varphi f(\varphi)\,d\varphi$, the Stieltjes integral reduces, as we have seen from Chapter I, to a Lebesgue integral, and we have

$$(B) \qquad u(r,\theta) = (1/2\pi)\int_0^{2\pi} \frac{(1-r^2)f(\varphi)\,d\varphi}{1+r^2-2r\cos(\varphi-\theta)}.$$

In (B), *the $f(\varphi)$ represents an arbitrary function summable in the Lebesgue sense,* and is consequently the unique derivative of $F(\varphi) = \text{const.} + \int_0^\varphi f(\varphi)\,d\varphi$ $\left(\text{or of } F(e) = \int_e f(\varphi)\,d\varphi\right)$ for almost all φ. It is worth while to restate the theorems already given for this case.

THEOREM 8. *If $u(r,\theta)$ is given by (B), it takes on the boundary values $f(\varphi)$ almost everywhere, in the wide sense; moreover its absolute value takes on the boundary values $|f(\varphi)|$ in the same sense; and the following equations are valid:*

(18) $$\lim_{r=1}\int_{\theta_1}^{\theta_2} u(r,\theta)\,d\theta = \int_{\theta_1}^{\theta_2} f(\theta)\,d\theta$$

(19) $$\lim_{r=1}\int_{\theta_1}^{\theta_2} |u(r,\theta)|\,d\theta = \int_{\theta_1}^{\theta_2} |f(\theta)|\,d\theta.$$

* Acta Mathematica, vol. 30 (1906), p. 345 and p. 357. Compare Kellogg's theorem on approach along curves with finite terminal curvature orthogonal to the boundary of a finitely connected region "subject to condition (A[(2)])" [Transactions of the American Mathematical Society, vol. 13 (1912), p. 128].

There is a further property of the function $F(r, \theta)$ $= \int_0^\theta u(r, \theta) \, d\theta$ which it will be shown later is characteristic for the equation (B), and which we now proceed to demonstrate.

THEOREM 9. *If $u(r, \theta)$ is given by (B), the absolute continuity of the function $F(r, \theta)$, is uniform for all values of r, $r < 1$.*

For the purpose merely of this theorem we shall define $u(1, \theta)$ to be $f(\theta)$. Since the uniform absolute continuity of the integral comes from the bounded character of the function when r is less than some constant r_1, r_1 being < 1, it is sufficient to consider r_1 in the closed interval $(r_1, 1)$. It is also sufficient to consider $f(\theta) \geq 0$ and therefore $u(r, \theta) \geq 0$, since $f(\theta)$ otherwise is merely the difference of two such non-negative functions. In fact if $f(\theta)$ is the difference $f_1(\theta) - f_2(\theta)$ of two non-negative functions, then the corresponding functions $u_1(r, \theta)$, $u_2(r, \theta)$ given by (B) are not negative, and

$$\int_e |u(r, \theta)| \, d\theta \leq \int_e u_1(r, \theta) \, d\theta + \int_e u_2(r, \theta) \, d\theta,$$

where e denotes any set consisting of a finite number of non-overlapping intervals. Let us then assume $f(\theta) \geq 0$.

Let r_0 be any value of r in the closed interval $(r_1, 1)$ and r_m any denumerable sequence of values of r which tend toward it as a limit. Then

(20) $$\lim_{m=\infty} \int_0^\theta u(r_m, \theta) \, d\theta = \int_0^\theta u(r_0, \theta) \, d\theta.$$

This property is obvious if $r_0 < 1$. If $r_0 = 1$ it is given by Theorem 8. From Theorem 8 we know also that $\lim_{m=\infty} u(r_m, \theta) = u(r_0, \theta)$ everywhere or almost everywhere. Hence the Theorem of De la Vallée Poussin tells us that the absolute continuity of $F(r_m, \theta)$ is uniform for all m.

Suppose now that the absolute continuity of $F(r, \theta)$ were not uniform for *all* r, $r_1 \leq r \leq 1$. Then we could find an ε such that no matter what $\delta_i > 0$ we chose there would be

FUNCTIONS HARMONIC WITHIN A CIRCLE

a value of $r = r_i$ and a finite number of intervals e_i of total measure $< \delta_i$, such that

$$(21) \qquad \int_{e_i} u(r_i, \theta)\, d\theta > \varepsilon.$$

Hence if we chose $\delta_1, \delta_2, \delta_3, \cdots$ such that $\lim_{i=\infty} \delta_i = 0$, we should have a sequence of values of r, namely $r_1, r_2, \cdots$ which would have at least one limiting point, say r', in the closed interval $(r_1, 1)$.* We could then choose a subsequence from them, say $r_{n_1}, r_{n_2}, \cdots$ which would approach r' as a true limit, and for which (21) would hold. But this is in contradiction with our previous result, taking

$$r_0 = r',$$
$$r_m = r_{n_m}.$$

Hence the theorem is proved.

* The case where there are only a finite number of different values of r_i is trivial.

CHAPTER III
NECESSARY AND SUFFICIENT CONDITIONS
THE DIRICHLET PROBLEMS FOR THE CIRCLE

17. Fundamental theorem and lemma. The principal object of the present chapter is to prove a theorem which gives sufficient conditions for the Poisson-Stieltjes and Poisson integrals. The necessity of them has already been proven, so that these conditions are both *necessary and sufficient*.

THEOREM 1. FUNDAMENTAL THEOREM. *Let $u(r, \theta)$ be harmonic inside the unit circle, denote as before $\int_0^\theta u(r, \theta)\, d\theta$ by $F(r, \theta)$, and let $r_1, r_2, \cdots$ be a sequence of values of r, $r < 1$, such that $\lim_{n=\infty} r_n = 1$. If now*

(i) $F(r_n, \theta)$ is of limited variation uniformly for all n *then $u(r, \theta)$ is given by a Poisson-Stieltjes integral* (A)

$$\text{(A)} \quad u(r, \theta) = \frac{1}{2\pi} \int_0^{2\pi} \frac{1-r^2}{1+r^2-2r\cos(\varphi-\theta)}\, dF(\varphi),$$

where $F(\varphi)$ is of limited variation and $F(2\pi+\varphi) = F(\varphi) + F(2\pi) - F(0)$; and if

(ii) $F(r_n, \theta)$ is absolutely continuous uniformly for all n *then $u(r, \theta)$ is given by a Poisson integral* (B)

$$\text{(B)} \quad u(r, \theta) = \frac{1}{2\pi} \int_0^{2\pi} \frac{1-r^2}{1+r^2-2r\cos(\varphi-\theta)} f(\varphi)\, d\varphi,$$

where $f(\varphi)$ is summable in the Lebesgue sense.

The condition (i) is equivalent to the condition

(i) $\qquad \int_0^{2\pi} |u(r_n, \theta)|\, d\theta < N, \qquad$ for all n.

The condition (ii) can also be stated in the form:

(ii) Given ε, arbitrarily, we can find δ_ε such that if e is a set consisting of a finite number of intervals of total measure $< \delta_\varepsilon$, then

$$\int_e |u(r_n, \theta)|\, d\theta \leq \varepsilon, \qquad \text{for all } n, e.$$

The condition (ii) obviously implies the condition (i).

NECESSARY AND SUFFICIENT CONDITIONS

If the function is harmonic within the unit circle, it is given by (2) Chap. II

$$u(r, \theta) = \frac{a_0}{2} + \sum_1^\infty r^n (a_n \cos n\theta + b_n \sin n\theta).$$

Since a constant $a_0/2$ satisfies the conditions (i), (ii), and is moreover given by a Poisson integral

$$\frac{a_0}{2} = \frac{1}{2\pi} \int_0^{2\pi} \frac{(a_0/2)(1-r^2)}{1+r^2-2r\cos(\varphi-\theta)} d\varphi,$$

we may omit it from consideration, writing $u(r, \theta) = 0$ when $r = 0$ and

(1) $\quad u(r, \theta) = \sum_1^\infty r^n (a_n \cos n\theta + b_n \sin n\theta), \qquad r < 1.$

The function $F(r, \theta)$ is then periodic as a function of θ with period 2π; in fact

$$F(r, \theta) = \sum_1^\infty \frac{r^n}{n} (a_n \sin n\theta + b_n \cos n\theta)$$
$$= F_1(r, \theta) - F_1(r, 0)$$

where

(2) $\quad \begin{cases} F_1(r, \theta) = \sum_1^\infty r^n (A_n \cos n\theta + B_n \sin n\theta), \\ n A_n = -b_n, \\ n B_n = a_n. \end{cases}$

LEMMA. *The series for $F_1(r, \theta)$ converges when $r = 1$ and represents a function of limited variation. It is the Fourier series for that function.*

In fact, $F_1(r, \theta)$ is of limited variation uniformly in r, since its total variation in θ is precisely the total variation of $F(r, \theta)$. Hence by the Corollary to Theorem 4, Chap. I,

$$r^n A_n \leq T/\pi n, \qquad r^n B_n \leq T/\pi n,$$

where T is independent of r, $r < 1$. Consequently we have

(3) $\qquad A_n \leq T/\pi n, \qquad B_n \leq T/\pi n.$

By Theorem 5, Chap. I, we know that the series

$$\sum_1^\infty A_n \cos n\theta + B_n \sin n\theta$$

converges almost everywhere in the interval $(0, 2\pi)$ and that it represents there a function $F_1(\theta)$ which with its square is summable over that interval; moreover that A_n, B_n are the Fourier coefficients for $F_1(\theta)$.

Now $F_1(\theta)$ is of limited variation on the set E where it is defined. The set E is dense on the circumference of the unit circle, for since the complementary set CE is of measure 0 there can be no arc without points of E. Hence by Corollary 2 of Theorem 1, Chap. I, $F_1(\theta)$ can be extended so as to be of limited variation over the closed interval $(0, 2\pi)$, periodic and with regular discontinuities; and since we are changing $F_1(\theta)$ only on a set of measure 0 we shall still have A_n, B_n as its Fourier coefficients. But now, since $F_1(\theta)$ is of limited variation, periodic and with regular discontinuities, its Fourier series converges everywhere to $F_1(\theta)$, and, moreover, everywhere

(4) $$\lim_{r=1} F_1(r, \theta) = F_1(\theta),$$

from Theorem 6, Chap. I. This proves the lemma and also the following corollary.

COROLLARY 1. *The function $F_1(\theta)$ has everywhere regular discontinuities, and* $\lim_{r=1} F_1(r, \theta) = F_1(\theta)$ *for all θ.*

18. Proof of fundamental theorem. We may now prove the first part of the fundamental theorem. Since $F(r, \theta) = F_1(r, \theta) - F_1(r, 0)$, if we define $F(\theta) = F_1(\theta) - F_1(0)$, we have for all θ

$$\lim_{n=\infty} F(r_n, \theta) = F(\theta).$$

But also the total variation of $F(r_n, \theta)$ is bounded for all n. Fix then an arbitrary point $M = (r, \theta)$, $r < 1$. We have

$$u(r, \theta) = \frac{1}{2\pi} \int_0^{2\pi} \frac{(r_n^2 - r^2)}{r_n^2 + r^2 - 2r_n r \cos(\varphi - \theta)} \, dF(r_n, \varphi), \ r_n > r.$$

Here however, since r is fixed,

$$\left| \frac{1-r^2}{1+r^2-2r\cos(\varphi-\theta)} - \frac{r_n^2-r^2}{r_n^2+r^2-2r_n r \cos(\varphi-\theta)} \right|$$

is uniformly small, for all φ, with $1-r_n$. Hence we have the hypotheses of Corollary 3 of Theorem 3, Chap. I, and

$$\lim_{n=\infty} \frac{1}{2\pi} \int_0^{2\pi} \frac{(r_n^2-r^2)\,dF(r_n,\varphi)}{r_n^2+r^2-2r_n r \cos(\varphi-\theta)}$$
$$= \frac{1}{2\pi} \int_0^{2\pi} \frac{(1-r^2)\,dF(\varphi)}{1+r^2-2r\cos(\varphi-\theta)}.$$

Consequently $u(r,\theta)$ is given by this last integral, which is (A). Thus the point is proved.

Corollary 2. *The conditions that $u(r,\theta)$ be the difference of two not negative harmonic functions within the circle is sufficient as well as necessary for the Poisson-Stieltjes formula*(A).

In fact, if $u(r,\theta) = u'(r,\theta) - u''(r,\theta)$ where u', u'' are not negative within the circle, and harmonic, the mean value theorem for harmonic functions tells us that

$$\int_0^{2\pi} |u(r,\theta)|\,d\theta \leq (u'_{r=0} + u''_{r=0})2\pi, \qquad r<1.$$

But this condition we have just proved to be sufficient for (A). The necessity of the condition in the corollary has already been proved. It is not assumed of course that $u(r,\theta)$ is bounded.

As an application of this corollary, consider the following case. Let there be an arbitrary distribution of positive and negative masses, each finite in total amount, outside or on the circumference of the unit circle, but in a finite region, where the law of attraction of point masses gives a force $-kmm'/r$, r being the distance between them. Since the potential due to the totality of positive masses is a constant plus a not negative function, and that due to the totality of negative masses is a constant plus a not positive function inside the circle, the conditions of the corollary are met with, and *the function inside the circle may be represented by a Poisson Stieltjes integral on the boundary.**

Let us now return to the second part of the fundamental theorem. Since (ii) implies (i), the $u(r, \theta)$ is given by a formula (A), in which the $F(\theta)$ may be chosen so as to have regular discontinuities if any. If by $f(\theta)$ we denote the unique derivative of $F(\theta)$ where it exists, $f(\theta)$ will be summable and the theorem of De la Vallée Poussin tells us that in virtue of (ii) we shall have

$$\lim_{n=\infty} \int_0^\theta u(r_n, \theta)\,d\theta = \int_0^\theta f(\theta)\,d\theta.$$

However, in virtue of (4),

$$\lim_{n=\infty} \int_0^\theta u(r_n, \theta)\,d\theta = F(\theta) - F(0).$$

Hence $F(\theta) - F(0) = \int_0^\theta f(\theta)\,d\theta$; $F(\theta)$ is absolutely continuous, and (A) reduces to (B). This is what we wished to prove.

We say that $f_r(x)$ converges in the mean of the first order to $f(x)$ in an interval (a, b) as r approaches 1 if

$$\lim_{r=1} \int_a^b |f(x) - f_r(x)|\,dx = 0.$$

In terms of this concept we have the following proposition.

COROLLARY 3. NOAILLON'S THEOREM.† *Given $f(\theta)$ summable in the Lebesgue sense, a necessary and sufficient condition that $u(r, \theta)$, harmonic inside the unit circle, be given by* (B) *is that $u(r, \theta)$, considered as a function of θ in the interval $(0, 2\pi)$ converge in the mean of the first order to $f(\theta)$ as r tends to* 1.

---

* The constant may be taken as $\pm (\log R)(\sum |m|)$ where R is the diameter of the set of points on which the mass is distributed.

† Comptes Rendus, t. 182 (1926), p. 1371. Compare Theorem XIV in Kellogg's paper [*Harmonic functions and Green's integral*, Transactions of the American Mathematical Society, vol. 13 (1912), p. 132], which in a sense is a special case of those given in this chapter.

That the condition is necessary follows from the uniform absolute continuity of the integral $\int |u(r,\theta) - f(\theta)| d\theta$ when $u(r,\theta)$ is given by (B). By De la Vallée Poussin's theorem, then, we have

$$\lim_{r=1} \int_0^{2\pi} |u(r,\theta) - f(\theta)| d\theta = 0.$$

That the condition is sufficient comes from the fact that

$$\int_0^{2\pi} |u(r,\theta)| d\theta \leq \int_0^{2\pi} |f(\theta)| d\theta + \int_0^{2\pi} |u(r,\theta) - f(\theta)| d\theta$$

and therefore remains bounded. Hence $u(r,\theta)$ is given by a formula (A) in which $F(\theta) = \int_0^\theta f(\theta) d\theta$, that is, by (B).

It is desirable to write this convenient condition in a form which does not employ the boundary values $f(\theta)$, but merely the values $u(r,\theta)$, $r < 1$. We say that $u(r,\theta)$ converges in the mean of the first order, as r approaches 1, in the interval $(0, 2\pi)$, if

$$\lim_{\substack{r'=1 \\ r''=1}} \int_0^{2\pi} |u(r'',\theta) - u(r',\theta)| d\theta = 0$$

however r' and r'' approach 1; that is, if given ε arbitrarily we can find $r_\varepsilon < 1$ so that

$$\int_0^{2\pi} |u(r'',\theta) - u(r',\theta)| d\theta < \varepsilon$$

for all values of r', r'' in the open interval $(r_\varepsilon, 1)$.*

We know however that in this case (as with convergence in the mean of the second order) that there is one and only one function $f(\theta)$, except for definition on an arbitrary set of values of θ, of zero measure, such that $u(r,\theta)$ converges in the mean of the first order to $f(\theta)$.† Moreover $f(\theta)$ is summable. Hence we have immediately

---

* This concept is used in even more fundamental fashion by H. E. Bray in a memoir, written in conjunction with the author of the present monograph, in press, on the generalized Dirichlet problems for the sphere.

† See Chap. VII, Art. 49.

COROLLARY 4. *A necessary and sufficient condition that $u(r, \theta)$, harmonic within the unit circle, be given by a Poisson integral* (B) *is that $u(r, \theta)$ converge in the mean of the first order as r approaches* 1.

19. Special cases of the Poisson integral (B). There are several particular classes of functions given by (B), which it is worth while to specify. In the first place we have *the class of functions harmonic and bounded, $r < 1$.*

In this case, if $|u(r, \theta)| \leq M$

(4') $$\int_e |u(r, \theta)|\, d\theta \leq M \cdot (\text{meas } e)$$

so that the absolute continuity of the integral $\int u(r, \theta)\, d\theta$ is uniform for all $r < 1$. Hence in this case $u(r, \theta)$ is given by a formula (B) where $|f(\varphi)|$ is bounded (Of course $f(\varphi)$ may be assigned arbitrarily on a set of values of φ of measure 0, without affecting the value of the integral (B)). Conversely, *if $f(\varphi)$ is bounded and integrable in the Lebesgue sense*, the $u(r, \theta)$ will be bounded, $r < 1$.

Secondly, we have the class of functions, which includes the former class, where $u(r, \theta)$, *harmonic within the circle, satisfies a condition of the form*

(5) $\qquad |u(r, \theta)| \leq \psi(\theta), \qquad r < 1,$

where $\psi(\theta)$ *is summable in the Lebesgue sense.*

In this case,

(6) $$\int_{\theta'}^{\theta''} |u(r, \theta)|\, d\theta \leq \int_{\theta'}^{\theta''} \psi(\theta)\, d\theta$$

and therefore it follows immediately that the absolute continuity of the integral $\int |u(r, \theta)|\, d\theta$ is uniform for all $r < 1$. Hence (5) is a sufficient condition for (B).

Consider as a third case the class of harmonic functions such that the numerical value of the derivative with respect to r is summable over the region bounded by the circle.

The quantity $(1/r)\dfrac{\partial u}{\partial r}$ will also be summable over the same region. Consider then regions $\sigma' = \sigma(r_0, r_1; \theta', \theta'')$ and

$\sigma = \sigma(r_0; \theta', \theta'')$ where σ' is the region $r_0 \leq r \leq r_1$; $\theta' \leq \theta \leq \theta''$, and σ is the region $r_0 \leq r < 1$; $\theta' \leq \theta \leq \theta''$. We have

$$\int_{\sigma'} \frac{1}{r} \frac{\partial u}{\partial r} \cdot d\sigma = \int_{\theta'}^{\theta''} d\theta \int_{r_0}^{r_1} \frac{\partial u}{\partial r} \cdot dr$$
$$= \int_{\theta'}^{\theta''} [u(r_1, \theta) - u(r_0, \theta)] \, d\theta$$

and therefore

$$\left| \int_{\theta'}^{\theta''} u(r_1, \theta) \, d\theta \right| \leq \int_{\theta'}^{\theta''} |u(r_0, \theta)| \, d\theta + \int_{\sigma'} \left| \frac{1}{r} \frac{\partial u}{\partial r} \right| d\sigma$$
$$\leq \int_{\theta'}^{\theta''} |u(r_0, \theta)| \, d\theta + \int_{\sigma} \left| \frac{1}{r} \frac{\partial u}{\partial r} \right| d\sigma.$$

A continuous function is positive or zero on at most a denumerable infinity of closed intervals. If accordingly we take the denumerable set of closed intervals in (θ', θ'') where $u(r_1, \theta) \geq 0$, constituting a set e', and the corresponding denumerable set of angular portions of σ constituting a set E', the above inequality yields the relation

$$\int_{e'} |u(r_1, \theta)| \, d\theta \leq \int_{e'} |u(r_0, \theta)| \, d\theta + \int_{E'} \left| \frac{1}{r} \frac{\partial u}{\partial r} \right| d\sigma.$$

In a similar fashion, for the set of open intervals e'' in (θ', θ'') where $u(r_1, \theta) < 0$, and the corresponding set E'', we have

$$\int_{e''} |u(r_1, \theta)| \, d\theta \leq \int_{e''} |u(r_0, \theta)| \, d\theta + \int_{E''} \left| \frac{1}{r} \frac{\partial u}{\partial r} \right| d\sigma.$$

But e'' and e' have no points in common, likewise E'' and E'. Therefore we have the inequality

$$(7) \quad \int_{\theta'}^{\theta''} |u(r_1, \theta)| d\theta \leq \int_{\theta'}^{\theta''} |u(r_0, \theta)| \, d\theta + \int_{\sigma} \left| \frac{1}{r} \frac{\partial u}{\partial r} \right| d\sigma.$$

But the integrals of the right hand member of (7) are absolutely continuous functions of θ'' which do not involve r_1. It follows that the absolute continuity of the left hand member of (7) is uniform for all $r_1 < 1$. Hence this third case also gives us functions represented by the formula (B).

EXERCISE. Show that the third case is included in the second.

An interesting type of functions covered by the third case is that where

$$\int \left[\left(\frac{\partial u}{\partial x} \right)^2 + \left(\frac{\partial u}{\partial y} \right)^2 \right] d\sigma$$

exists when the region of integration is the circle. In fact

$$\int \left[\left(\frac{\partial u}{\partial x} \right)^2 + \left(\frac{\partial u}{\partial y} \right)^2 \right] d\sigma = \int \left[\left(\frac{\partial u}{\partial r} \right)^2 + \frac{1}{r^2} \left(\frac{\partial u}{\partial \theta} \right)^2 \right] d\sigma$$
$$\geq \int \left(\frac{\partial u}{\partial r} \right)^2 \partial \sigma,$$

and since the last integral exists, then $\int \left| \frac{\partial u}{\partial r} \right| d\sigma$ also exists.

20. The Dirichlet problem and its extension. The formula (B) yields the solution of the Dirichlet problem.

THEOREM 2. *Given the function $f(\theta)$, summable in the Lebesgue sense there is one and only one function $u(r, \theta)$ harmonic within the circle, of the class defined by* (ii), *such that* $\lim_{r=1} u(r, \theta) = f(\theta)$ *almost everywhere. In particular, if $f(\theta)$ is bounded $u(r, \theta)$ is bounded, and conversely.*

The formula (A) yields the solution of a generalized Dirichlet problem.

THEOREM 3. *Given the function $F(\theta)$, of limited variation, there is one and only one function $u(r, \theta)$ harmonic within the circle, of the class defined by* (i), *such that* $\lim_{r=1} \int_{\theta_1}^{\theta_2} u(r, \theta) d\theta = F(\theta_2) - F(\theta_1)$, *for θ_1, θ_2 in a set of values of θ dense on the circumference. In fact this boundary condition is satisfied except for values of θ_1 or θ_2 which correspond to discontinuities of $F(\theta)$, at most denumerable in number, when these are not regular.*

EXERCISE. State the solution of the Dirichlet problem in terms of Noaillon's theorem.

CHAPTER IV

POTENTIALS OF A SINGLE LAYER AND THE NEUMANN PROBLEM

21. The Stieltjes integral for potentials of a single layer.* Consider the integral

(C) $\quad v(M) = \dfrac{a}{\pi} \displaystyle\int_0^{2\pi} \log \dfrac{1}{MP}\, dF(\varphi_P) + A,$

or, what is the same thing,

(1) $\quad v(r, \theta) = -\dfrac{a}{2\pi} \displaystyle\int_0^{2\pi} \log\left(a^2 + r^2 - 2ar\cos(\varphi - \theta)\right)$
$\qquad\qquad \times dF(\varphi) + A,$

where A is a constant. This represents the potential due to the most general distribution of simple masses on the circumference of a circle of radius a if $F(\varphi)$ is the most general function of limited variation. We notice however, that for the particular case $F(\varphi) = c\varphi$ the potential (C) reduces to a constant.

In fact, for $a = 1$, $r = \varrho < 1$,

$$\log \sqrt{1 + \varrho^2 - 2\varrho\cos\varphi} = -\varrho\cos\varphi - \varrho^2 \frac{\cos 2\varphi}{2}$$
$$-\varrho^3 \frac{\cos 3\varphi}{3} - \ldots$$

and so for $r < a$

$$\log \sqrt{a^2 + r^2 - 2ar\cos(\varphi - \theta)}$$
$$= \log a - \frac{r}{a}\cos(\varphi - \theta) - \frac{r^2}{a^2}\cos\frac{2(\varphi-\theta)}{a} - \ldots.$$

Hence

$$-\frac{a}{2\pi}\int_0^{2\pi} \log\left(a^2 + r^2 - 2ar\cos(\varphi - \theta)\right) dc\varphi$$
$$= -ac\log a^2, \quad r < a,$$

which we wished to prove.

* **My** predecessor in this point of view is Plemelj, *Potentialtheoretische Untersuchungen*, Leipzig (1911).

EXERCISE. Deduce the above development in terms of ϱ and φ, as the real part of a function of a complex variable.

The constant which we have just found may be included in the constant A, and therefore there is no loss of generality in the form (C) if we suppose that $F(\varphi)$ is periodic, of period 2π. Moreover the value of the Stieltjes integral for $r < 1$ does not depend on the particular values which $F(\varphi)$ has at discontinuities, provided it remains of limited variation. We may accordingly regard the discontinuities as *regular*, that is, of the type of the Fourier representation of the function. In this way the function $F(\varphi)$ will be uniquely determined by the values of $v(r, \theta)$ at interior points of the circle, as we shall see.

22. Necessary and sufficient condition. THEOREM 1. *A necessary and sufficient condition in order that $v(r, \theta)$, harmonic inside the circle of radius a, be given by an integral of type (C) is that there should be a constant N, such that*

$$(2) \qquad \int_0^{2\pi} \left| \frac{\partial v(r, \theta)}{\partial r} \right| d\theta < N, \qquad r < a.$$

Since the $F(\varphi)$ is periodic, (1) is equivalent, by an integration by parts, to the following:

$$v(r, \theta) = \frac{a}{2\pi} \int_0^{2\pi} \frac{2ar \sin(\varphi - \theta)}{a^2 + r^2 - 2ar \cos(\varphi - \theta)} F(\varphi) \, d\varphi + A,$$

which represents the function conjugate to the function

$$u(r, \theta) = -\frac{a}{2\pi} \int_0^{2\pi} \frac{a^2 - r^2}{a^2 + r^2 - 2ar \cos(\varphi - \theta)} F(\varphi) \, d\varphi + B,$$

where B is a constant.* Perform now on this integral a differentiation and follow it by an integration by parts,

---

* Osgood, *Funktionentheorie*, Leipzig (1912), vol. 1, p. 634. The formula is obtained by finding the imaginary part of $\frac{1+z}{1-z}$ (See Art. 11).

keeping account of the periodicity of $F(\varphi)$. We obtain the following equation

$$(3) \quad r\frac{\partial v}{\partial r} = -\frac{\partial u}{\partial \theta} = \frac{a}{2\pi}\int_0^{2\pi}\frac{(a^2-r^2)\,dF(\varphi)}{a^2+r^2-2ar\cos(\varphi-\theta)}.$$

Conversely, from (3) we have (1), since $v(r,\theta)$ is a single valued harmonic function within the circle.

But we have already seen that a necessary and sufficient condition for a representation of type (3) is that the harmonic function to be so represented satisfy a certain integral condition—the condition (i), Chap. III,—which in this case becomes

$$\int_0^{2\pi}\left|r\frac{\partial v}{\partial r}\right|d\theta < N', \qquad N' \text{ independent of } r.$$

And this is equivalent to the condition of the theorem to be proved.

23. Further properties. If we proceed to apply the results already obtained for (3) to the problem in hand, we obtain the following facts:

$$(4) \qquad \lim_{r=a}\frac{\partial v(r,\theta)}{\partial r} = F'(\theta),$$

for every θ for which $F'(\theta)$ exists and is unique, and that is almost everywhere on the circumference of the circle. Here the manner of approach of (r,θ_r) to (a,θ) is in the wide sense of Theorem 7, Chap. II, which includes the particular case of $\theta_r = \theta$, where the point approaches the boundary along a radius.

$$(5) \qquad \lim_{r=a}\int_0^\theta\frac{\partial v}{\partial r}\,d\theta = F(\theta) - F(0),$$

for every θ.

(j) A necessary and sufficient condition for (1) is that $v(r,\theta)$ be the difference of two functions each harmonic at interior points of the circle, except at the center, where both functions have the same logarithmic singularity, and each is non-decreasing as a function of r.

(jj) A necessary and sufficient condition that $F(\varphi)$ be absolutely continuous, that is, the integral of its derivative $F'(\varphi) = f(\varphi)$, is that the absolute continuity of the integral $\int \dfrac{\partial v}{\partial r} d\theta$ be uniform, $r < a$.

THEOREM 2. *Under the condition* (jj), *the* $v(r, \theta)$ *is given by the formula*

(D) $\quad v(r, \theta) = \dfrac{a}{2\pi} \displaystyle\int_0^{2\pi} \log \dfrac{1}{a^2 + r^2 - 2ar\cos(\varphi - \theta)} f(\varphi) d\varphi + A,$

where $\int_0^{2\pi} f(\varphi) d\varphi = 0$, *by means of the proper choice of* A. *In fact*, A *is the value of* $v(r, \theta)$ *when* $r = 0$.

24. The Neumann problems. Equation (D) gives the solution of the Neumann problem for the class of functions (jj), the values $f(\varphi)$ which the normal derivative is to take on at the boundary almost everywhere being arbitrary provided that they are summable in the Lebesgue sense and such that their integral around the boundary is zero (if the integral of the boundary values of $\dfrac{\partial v}{\partial r}$ around the circumference were not zero, the function v could not be finite inside). A particular class of the functions (jj) are those for which $\left|\dfrac{\partial v}{\partial r}\right|$ remains bounded, $r < a$. For these functions $f(\varphi)$ is also bounded, except of course on an arbitrary set of values of φ of zero measure.

Equation (C) solves a generalized Neumann problem. This rather than the other is essentially the physical problem of the Neumann type.

THEOREM 2a. *Among the functions whose total absolute flux* $\int_0^{2\pi} \left|\dfrac{\partial v}{\partial r}\right| d\theta$ *is bounded*, $r < a$, *there is one and only one, except for an arbitrary additive constant, which satisfies the equation* (5),— *given* $F(\theta)$ *of limited variation, with regular discontinuities, and periodic with period* 2π.

25. General points of view. It is now an interesting question to ask if the class of functions of the theorem of

Art. 11, which gives the solution of the generalized Neumann problem is more or less general than the class of functions of condition (i) which gives the solution of the generalized Dirichlet problem. The question is answered at once. If $\int_0^{2\pi} \left|\frac{\partial v}{\partial r}\right| d\theta < N$, it follows that $\left|\frac{\partial v}{\partial r}\right|$ is summable on the region defined by the circle $r = a$, and therefore (see Art. 19) that $v(r,\theta)$ is given by a formula

$$(6) \quad v(r,\theta) = \frac{1}{2\pi} \int_0^{2\pi} \frac{a^2 - r^2}{a^2 + r^2 - 2ar\cos(\varphi - \theta)} \cdot g(\varphi) d\varphi,$$

where $g(\varphi)$ is some function summable (L). But (6) is merely a special case of the Poisson-Stieltjes integral of Chapter II.

We are now faced with another question. Are the harmonic functions of what we might call *physical character* the only possible ones? Are functions which can be written inside the circle as the difference of two not negative harmonic functions the only ones possible which are harmonic within the circle? Again the answer is in the negative. There are for example those which arise from the class considered by differentiation one or more times with respect to θ. Thus there are the functions

$$(7) \quad u(r,\theta) = a_0/2 + \sum_1^\infty r^n (a_n \cos n\theta + b_n \sin n\theta)$$

where a_n, b_n become infinite like a positive power of n. But there are also functions which are not derivatives with respect to θ, of any finite order, of functions of *physical character*. Consider for example a function given by (7) when the a_n, b_n become infinite like the quantity $a^{\sqrt{n}}$, $a > 1$.*

26. Digression. Physical interpretation of a general distribution of mass. It has already been pointed out in

---

* The extension of this treatment to spaces of dimension > 2 constitutes one of the most interesting applications of Volterra's theory of functions of curves and integral equations. In fact the limits of the absolutely continuous functions which must be considered are functions of curves which are additive, but which are not additive functions of point sets.

CHAPTER IV

Art. 13 that the function $F(\varphi)$ corresponds to an arbitrary distribution of positive and negative mass, $\Phi(e)$, on the circumference of a circle, finite in total absolute amount. In Chapters II and III we were dealing with distributions of a double layer and in the present chapter we deal with distributions of a single layer. But we have not considered any method of constructing or laying down such a distribution $\Phi(e)$, and this is a process which is not easy to imagine. In this section we shall consider simple approximations to such arbitrary distributions, the approximations being "as close" as desired.

When we speak of an arbitrarily close approximation in a physical sense, we mean not merely that the given one is a limiting one for the approximation, but also that the physical measurements — attraction, potential, etc., from which one infers a distribution of matter, — which are given in terms of the approximation differ by as little as we please from those which belong to the given distribution.

Let P be a generic point of the circumference and M a point not on the circumference; and let $h(M, P)$ be a continuous function of M and P. The quantities which are measured depend on the whole distribution and will therefore be given by Stieltjes integrals of the form

(8) $$\xi(M) = \int_0^{2\pi} h(M, P) \, dF(\varphi_P).$$

If then $F_n(\varphi)$ is the function of limited variation which corresponds to an approximate distribution, the corresponding measurement function will be

(8') $$\xi_n(M) = \int_0^{2\pi} h(M, P) \, dF_n(\varphi_P).$$

Consequently sufficient conditious for approximate measurement,
(8'') $$\lim \xi_n(M) = \xi(M),$$

are given by the hypotheses of the Helly-Bray theorem.

THEOREM OF APPROXIMATE MEASUREMENT. *If* $\lim_{n=\infty} F_n(\varphi)$ $= F(\varphi)$ *on a set of values of φ dense on the circumference, and $F_n(\varphi)$ is of uniformly limited variation, then equation* (8″) *is valid*.

Incidentally, all functions of limited variation which correspond to the same additive function of point sets $\Phi(e)$ (that is, to the same distribution of matter) yield the same value of the integral $\mathfrak{F}(M)$, for any two such functions are extensions of each other as defined on a common set E where both are continuous, except for an arbitrary additive constant. Thus if there is a point charge at φ_0 its value is always $F(\varphi_0+0)-F(\varphi_0-0)$.

Any function of limited variation $F(\varphi)$ may be written as the sum of three terms of different kinds

$$(9) \qquad F(\varphi)-F(0) = \alpha(\varphi)+\int_0^\varphi f(\varphi)\,d\varphi + \lambda(\varphi).$$

In this representation $\alpha(\varphi)$ represents the function of discontinuities; i. e., if the discontinuities are at points $\varphi_1, \varphi_2, \cdots$, taken in some denumerable order,

$$(10) \qquad \begin{aligned} \alpha(\varphi) &= \sum_{0 < \varphi_n \leq \varphi} \{F(\varphi_n) - F(\varphi_n - 0)\} \\ &+ \sum_{0 \leq \varphi_n < \varphi} \{F(\varphi_n+0) - F(\varphi_n)\}, \text{ for } 0 < \varphi \leq 2\pi, \\ &= 0, \text{ for } \varphi = 0, \end{aligned}$$

where the summations are extended over the discontinuities φ_n which are indicated. We then write $f(x)$ as the derivative, where it exists, of the continuous function $F(\varphi)-F(0)-\alpha(\varphi)$, and define it arbitrarily otherwise; in fact, $f(\varphi) = F'(\varphi)$ almost everywhere. What is left, namely

$$(11) \qquad \lambda(\varphi) = F(\varphi)-F(0)-\alpha(\varphi)-\int_0^\varphi f(\varphi)\,d\varphi,$$

is a continuous function of limited variation with a zero derivative almost everywhere.

Vitali has shown that $\lambda(\varphi)$ may be written in the form

(12) $$\lambda(\varphi) = \sum_1^\infty k_i \psi_i(\varphi),$$

where $\sum |k_i|$ is convergent, and each of the functions $\Psi(\varphi) = \psi_i(\varphi)$ (called an *elementary discard*) is continuous, non-decreasing (with $\Psi(0) = 0$, $\Psi(2\pi) = 1$) and constant on each interval of a denumerable sequence of intervals $s_1, s_2, \cdots$ (called a *plurisegment*) of which the complementary set E_Ψ is perfect and of zero measure.* According to Vitali a denumerable set of values y_i can be found, dense in $(0, 1)$ such that $\Psi(\varphi) = y_i$ on the interval s_i, and $y_k > y_m$ if s_k is to the right of s_m. We might call the distribution which is given by $\Psi(\varphi)$ an *elementary discard distribution*.

The distribution given by $F(\varphi) - F(0)$ is the sum of those given by its three parts. In the case of $\alpha(\varphi)$ the corresponding distribution may be interpreted as the sum of those represented by its separate terms; in fact, the function corresponding to the sum of the first p terms (following a denumerable order) has $\alpha(\varphi)$ as its limit for all values of φ, and its total variation is uniformly bounded, for all p, not exceeding the total variation of $\alpha(\varphi)$. Hence the requirements for the theorem of approximate measurement are fulfilled. Each of the single terms represents a point charge.

Similarly the distribution represented by $\lambda(\varphi)$ may be regarded as formed by successively adding the distributions corresponding to the separate terms $k_i \psi_i(\varphi)$. In fact the function

(13) $$\lambda_p(\varphi) = \sum_1^p {}_i k_i \psi_i(\varphi)$$

has $\lambda(\varphi)$ for its limit for all values of φ; moreover the total variation of $\lambda_p(\varphi)$ is $\leq \sum_1^p |k_i| \leq \sum_1^\infty |k_i|$, and is therefore uniformly bounded. Hence again the conditions of the measurement theorem are satisfied. We are thus left with three types of distribution to interpret, viz., the elementary discard

* Rendiconti del Circolo Matematico di Palermo, vol. 46 (1922), p. 388. The illustration in Art. 2 above is the graph of an elementary discard.

distribution, the point charge and the absolutely continuous function, the last being a charge which is the integral of its density over any interval, and therefore the kind of charge usually considered in integral formulae.

As a first interpretation let us consider the elementary discard as the limiting form of a distribution spread uniformly (that is, with constant density) over each of a finite number of intervals. To be precise, let $\Psi_n(\varphi)$ be a polygonal approximation to $\Psi(\varphi)$, formed by joining successively with straight lines a finite number of points on the graph of $\Psi(\varphi)$, including for convenience those for 0 and 2π. The approximation $\Psi_{n+1}(\varphi)$ is formed by retaining the vertices of $\Psi_n(\varphi)$ and inserting new ones between them, in such a way that the vertices of all the polygons taken together form a set E of values of φ dense in $(0, 2\pi)$. In this way the approximate distribution consists of uniform patches on the circumference of the circle, and the successive approximations subdivide these into smaller patches, only those which contain points of the perfect set E_Ψ possessing any charge. The perfect set E_Ψ is in fact obtained by cutting out intervals.

The total variation of $\Psi_n(\varphi)$ obviously is precisely unity, and is therefore bounded uniformly, and $\lim\limits_{n=\infty} \Psi_n(\varphi) = \Psi(\varphi)$ for the points of E. The equation (8″) is therefore valid for the approximation.

Instead of a polygonal approximation we may use a step-function as $\Psi_n(\varphi)$, based on the same points of division, the graph now being composed of a finite number of horizontal lines, and discontinuous. The corresponding approximate distribution consists of a finite number of point masses, and for further approximations these are successively subdivided and shifted (the portions of any mass for a given approximation always remaining, however divided, within two contiguous intervals of that approximation) until they are located on the set E_Ψ of zero measure.

The absolutely continuous function $\int_0^\varphi f(\varphi)\,d\varphi$ may similarly be interpreted either with the help of a polygonal approximation as the limit of a distribution uniform in patches, or with the help of a step function as the limit of a distribution of point charges. The point charge may also be regarded as the limit of a uniform patch. In fact there is nothing to prevent us from applying these methods directly to the function $F(\varphi)$ itself; we know that as long as we form a limiting distribution which agrees with $\Phi(e)$ on all intervals with end points in E, it must be identical with $\Phi(e)$ itself. This is of course not the same as saying that $\lim_{n=\infty} \Phi_n(e) = \Phi(e)$ for all measurable sets e. It may be remarked however that in dealing directly with $F(\varphi)$ we mask the nature of the particular kinds of distributions (like elementary discards) out of which $\Phi(e)$ is built.

Other kinds of approximations are also available. We might for instance have used the Fejér trigonometric sums, which converge to $F(\varphi)$ except at its discontinuities and are of uniformly limited variation, as was seen in Chap. I. They yield approximate distributions in which the density varies continuously, but lack the concreteness of the other interpretations suggested.

We are not able, merely by describing possible distributions of mass, to infer their physical existence. On the other hand, apparent resolution of matter into components of one kind does not preclude the existence of other kinds. To take a concrete example in the field of energy, it may be remarked that although band spectra seem to be resolvable indefinitely into sequences of fine lines, it may be that physicists, by means of that resolution, are dealing with an approximation by points to distributions of energy which would ultimately be more conveniently interpreted in terms of elementary discards. Whether we arrive at point distributions as a theoretical explanation or elementary discards, apparently hinges on the question as to whether it is the intervals of wave length in which distributions occur, which are most important, or the intervals from which they are excluded.

The results obtained in this section have been stated in physical terms. They may also, if we take for the function

$h(M, P)$ the two special forms which we have used in the last three chapters, be regarded as the statement of closure properties for certain classes of harmonic functions.

27. Cauchy's integral formula. Let $u(r, \theta)$ be a harmonic function of class (ii) inside the circle of radius a, so that it is given by a Poisson integral of type (B)

$$u(M) = u(r, \theta) = \frac{1}{2\pi} \int_0^{2\pi} \frac{(a^2 - r^2) f(\varphi)}{a^2 + r^2 - 2ar \cos(\varphi - \theta)} d\varphi$$

in which $f(\varphi)$ is summable in the Lebesgue sense. This function is the real part (see Art. 11, page 27) of the function

$$\begin{aligned} w(z) &= u + iv \\ &= \frac{1}{2\pi} \int_0^{2\pi} \frac{t+z}{t-z} f(\varphi) d\varphi + \text{const.}, \\ &= -\frac{1}{2\pi} \int_0^{2\pi} f(\varphi) d\varphi + \frac{1}{\pi} \int_0^{2\pi} \frac{t}{t-z} f(\varphi) d\varphi + \text{const.}, \end{aligned}$$

where we write $t = ae^{i\varphi}$, $z = re^{i\theta}$. If we put $z = 0$ in this equation we find the value of the constant to be $iv(0)$, so that the equation becomes the following, when we write $dt = it\, d\varphi$ and $f(t)$ for the corresponding $f(\varphi)$,

$$(14) \qquad w(z) = u(0) + iv(0) + \frac{1}{\pi i} \int_c \frac{f(t)\, dt}{t-z},$$

with

$$(14') \qquad u(0) = \frac{1}{2\pi} \int_0^{2\pi} f(\varphi) d\varphi.$$

We are able therefore to state the following lemma.

LEMMA. That $u(r, \theta)$ be a harmonic function of class (ii) inside the circle of radius a is a necessary and sufficient condition that $w(z)$ satisfy an equation of form (14), where $f(t)$ is a real-valued function summable on the circumference and $u(0)$ is given by (14'). The real part of $w(z)$ takes on the boundary values $f(t)$ almost everywhere, as z approaches t in the wide sense, and $w(z)$ is uniquely determined by those values except for the arbitrary imaginary constant $iv(0)$.

Suppose now that both u and v are of class (ii), for which a necessary and sufficient condition is that $|u+iv|$ be of class (ii). In this case we shall say that $w(z)$ is of class (ii). Then (14) holds; but also, similarly,

$$-iw(z) = v(r, \theta) - iu(r, \theta)$$
$$= -v(0) - iu(0) + \frac{1}{\pi i}\int_c \frac{g(t)\,dt}{t-z}$$

or

(15) $$w(z) = u(0) - iv(0) + \frac{1}{\pi i}\int_c \frac{ig(t)\,dt}{t-z}$$

where $g(t)$ is a summable real-valued function. From (14) and (15) we deduce the equation

(16) $$w(z) = \frac{1}{2\pi i}\int_c \frac{f(t)+ig(t)}{t-z}\,dt.$$

In other words, $w(z)$ may be given by a Cauchy integral formula (16) where $\lim_{z=t} w(z) = f(t)+ig(t)$ almost everywhere, in the wide sense.

But the real-valued functions f and g, as developed in (16), are not independent. In order to state the relations between them in convenient form let us introduce the Fourier coefficients a'_m, b'_m of $f(\varphi)$ and a''_m, b''_m of $g(\varphi)$, quantities which we know to exist since the two functions are summable. In fact, from the convergence of the Fejér summation method, we know that these summable functions are determined almost everywhere by their Fourier coefficients. We may state then the following theorem.

THEOREM 3. *A necessary and sufficient condition that $w(z)$ may be written in terms of a Cauchy integral formula* (16), *with summable real-valued functions $f(t)$, $g(t)$, is that $w(z)$ be analytic inside the circle and of class* (ii). *In order that $w(z)$, so given, take on the boundary values $f(t)+ig(t)$ almost everywhere in the wide sense, it is necessary and sufficient that*

(17) $$a'_m = b''_m, \qquad b'_m = -a''_m, \qquad m = 1, 2, \cdots,$$

where a'_m, b'_m are Fourier coefficients of $f(\varphi)$ and a''_m, b''_m are Fourier coefficients of $g(\varphi)$, φ and θ *being held concentric.*

The functions $f(\varphi)$, $g(\varphi)$ are therefore determined in terms of each other except for arbitrary additive constants.

It is only the second part of the theorem which requires further proof. We write $w(z) = w_1(z) + w_2(z)$ where

$$(18) \quad \begin{aligned} w_1(z) &= u_1(r,\theta) + iv_1(r,\theta) \\ &= \frac{1}{\pi i} \int_c \frac{\frac{1}{2}f(t)}{t-z} dt - \frac{1}{2\pi} \int_0^{2\pi} \frac{1}{2} \{f(\varphi) - ig(\varphi)\} d\varphi \end{aligned}$$

$$(19) \quad \begin{aligned} w_2(z) &= u_2(r,\theta) + iv_2(r,\theta) \\ &= \frac{1}{\pi i} \int_c \frac{\frac{i}{2}g(t)}{t-z} dt + \frac{1}{2\pi} \int_c^{2\pi} \frac{1}{2} \{f(\varphi) - ig(\varphi)\} d\varphi, \end{aligned}$$

so that (18) and (19) become special cases of (14) and (15) respectively, and, almost everywhere,

$$\lim_{r=1} u_1(r,\theta) = \frac{f(\theta)}{2}, \quad \lim_{r=1} v_2(r,\theta) = \frac{i}{2}g(\theta),$$

Hence, by the lemma,

$$w(z) = 2w_1(r,\theta) + Ai = 2w_2(r,\theta) + B$$

where A and B are real constants. But putting $z = 0$ in (18) and (19) we have

$$w(0) = \begin{aligned} 2w_1(0) + Ai &= \frac{1}{2\pi} \int_0^{2\pi} \{f(\varphi) + ig(\varphi)\} d\varphi + Ai \\ 2w_2(0) + B &= \frac{1}{2\pi} \int_0^{2\pi} \{f(\varphi) + ig(\varphi)\} d\varphi + B, \end{aligned}$$

so that $A = B = 0$, and $w_1(z) \equiv w_2(z)$.

This last condition may be written in the form

$$(20) \quad \frac{1}{2\pi i} \int_c \frac{f(t) - ig(t)}{t-z} dt + \frac{1}{2\pi} \int_0^{2\pi} \{-f(\varphi) + ig(\varphi)\} d\varphi = 0,$$

and by substition in (18) and (19) is seen to be sufficient, as well as necessary, that $w(z)$ take on the required boundary values. The left hand member of (20) however represents an analytic function of z, and for this function to be identically zero it is necessary and sufficient that it should vanish with all its derivatives, at the point 0.

The equation is already satisfied when $z = 0$. Hence the required conditions are merely the following,

$$\int_c \frac{f(t)-ig(t)}{t^k}\,dt = 0, \quad k = 2, 3, \cdots,$$

or, finally, when we have written $t = e^{\varphi i}$, $dt = it\,d\varphi$,

$$\int_0^{2\pi} \{f(\varphi)-ig(\varphi)\}\{\cos m\varphi - i\sin m\varphi\}\,d\varphi = 0, \quad m = 1, 2, \cdots,$$

which are precisely the equations (17), Hence the theorem is proved.

COROLLARY 1. Given the real function $f(\varphi)$ summable with its square, whose Fourier coefficients a'_m, b'_m, are such that the series

$$\sum_1^\infty (a'^2_m + b'^2_m)(\log m)^2$$

converges, there is a real function $g(\varphi)$, also summable with its square, uniquely determined by (17) except for an arbitrary constant. These two functions are given by their Fourier series, and the function $w(z)$, given by (16) takes on almost everywhere the boundary values $f(t)+ig(t)$ in the wide sense.

This corollary is verified at once by using Theorem 5, Chap. I. With the aid of the Fischer-Riesz theorem we have a slightly more general result.*

COROLLARY 2. Given $f(\varphi)$, real and summable with its square, $g(\varphi)$ and $w(z)$ are determined so that the conclusions of Corollary 1 follow, except that $f(\varphi)$ and $g(\varphi)$, instead of

---

* A proof of the Fischer-Riesz theorem is given in Hobson, *Theory of functions of a real variable*, vol. II, Cambridge (1926), p. 576.

POTENTIALS OF A SINGLE LAYER

being given by their Fourier series, are given by the Fejer summation process.

EXERCISE 1. With the help of (18) and (19) discuss the integral (16) when $f(t)+ig(t) = 1/t$.*

EXERCISE 2. We shall say that $w(z)$ converges in the mean of the first order as z approaches the boundary if $\lim_{r=1} \int_c |w(z')-w(z'')| \, |dz| = 0$, where z' denotes the point (r', θ), z'' the point (r'', θ) and $r \leqq r' < r'' < 1$. Show that this condition is necessary and sufficient that $w(z)$, analytic inside the circle, may be given by (16) (see Art. 18).

EXERCISE 3. Discuss the equation

$$\frac{w(z)-w(0)}{z} = \frac{1}{2\pi} \int_c \frac{d\Psi(t)}{t-z},$$

where $\Psi(t) = F(\varphi) + iG(\varphi)$, these functions of φ being real and of limited variation.

EXERCISE 4. Show that if $u(r, \theta)$ is given, harmonic within the circle, such that $\left(\frac{\partial u}{\partial x}\right)^2 + \left(\frac{\partial u}{\partial y}\right)^2$ is summable over the circle, the analytic function $u+iv = w$ may be given by a formula (16).

EXERCISE 5. If $u(r, \theta)$ is given by a formula (C), Art. 21, and is of uniformly limited variation in θ for all $r < 1$, show that $w = u+iv$ may be given by a formula (16). The circle is transformed by the corresponding conformal transformation into a region with rectifiable boundary.†

EXERCISE 6. If $w(z)$ is given by a formula (16) in which $f(t)$ and $g(t)$ are real and summable on the circumference, but do not satisfy (17), it may also be given by such a formula where (17) will be satisfied.

* See Osgood, *Funktionentheorie*, Leipzig (1912), p. 297.

† It is clear that the boundary functions $f(\varphi)$ and $g(\varphi)$ are of limited variation. It is desired to show that they are continuous. Suppose that $f(\varphi)$ has a positive jump at $\varphi = 0$, such that $f(0+)-f(0-) = A$. Then, since the discontinuity is regular, we have for the positive and negative variation functions, $\varphi(0+)-\varphi(0-) = A$, $\psi(0+)-\psi(0-) = 0$.

Now, $v(r, \theta)$ being conjugate to $u(r, \theta)$, we have

$$v(M) = \frac{1}{\pi} \int_{-\pi}^{\pi} \log \frac{1}{MP} d\varphi(P) + \frac{1}{\pi} \int_{-\pi}^{\pi} \log \frac{1}{MP} d\psi(P) + \text{const.}$$

$$= \frac{1}{\pi} \int_{-\varepsilon}^{\varepsilon} \log \frac{1}{MP} d\varphi(P) + \frac{1}{\pi} \int_{-\varepsilon}^{\varepsilon} \log \frac{1}{MP} d\psi(P) + V(M),$$

where $M = (r, 0)$, $P = (1, 0)$ and $V(M)$ remains bounded as M approaches P along the radius. But

$$v(r, 0) \geq \frac{A}{\pi} \log \frac{1}{1-r} - \frac{1}{\pi} \left(\log \frac{1}{1-r} \right) |\psi(\varepsilon) - \psi(-\varepsilon)| + v(r, 0),$$

the first term being the contribution to the first integral at $\varphi = 0$. But this whole expression becomes positively infinite as r tends to 1, which is contrary to the hypothesis; for $v(r, 0)$ tends to $g(0)$ which is finite. Hence $f(\varphi)$ is continuous at $\varphi = 0$, and similarly for any other φ.

The condition of Exercise 5, that $u(r, \theta)$, $v(r, \theta)$ be of limited variation as functions of θ uniformly for all r, or that $\int_0^{2\pi} |\partial u/\partial r| d\theta$, $\int_0^{2\pi} |\partial v/\partial r| d\theta$ be bounded, is necessary and sufficient that the boundary of the transformed region be a rectifiable (not necessarily simple) continuous closed curve.

CHAPTER V

GENERAL SIMPLY CONNECTED REGIONS AND THE ORDER OF THEIR BOUNDARY POINTS BOUNDARY VALUE PROBLEMS

28. Conformal transformations and general regions. Since a conformal transformation of the plane carries a harmonic function into another harmonic function, the results so far obtained will apply to regions which can be obtained from a circle by conformal transformation, in so far as those results can be stated in a form which is invariant of such transformations.

Let T be an arbitrary simply connected open region, for simplicity, in the finite plane, let C be its frontier or boundary, and let g, h be the Green's function and its conjugate respectively for T.

Then the function
$$(1) \qquad w = f(z) = e^{-g-hi}$$

transforms the region T in a one-one manner and conformally on the interior of the unit circle of the w plane.

In particular, if T is bounded by a finite number of pieces of analytic curves, the transformation is also one-one for points on the boundary, and, except at the vertices, conformal.[*]

29. Invariant forms of conditions (i), (ii), **etc.** In the circle of radius 1, the Green's function for which the pole is at the center 0 is given by $\log 1/r$, simply, and its conjugate function is $-\theta$. Hence if we denote this Green's function for the circle by $g_0(O|M)$, and its conjugate by $h_0(O|M)$, the conditions (i), (ii) may be restated as the following ones, the integration being extended along curves $g_0(O|M) = \text{const.} = g$:

[*] Osgood, *Funktionentheorie*, Leipzig (1912), vol. 1, p. 682. The existence of the Green's function is an application of the theory of increasing sequences of functions to an open set. (Ibid, p. 701.)

(i) $\int_0^{2\pi} |u(M)|\, dh_0(O|M) < N, \quad g > 0,$

(ii) The absolute continuity of $\int_0^h |u(M)|\, dh_0(O|M)$ is uniform for $g > 0$.

EXERCISE. Show by means of a linear transformation that O may be replaced by an arbitrary interior point A' of the circle.

The conditions (i), (ii) are now obviously in a form independent of a conformal transformation. For we get equivalent conditions for the general region T merely by replacing $g(O|M)$ by $g(A|M)$ and $h(O|M)$ by $h(A|M)$, which are the Green's function and its conjugate, respectively, for T, the point A being an arbitrary fixed point of T. We have, in T, the conditions

(i) $\int_0^{2\pi} |u(M)|\, dh(A|M) < N, \quad g > 0,$

(ii) The absolute continuity of $\int_0^h |u(M)|\, dh(A|M)$ is uniform for $g > 0$,

as conditions to describe the classes of functions which arise by means of conformal transformations from the corresponding classes in the circle. This is merely an application of equation (1).

We verify at once then, also by (1), that *the class* (i) *in T consists of the functions which may be written as the difference of two not negative functions, harmonic in T, and that the class* (ii) *contains in particular the class of bounded functions.*

EXERCISE. Show that the integral

$$\int_T \left\{ \left(\frac{\partial u}{\partial x}\right)^2 + \left(\frac{\partial u}{\partial y}\right)^2 \right\} d\sigma$$

if it exists, is invariant of a conformal transformation, and thus can be used to describe an important class of harmonic functions. Write the condition that $\partial u/\partial r$ is summable over the circle in a form which is invariant of conformal transformations.

30. **Invariant forms of conclusions.** In order to make our theorems general, the conclusions also must be given a form which is invariant of a conformal transformation, i. e., in terms of the Green's function and its conjugate. An application of the theorems of Chapter II gives us at once the following theorems.

THEOREM 1. *Under* (i), *the function* $u(M)$ *has a limit as* M *approaches the frontier* C *along almost all curves* $h = const$; *moreover, if we integrate along curves* $g(AM) = const$, *we have*

(2) $$\lim_{g=0} \int_{h_1}^{h} u(M)\, dh \text{ exists, } = F(h),$$

for every value of h, *with* $F(h)$ *a function of* h *of limited variation with regular discontinuities. The function* $u(M)$ *of class* (i) *is uniquely determined by this frontier function* $F(h)$.

We may say that $u(M)$ takes on boundary values for almost all h if it behaves as in the first clause of Theorem 1.

THEOREM 2. *Under* (ii), *the function* $F(h)$ *is absolutely continuous as a function of* h, *and if we write* $f(h) = F'(h)$, *where the latter exists, and* $f(h)$ *arbitrary otherwise, the function* $u(M)$ *takes on the boundary values* $f(h)$, *as* M *approaches* C *along a curve* $h = const.$, *for almost all* h.

Moreover if $f(h)$ *is an arbitrary summable function of* h, *there is one and only one* $u(M)$ *of class* (ii) *which takes on these boundary values. A boundary value* $f(h)$ *will be attained by* $u(M)$ *whenever the condition*

$$f(h) = \frac{d}{dh} \int f(h)\, dh$$

is satisfied.

The second theorem itself yields a solution of the Dirichlet problem for the general simply connected region, and the first theorem gives a solution of a generalized problem of nature similar to that of Dirichlet. In order to make these theorems more useful, however, the functions $f(h)$ and $F(h)$ should be defined not merely in terms of limits of functions of h when g approaches 0, but rather, directly in terms of the boundary points themselves. This problem is primarily that of the order of boundary points; and we find that for a solution of it, consideration of the so-called "accessible points" of C will suffice.

31. Order of boundary points. A point P of C is *accessible* if it can be joined to an interior point M of T by a simple continuous curve γ. The order of these accessible points is given by means of the behavior of the conformal transformation already mentioned, with relation to the points of C.* We quote as lemmas two theorms of Osgood.†

LEMMA 1. Let T be a simple connected plane region whose boundary consists of more than one point, and let the interior of T be mapped conformally on the interior of a circle S. Let P be an accessible boundary point of T, and let γ be a curve lying within T (except for one extremity) and leading to P. Then the image of γ in S is a curve γ' with a single limiting point P' on the circumference of S; so that if a point M approach P along γ its image will approach P' as a limit.

LEMMA 2. Let γ_1 be a second curve of T also leading from a point A of γ to P and meeting γ only in A and P. Let γ_1' be the image of γ_1 in S, P_1' its limiting point on the boundary of S. The necessary and sufficient condition that P_1' coincide with P' is that the simple closed curve $\overline{\gamma}$ consisting of γ and γ_1 may be drawn together continuously to the point P without passing out of T; or in other words, that $\overline{\gamma}$ shall contain in its interior only interior points of T.

In particular, the proof of Osgood's theorem shows that if P is an accessible point there is a definite value of $h(A|P) = h_0(O|P')$ that corresponds to the limiting position of M as M approaches P along γ; moreover P is accessible along that $h_A = $ const. curve or is a limit point of that curve. The same remark also applies to some value $h(B|P)$, and

---

* The properties of the boundary points are investigated by Osgood, Carathéodory and Courant. [See Hurwitz-Courant, *Funktionentheorie*, Berlin (1922), p. 349.]

† Osgood and Taylor, *Conformal transformations on the boundaries of their regions of definition,* Transactions of the American Mathematical Society, vol. 14 (1913), pp. 277-298.

to accessibility along $h_B = $ const., where B is any other pole; for we might have made a conformal transformation with B going into O. For convenience let us refer to these two h curves from A and B, respectively, determined by P, as λ and λ_1. The curve λ (or λ_1) cannot itself lead to two different accessible positions on C, as an immediate consequence of Lemma 1.

As is indicated in Lemma 2, however, two different h curves from A may belong to the same position P on C. Such positions we describe as *multiple points* and we define each accessible part of the position which belongs to an h curve from A as a separate accessible point P, determined by a specific value of h. So far the identity of a point refers to a single pole of reference A, and the curve λ which belongs to it. But if B is a second pole in T, the preceding paragraph tells us that we can set up a one to one correspondence between curves λ from A and curves λ_1 from B which belong to accessible points; for the relation described is a symmetric one. The curves λ and λ_1 have images in each conformal transformation, λ', λ_1' and λ'', λ_1'' respectively, and λ' and λ_1' lead to the same point P' on the circumference of S while λ'' and λ_1'' lead to the same point P'' on the circumference of S. As we see from Lemma 2, a necessary and sufficient condition that λ and λ_1 correspond if they actually lead to points of C is that the closed curve formed by λ, λ_1 and $h(A|B)$ contain in its interior no points of C. The individuality of an accessible point of C, as now defined, is thus independent of the pole A. Moreover Lemma 2 enables us to decide whether any curve, not necessarily an h curve, which leads to a point C, leads or does not lead to the particular point P if P is accessible along an h curve.*

We now assign to the accessible points of C an order. It will be defined merely as the circular order of the values $-h(A|P)$. The order is thus defined with reference again

---

* Carathéodory gives an example of an accessible boundary point, of a simply connected region, of non-denumerable multiplicity, Math. Annalen, vol. 73 (1913), p. 363.

to a particular pole A. But this order is independent of the pole which is chosen, or what amounts to the same thing, of the conformal transformation which identifies the order of the points of C with the counter clockwise order of boundary points of the circle. For any conformal transformation of S into T is merely one of the interior of the circle into itself, which does not change the order of points on the circumference, followed by the particular conformal transformation in which $-\theta$ corresponds to the values of $h(A|P)$. The order of the accessible boundary points as above defined is thus an intrinsic property of the region T.

If we have a family of curved rays γ leading from an arbitrary point A of T to points of C, such that one goes to every accessible point as above defined, and no two have common points in T except at A, the order of the points of C will be the same as the counter clockwise order of the rays from A. In fact, the images of these rays in S make a similar family of rays γ' with the same order about O, and this again is the counter clockwise order of their end points on the circumference of the circle. This is evident if we remember that having drawn one ray γ'_α, any second ray γ'_β forms a cross-cut in the resulting simply connected region. If these are divided by another pair of rays γ'_μ, γ'_ν, their end points will be similarly divided by the end points of the latter pair of curves.

We have stated that it is sufficient to consider the accessible points of C.* This is so, because we can prove the following theorem:

THEOREM 3. *Almost all the h curves issuing from A lead to accessible points of C.*†

The truth of the theorem follows at once from the lemma

* In fact, as follows from Theorem 3, it would be sufficient to consider the points accessible along curves $h_s = $ const., for some pole A.

† Osgood shows that the set of values of h corresponding to accessible points of C is dense [loc. cit.]. Carathéodory [loc. cit. p. 365] proves the theorem given above by a different method. The method given above is taken from our article on Fundamental Points of Potential Theory, already referred to.

that *almost all the h curves issuing from A are finite in length.*
We do not mean of course that their length is bounded for
this set of values.

In order to prove this lemma we consider first the part
of T contained between two curves $g(A|M) = k$ and
$g(A|M) = k'$, and form the integral

$$\int_\sigma \left\{ \left(\frac{\partial g}{\partial x}\right)^2 + \left(\frac{\partial g}{\partial y}\right)^2 \right\} d\sigma = -\int_s g \frac{\partial g}{\partial n} ds = (k-k')\, 2\pi,$$

where s denotes the complete boundary of σ and n the interior
normal. But this integral has the upper bound $2\pi k$ for
all k', $0 < k' < k$, and since the integrand is not negative it
follows that the squares of the partial derivatives of $g(A|M)$
are summable over T when the neighborhood of the pole A
is excluded.

Now let ds_g be the element of arc of a curve $h = \text{const.}$
in the direction of increasing g, and let ds_h be similarly
defined for h. Then we may write

$$I_h = \int_0^1 ds_g$$

as the length of the portion of the curve $h = \text{const.}$, finite
or infinite, outside the contour $g = 1$. Now everywhere in T

$$0 \leq \frac{dg}{ds_g} \leq \left(\frac{\partial g}{\partial x}\right)^2 + \left(\frac{\partial g}{\partial y}\right)^2 + 1$$

so that the integral

$$\int_{\sigma'} \frac{dg}{ds_g} d\sigma$$

exists, if σ' is the portion of T outside $g = 1$. But we have

$$\int_{\sigma'} \frac{dg}{ds_g} d\sigma = \int_{(g<1)} \frac{dg}{ds_g} \frac{ds_g}{dg} \frac{ds_h}{dh} dg\, dh = \int_{(g<1)} \frac{ds_h}{dh} dg\, dh$$
$$= \int_{(g<1)} \frac{ds_g}{dg} dg\, dh,$$

and the integral may be written as an iterated integral

$$\int_0^{2\pi} dh \int_0^1 \frac{ds_g}{dg} dg = \int_0^{2\pi} I_h\, dh$$

from which follows the summability of I_h.* To say that I_h is summable however is to say that it is finite *almost everywhere*.

This completes the study of the order of the accessible points of C. The inaccessible points may however also be treated by blocking them off in subsets, yielding elements called *Primenden* by Carathéodory, in the memoir to which references have been given. He establishes the remarkable theorem that the *Primenden* (including accessible points among them) may be put into one to one correspondence with the values of h. But this complete order is not necessary for us.

32. Integrals on the boundary and the Dirichlet problems. We shall say that a function $f(P)$ is defined *almost everywhere on the frontier* (or boundary) *of T* if it is defined on accessible points of the frontier which correspond to *allmost all* values of h, and that it is *summable with respect to h* if when defined on the rest of the interval from 0 to 2π in h it becomes a summable function of h. We notice that if $f(P)$ is summable with respect to $h(A, P)$, A being a point of T it is also summable with respect to $h(B, P)$, B being any other point of T. In fact the mapping on S shows that $h(B, P)$ is a continuous function of $h(A, P)$ with continuous derivative, on the set of values of $h(A, P)$ which belong to accessible points of C, and that it can be defined so as to be continuous with its derivative for all values of $h(A, P)$.

If a function $F(P)$ is defined almost everywhere on C, and is of limited variation with respect to h on the set of values of h for which it is defined, it may by Theorem 1, Chap. I, be extended so as to be of limited variation for all values of h in the interval $(0, 2\pi)$. Moreover the property is

*De la Vallée Poussin, *Cours d'Analyse Infinitésimale*, vol. II (1912), p. 121.

independent of the pole A to which the function h refers. We shall therefore speak of $F(P)$ as being of *limited variation on C*. If P is an accessible point of C the quantities $F(P+0)$ and $F(P-0)$ (which refer to the order of points of C) both exist, and if $F(P) = \frac{1}{2}\{F(P+0)+F(P-0)\}$ the discontinuities may be regarded as regular. In fact the function may be extended, as we have seen, so as to have regular discontinuities as a function of h for all h. The function $F(P)$ need not be single valued on C (i. e., periodic as a function of h). We shall consider however merely functions $F(P)$ whose successive multiple values at a generic point of C differ by a single constant value (i. e., $F(h+2\pi) = F(h)+F(2\pi)-F(0)$).

We are now in a position to define the two integrals

(E) $$u(M) = \frac{1}{2\pi} \int_C \frac{dh(M, P)}{dh(A, P)} dF(P),$$

(F) $$u(M) = \frac{1}{2\pi} \int_C f(P) dh(M, P),$$

the integration being extended over values of h from 0 to 2π, A being a fixed and M a variable point of T, and P being a point of C. In particular, the integral (E) may, as is easily verified, be formed or defined directly in terms of a Riemann or Cauchy sum on the accessible points of C. Moreover if we have one class of functions on C which are summable with respect to h, we may define the integral (F), directly on C, for further classes of functions by the method of increasing and decreasing sequences of functions.* The integral may also be regarded from the point of view of the general integral of Daniell, which in effect amounts to the same thing. Possible primary classes of functions will be given shortly. The relation of the secondary functions, defined as limits, to the Dirichlet problem is stated in Theorem 6, below.

A conformal transformation which carries T into S and A into the center of S, transforms (E) into the Poisson-Stieltjes

* Cf. Evans, *Fundamental points of potential theory*, Rice Institute Pamphlets, (1920), p. 322 and 329. Wiener, Trans. Amer. Math. Soc., vol. 25 (1923), p. 307.

integral (A) of Ch. II, and (F) into the Poisson integral (B) of Ch. II, by the corresponding transfer of values, and the inverse transformation performs the reverse transfer; hence we have the following theorems on the generalization of the Dirichlet problem and the problem itself.

THEOREM 4. *A necessary and sufficient condition for the representation* (E) *is given by* (i), *or one of its equivalents. Equation* (2) *holds except for an additive constant for all accessible points of C except the denumerable infinity where $F(P)$ is discontinuous, and for these also if the discontinuities are made regular; moreover if $u(M)$ is of the class* (i) *it is uniquely determined by $F(P)$ and equation* (2).

THEOREM 5. *A necessary and sufficient condition for the representation* (F) *is given by* (ii). *The $u(M)$ takes on the boundary values $f(P)$ almost everywhere on C and is uniquely determined by them, if of the class* (ii). *The function $f(P)$ is arbitrary provided that it is summable with respect to $h(A|P)$, and the boundary value is taken on wherever $f(P)$ is the derivative of its integral with respect to $h(A|P)$ both of these characterizations being independent of the choice of the point A.*

Consider denumerable sequences of not negative functions, for which a limiting function exists. Since in this case, a necessary and sufficient condition for the limit of the integral to be the integral of the limit, is the uniform absolute continuity of the integral of the function of the sequence, we have at once the following theorem on the limits of Dirichlet problems:

THEOREM 6. *Let $f_m(P)$ be a denumerable sequence of functions defined almost everywhere on C and not negative. Let $f_m(P)$ have a limit $f(P)$ almost everywhere on C, and let $f_m(P)$ and $f(P)$ be summable with respect to h and such that $\lim_{m=\infty} \int f_m(P) \, dh = \int f(P) \, dh$. Then, if $u_m(M)$ and $u(M)$ are the harmonic functions of class* (ii) *determined respectively by the boundary values $f_m(P)$ and $f(P)$, we shall have*

(3) $\qquad \lim u_m(M) = u(M), \quad M \text{ in } T.$

COROLLARY. *Equation* (3) *applies if the* $f_m(P)$, *instead of being not negative, form an increasing sequence of summable functions with summable limit.*

In fact the integral of the limit is the limit of the integral, and since the functions $f_m(P) - f_1(P)$ form a not negative sequence the theorem on uniform absolute continuity may again be applied.

It may be opportune to emphasize that in this theorem we are not merely establishing a one-one correspondence between limits of functions on the boundary and limits of functions harmonic in T. Since the limit of the harmonic functions is of class (ii), being given by (F), it follows that it actually takes on almost everywhere the limit of the $f_m(P)$ and is determined by those boundary values — a fact which would not be true if we did not limit the $u_m(M)$ to the class (ii).

33. Special cases of the condition (ii). **The continuous boundary value problem.** A more usual statement of the Dirichlet problem is solved by the following theorem:

THEOREM 7. *Let* $f(M)$ *be continuous in a region* $\sum$ *which contains* T *and its boundary* C. *There is one and only one function bounded and harmonic in* T *which takes on the values* $f(P)$ *almost everywhere on* C.

In fact $f(P)$ is bounded on C. It is also integrable with respect to h; for almost everywhere on C it is the limit of a continuous function of h, $0 \leq h \leq 2\pi$, namely the limit of the function $f(M)$ formed on the analytic curve $g(A, M) = \text{const} = c$ as c approaches zero and M moves out along curves $h = \text{const}$.

If we demand merely that $f(M)$ be continuous in T (instead of $\sum$) and "semi-continuous" on C (that is, the value at an accessible point P of C is the limit of the values of $f(M)$ as M approaches P along every arc in T leading to P), and if $f(M)$ is bounded, then the conclusion of Theorem 7 still applies, for it is sufficient to consider merely points of C which are accessible along curves $h = \text{const}$. This is a more general type of continuous boundary value problem than the preceding one; in fact $f(P)$ may have different values on the various accessible points P that make up the whole or a part of a multiple point of C.

34. A new continuous boundary value problem.

These special problems which we have just been describing discuss the application of a harmonic function to a continuous function as a sort of plaster glued around the edges. An interesting class of continuous boundary value problems may be described directly in terms of the boundary points, without the use of any function $f(M)$ on neighboring points of T, and solved in terms of the methods already developed.

Let P_0 be an accessible boundary point, that is, a simple accessible point or a simple accessible part of a multiple point; let s be a circumference with center P_0 and s_{P_0} the arc-cross-cut of it which divides T into two simply connected parts, of which one contains P_0. Let σ_{P_0} be this part of T. Then we say that $f(P)$, given almost everywhere on C, is continuous at P_0 if given ε we can take the radius r of s sufficiently small so that

$$|f(P)-f(P_0)|<\varepsilon,$$

provided P is any accessible boundary point of T in σ_{P_0}.

We say that $f(P)$ is continuous on the accessible points of the boundary of T if it is continuous at every accessible point of the boundary.

LEMMA. A function $f(P)$ continuous at the accessible points of the boundary of T is measurable with respect to $h(A, P)$.

Osgood has shown that to the ends of s_P, namely P_1 and P_2 correspond two values of $h(A, P)$, say h_1 and h_2, which are unequal and unequal to h_P, such that

$$|h_1-h_P|<\eta$$
$$|h_2-h_P|<\eta$$
(η arbitrary),

provided we approach the boundary of T along the two directions of s_P, if the radius of the circle s is taken small enough, equal say to r'. All the intermediate values of h,

$h_1 \leq h \leq h_2$, which correspond to curves $h = $ const. which lead to points of C, are such that those curves cut $s_P = s'_P$, and ultimately remain within σ'_P, or reach the points P_1 and P_2, since no two curves $h = $ const. may intersect at an interior point of T. If then, we take any value of $r > r'$ all the above values of h give curves that lead to points of σ_P which are boundary points of T.

In other words, if θ_0 is the value on the unit circle which corresponds to P_0 by the conformal transformation, we can, given ε, find δ (equal to the smaller of two values $|h_1 - h_{P_0}|$, $|h_2 - h_{P_0}|$) such that if $|\theta - \theta_0| < \delta$ we shall have $|f(P) - f(P_0)| < \varepsilon$ for all the corresponding boundary points of T which are accessible along curves $h = $ const. In fact these curves $h = $ const. lead to boundary points of σ_{P_0} which are boundary points of T.

We return now to the lemma to be proved. It suffices to show that the set of values of θ corresponding to accessible points of C for which $f(\theta) > b$, when b is given arbitrarily, is measurable; we denote this set by E_b. Let F be the set of values of θ, for which the corresponding h curve does not lead to an accessible point; F is of measure 0. Now if θ_1 is a value of θ for which $f > b$ there will be, as we have just shown, an open interval ω_1, with center at θ_1, such that

$$CF \cdot \omega_1 < E_b.$$

Hence the set E_b is identical except for a set of measure 0, with the set

$$\sum (CF) \cdot \omega_1 = CF \cdot \sum \omega_1$$

where the intervals ω_1 are formed for all points of E_b. But the sum of an infinity of open sets, denumerable or not, is an open set. Accordingly $\sum \omega_1$ is an open set and therefore measurable. Also CF is measurable, since F is measurable. It follows then that E_b which is, except for a set of zero measure, the product of two measurable sets, is measurable.

From Theorem 5 we have now at once the following one.

THEOREM 8. *If $f(P)$ is continuous on the accessible points of C, and bounded, there is one and only one function $u(M)$,* harmonic in T, and bounded, which takes on almost everywhere on C the given values $f(P)$. The function $u(M)$ takes on the value $f(P)$ at every accessible point.

In fact, $f(P)$ being measurable with respect to h, and bounded, is summable with respect to h. This proves the first part of the theorem.

In order to prove the second part it is sufficient to show that at any accessible point P_0, $f(P_0)$ is the derivative of its integral with respect to h. Let $\overline{f}(\theta)$ be the corresponding function on the circumference of S, and θ_0 the value which corresponds to P_0. For the moment we define $\overline{f}(\theta) = f(P_0)$ for all the points of the set F of the lemma, which does not affect the value of the integral with respect to h or θ, since F is of measure 0. Hence by taking δ small enough there will be an ε such that

$$f(P_0) - \varepsilon < \overline{f}(\theta) < f(P_0) + \varepsilon, \qquad |\theta - \theta_0| < \delta,$$

and for any set e in this neighborhood it will follow that

$$(f(P_0) - \varepsilon) \cdot \operatorname{meas} e < \int_e \overline{f}(\theta) \, d\theta$$
$$= \int f(P) \, dh < (f(P_0) + \varepsilon) \cdot \operatorname{meas} e.$$

Hence the derivative of the integral of $f(P)$, with respect to h, at P_0 lies between $f(P_0) - \varepsilon$ and $f(P_0) + \varepsilon$. But ε is arbitrarily small.

EXERCISE. Consider the extension of this theorem when the transformation is conformal at P on the boundary. Hence state a theorem which applies to a region bounded by a finite number of branches of analytic curves.

35. The generalized Neumann problem in the general region. For problems which refer to boundary values of the derivative we make use of the invariants

$$\int_{M_1}^{M_2} \frac{\partial u}{\partial n} ds = -\int_{h_1}^{h_2} \frac{\partial u}{\partial g} dh,$$
$$\left| \int_{M_1}^{M_2} \left| \frac{\partial u}{\partial n} \right| ds \right| = \left| \int_{h_1}^{h_2} \left| \frac{\partial u}{\partial g} \right| dh \right|,$$

where the integration is extented along curves $g = $ const., where n denotes the interior normal and ds is (for the purpose of this section) an algebraic magnitude counted positive in the direction opposite to increasing h.

We shall say that $u(M)$ harmonic in T, belongs to the class (j) if its total absolute flux on the level curves $g(A|M) = $ const. is bounded:
$$\int_{g=\text{const}} \left| \frac{\partial u(M)}{\partial n} \right| ds < N.$$

We shall say that the flux takes on the boundary values given by $F(P)$, where $F(P)$ is periodic, if
$$\lim_{g=0} \int_{M_1}^{M_2} \frac{\partial u}{\partial n} ds = F(P_2) - F(P_1)$$

where the integration is extended along curves $g = $ const., and M_1 and M_2 approach C respectively along curves $h = $ const. which belong to accessible points P_1 and P_2. We need not, as a matter of fact, restrict the path of integration in this integral to curves $g = $ const., since the value of the integral is the same over any other rectifiable path in T joining M_1 and M_2. We must however take $F(P)$ as single valued, since the integral of $\partial u/\partial n$ over a closed curve in T is 0.

The conformal transformation yields at once the following theorem.

THEOREM 9. *Given $F(P)$, single valued and of limited variation on C, there is one and only one function of class (j), harmonic in T, whose flux takes on the boundary values given by $F(P)$ except at the points of discontinuity of $F(P)$.*

* For a detailed study of a more special case of the discontinuous boundary value problems of the first and second kinds, see Lichtenstein, Journal für Math. vols. 141, 142, and 143 (1912–14). He treats also the extension to partial differential equations of elliptic type.

CHAPTER VI

PLANE REGIONS OF FINITE CONNECTIVITY

36. Functions harmonic outside a circle. Consider a circle of radius a and center O, and let $u(r, \theta)$ be harmonic, $r > a$. We say that $u(r, \theta)$ is regular at ∞ if $u(r, \theta)$ is bounded outside a circle of radius R, $R > a$.

If we make a transformation of the plane by inversion, say in a circle of radius 1 and center O, the transferred values of the given function $u(r, \theta)$, regular at ∞, yield a function $U(r', \theta)$ which is harmonic inside the circle of center O and radius $a' = 1/a$, except possibly at O, in the neighborhood of which point the function remains bounded. It is not difficult to show that the function either is harmonic at O also, or else merely has an unnecessary discontinuity, so that it becomes harmonic at O by a proper definition at that point. The fact becomes evident from the following theorem due to Lebesgue.*

LEMMA. A THEOREM OF LEBESGUE. Let $v(r, \theta)$ be harmonic and bounded in the neighborhood of a point O Then it is also harmonic at O or else has an unnecessary discontinuity at that point.

In order to prove this lemma, draw a circle with center O and radius b, small enough to lie entirely within the given neighborhood, and draw a smaller concentric circle of radius ϱ. Let $v_1(r, \theta)$ be the function which is harmonic within the first circle, bounded, and reduces to $v(b, \theta)$ when $r = b$; this function is in fact given by Poisson's integral. The difference, $V(r, \theta) = v(r, \theta) - v_1(r, \theta)$, vanishes as r approaches b and is bounded, say

$$M > V(r, \theta) > m.$$

Within the region between the circles ϱ and b, the function $V(r, \theta)$, being harmonic, satisfies the inequality

$$M \frac{\log b/r}{\log b/\varrho} \geqq V(r, \theta) \geqq m \frac{\log b/r}{\log b/\varrho}.$$

* Comptes Rendus t. 176 (1923), p. 1097.

In fact, otherwise, the difference between $V(r, \theta)$ and one of the other two members of the inequality would have a maximum or a minimum at an interior point of the region.

But now if we hold r fixed in this region, and let ϱ approach zero, both extreme members of the inequality become arbitrarily small. Hence $V(r, \theta) = 0$. Hence $v(r, \theta) = v_1(r, \theta)$, $0 < r < b$, and $v(r, \theta) = v_1(r, \theta)$ at the origin or else can be made harmonic within the circle by defining it that way at the origin.

The function $U(r', \theta)$, to which we return, will consequently be made harmonic within a circle of radius a', and will therefore be given in terms of its values on the circumference of any circle of radius $R' < a'$ by Poisson's integral

$$U(r', \theta) = \frac{1}{2\pi} \int_0^{2\pi} \frac{(R'^2 - r'^2) U(R', \varphi)}{R'^2 + r'^2 - 2R'r' \cos(\varphi - \theta)} d\varphi.$$

If therefore we return to the original function $u(r, \theta)$ by inverting back again, we find

$$(1) \quad u(r, \theta) = \frac{1}{2\pi} \int_0^{2\pi} \frac{(r^2 - R^2) u(R, \varphi)}{r^2 + R^2 - 2Rr \cos(\varphi - \theta)} d\varphi$$

where $R = 1/R' > a$. Thus we have

THEOREM 1. *If $u(r, \theta)$ is harmonic when $r > a$, and regular at ∞, it satisfies the identity* (1) *for $r > R > a$.*

We have also the following results of which the proofs are evident.

COROLLARY. Under the hypotheses of Theorem 1, we have

$$(2) \quad \begin{cases} \lim_{r=\infty} u(r, \theta) = \frac{1}{2\pi} \int_0^{2\pi} u(R, \varphi) d\varphi, \\ \lim_{r=\infty} r \frac{\partial u}{\partial \theta} = \frac{R}{\pi} \int_0^{2\pi} u(R, \varphi) \sin(\varphi - \theta) d\varphi, \\ \lim_{r=\infty} r^2 \frac{\partial u}{\partial r} = -\frac{R}{\pi} \int_0^{2\pi} u(R, \varphi) \cos(\varphi - \theta) d\varphi, \end{cases}$$

all uniformly in θ.

It follows that $\lim_{r=\infty} r \frac{\partial u}{\partial x} = 0$, no matter what the direction of x, and uniformly; a property which is included usually in the definition of regularity.

An inversion in the unit circle also gives us the following theorems.

THEOREM 2. *A necessary and sufficient condition that $u(r, \theta)$ be given by a Poisson-Stieltjes integral*

(A') $\quad u(r, \theta) = \frac{1}{2\pi} \int_0^{2\pi} \frac{(r^2-a^2)\,dF(\varphi)}{r^2+a^2-2ar\cos(\varphi-\theta)}, \quad r > a,$

where $F(\varphi)$ is of limited variation, with $F(2\pi+\varphi) = F(\varphi) + F(2\pi) - F(0)$, is that $u(r, \theta)$ be harmonic for $r > a$, regular at ∞, and satisfy what corresponds to the condition (i) *for the exterior region*:

(i') $\int_0^{2\pi} |u(r, \theta)|\,d\theta$ *is bounded in an exterior neighborhood of $r = a$.*

If in this theorem we replace (A') *by the Poisson integral*

(B') $\quad u(r, \theta) = \frac{1}{2\pi} \int_0^{2\pi} \frac{(r^2-a^2)f(\varphi)\,d\varphi}{r^2+a^2-2ar\cos(\varphi-\theta)}, \quad r > a,$

with $f(\varphi)$ summable, we replace (i') *by*

(ii') $\int_0^\theta u(r, \theta)\,d\theta$ *has uniform absolute continuity in an exterior neighborhood of $r = a$.*

A necessary and sufficient condition for (A'), as in Theorem 2, is that $u(r, \theta)$ be the difference of two not negative functions, harmonic outside the circle and regular at infinity.

If $u(r, \theta)$ is given by (A') it takes on the boundary values $F'(\varphi)$ almost everywhere, and if given by (B') the boundary values $f(\varphi)$ almost everywhere, provided $M = (r, \theta)$ approaches $P = (a, \varphi)$ in the wide sense, that is to say, in any way so that it remains in the angle formed by two rays directed outward from P, neither ray being tangent to the circle. The other properties of the integrals discussed in Chap. II have their obvious analogs

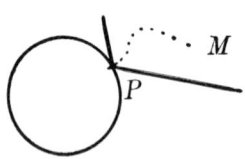

for (A′) and (B′) and it is unnecessary to state them in detail.

EXERCISE 1. Obtain values for u_∞, $\left[r\dfrac{\partial u}{\partial \theta}\right]_\infty$ and $\left[r^2\dfrac{\partial u}{\partial r}\right]_\infty$ when $u(r, \theta)$ is given by (A′).

EXERCISE 2. Show that a necessary and sufficient condition that
$$v(r, \theta) = \frac{a}{2\pi}\int_0^{2\pi} \log(a^2+r^2-2ar\cos(\varphi-\theta))dF(\varphi)+A, \quad r>a,$$
where A is a constant, and $F(\varphi)$ is of limited variation, periodic and with regular discontinuities, is that $v(r, \theta)$ be harmonic outside the circle of radius a and regular at ∞, and that the total absolute flux
$$\int_0^{2\pi} \left|\frac{\partial v}{\partial r}\right| d\theta$$
be bounded for the exterior neighborhood of $r = a$.

Show that $\lim\limits_{r=\infty} v(r, \theta) = A$ uniformly, that $\lim\limits_{r=a} \int_0^\theta \dfrac{\partial v}{\partial r} d\theta = F(\theta) - F(0)$, and that $\lim\limits_{r=a} \partial v/\partial r = F'(\varphi)$ almost everywhere, in the wide sense.

EXERCISE 3. Describe the special case when $F(\varphi)$ is absolutely continuous.

37. The multiply connected region bounded by $n+1$ distinct circles. Let S be an open region of finite connectivity whose boundary consists of non-intersecting circles, an external one s_0 and n internal ones $s_1, s_2, \cdots, s_n$. Let M be a point of S with polar coördinates r_i, θ_i referred to the center of the circle s_i of radius a_i, and let $u(M)$ be a function harmonic in S. Such a function may be written as the sum of $n+1$ functions $u_i(M)$,

(3) $$u(M) = u_0(M) + \sum_1^n u_i(M),$$

one of them, namely $u_0(M)$, being harmonic in the open region inside s_0, and each of the others being given by a formula

(3′) $u_i = w_i + m_i \log r_i,$ $i = 1, 2, \cdots, n,$

where w_i is harmonic outside s_i, is regular at ∞ and vanishes there.* We wish to find necessary and sufficient conditions on $u(M)$ in order that $u_0, w_1, \cdots, w_n$ may be expressed by Poisson-Stieltjes integrals of the form (A) and (A′). But a function $m_i \log r_i$ is continuous on and in the neighborhood of any contour s_j, and a function u_i is continuous on and in the neighborhood of any contour s_j, $i \neq j$, so that we have at once the theorem:

THEOREM 3. *A necessary and sufficient condition that $u(M)$ be given by (3), (3′) where the u_0 is given by* (A)

(A) $u_0(r_0, \theta_0) = \dfrac{1}{2\pi} \displaystyle\int_0^{2\pi} \dfrac{(a_0^2 - r_0^2)\, dF_0(\varphi)}{a_0^2 + r_0^2 - 2 a_0 r_0 \cos(\varphi - \theta_0)},$

with $F_0(\varphi)$ of limited variation and $F_0(\varphi + 2\pi) = F_0(\varphi) + F_0(2\pi) - F_0(0)$, and the w_i are given by (A′)

(A′) $w_i(r_i, \theta_i) = \dfrac{1}{2\pi} \displaystyle\int_0^{2\pi} \dfrac{(r_i^2 - a_i^2)\, dF_i(\varphi)}{a_i^2 + r_i^2 - 2 a_i r_i \cos(\varphi - \theta_i)}$

with $F_i(\varphi)$ of limited variation and periodic, with period 2π, is that $u(M)$ be harmonic in the open region S and

(4) $\displaystyle\int_0^{2\pi} |u(M)|\, d\theta_i < K,$ *K constant,*

when the integration is extended over all circles in S concentric with a boundary circle and in its neighborhood.

COROLLARY 1. A necessary and sufficient condition that u_0 or a particular $w_i(M)$ should reduce to the corresponding integral (B) or (B′) is that, in addition, the absolute continuity of $\displaystyle\int_0^\theta u(r_i, \theta_i)\, d\theta_i$ be uniform, when the integration is extended over all circles in S concentric with the particular s_i and in its neighborhood.

EXERCISE. Show that a necessary and sufficient condition that u_0 or a particular w_i should be given by the corresponding integral (C) or (C′) is that in addition to the

* Osgood, *Funktionentheorie*, Leipzig (1912), p. 643.

hypothesis of Theorem 3 the integral $\int_0^{2\pi} \left|\frac{\partial u}{\partial r_i}\right| d\theta_i$ should be bounded for all circles in S concentric with the particular boundary circle s_i and in its neighborhood. This condition of course implies the condition (ii) or (ii') for the particular circle.

COROLLARY 2. If $u(r, \theta)$ is harmonic in S and is either bounded or such that $\left\{\left(\frac{\partial u}{\partial x}\right)^2 + \left(\frac{\partial u}{\partial y}\right)^2\right\}$ is summable over S, then all the boundary integrals reduce to Poisson integrals.

We shall now prove the following theorem, of which the importance from the physical point of view hardly needs emphasis.

THEOREM 4. *A necessary and sufficient condition for the situation of Theorem 3 is that $u(M)$ be the difference of two not negative functions, each harmonic in the open region S.*

The condition is obviously necessary, since the $F_i(\varphi)$ are of limited variation. In order to prove the sufficiency we establish first two simple lemmas.

LEMMA 1. For the purpose of this lemma let $\overline{S}$ be a simply connected open region, bounded internally or externally by a circle $\overline{s}$ of radius $\overline{a}$, and let $u(M)$ be the difference of two not negative harmonic functions $u'(M)$, $u''(M)$ in an annular neighborhood in $\overline{S}$ of the boundary $\overline{s}$. Then $\int |u| d\theta$ is bounded for the circles of that neighborhood concentric with $\overline{s}$.

For convenience let $\overline{S}$ be an interior region. Let s' of radius r' be a fixed circle in the annular neighborhood and let there be a circle of radius r, $r' \leq r < a$, all these circles having the same center O. The lemma follows immediately from Bôcher's device, evaluating the area integral of $(1/r)\,\partial u/\partial r$. In fact by integrating this expression in u' over an annular region, and equating its two evaluations as an iterated integral, we have the equation

(5) $\quad \left(\int_0^{2\pi} u'\, d\theta\right)_r = \left(\int_0^{2\pi} u'\, d\theta\right)_{r'} + a_1 \log r - a_1 \log r'$

where α_1 is a constant. Hence

$$\int_0^{2\pi} u' d\theta = \int_0^{2\pi} |u'| d\theta$$

is bounded. Similarly $\int |u''| d\theta$ is bounded, and therefore $\int |u| d\theta$ is bounded.

LEMMA 2. In the open region S let $u(M)$ be the difference of two functions $u'(M)$, $u''(M)$, harmonic and not negative in S. Then $u(M)$ is the sum of $n+1$ functions $u_i(M)$, each harmonic except for possible logarithmic singularities at ∞, in the simply connected open region S_i bounded by s_i, and each satisfying in its region S_i the conditions of Lemma 1.

The first part of the lemma is the theorem already expressed equations (3), (3'). The last part follows immediately by writing, in the neighborhood of s_i, $u_i(M)$ as the difference $u(M) - \bar{u}(M)$ where

$$\bar{u}(M) = \sum_j' u_j(M),$$

the summation not including $j = i$.

It can now be shown that the condition of Theorem 4 is sufficient. In fact, this condition is the hypothesis of Lemma 2, and therefore by Lemma 1 the quantity $\int |u_i| d\theta_i$ is bounded for circles concentric with s_i and in its neighborhood. The same is true for $\int |u_j| d\theta_i$, $j \neq i$, since $u_j(M)$ is harmonic on s_i itself if j is unequal to i. But

$$\int_0^{2\pi} |u| d\theta_i \leq \sum_{j=0}^n \int_0^{2\pi} |u_j| d\theta_i.$$

Hence the condition (4) of Theorem 3 is satisfied.

38. Representation in terms of the Green's function. The class of functions so far considered may be given a representation directly in terms of the boundary values, rather than in terms of the functions $F_i(\theta)$. But for this we need one or two preliminary facts about the Green's function for S.*

LEMMA 1. Let n denote the interior normal at the boundary of S. If $g(Q, M)$ is the Green's function for S with pole at Q, there will be two positive numbers l, L, with $0 < l < L$, such that the relation

$$(6) \qquad l < \frac{\partial g}{\partial n} < L,$$

is satisfied for every point P of the boundary; i. e., of $s_0, s_1, \cdots, s_n$.

The fact is established immediately for any particular boundary s_i (whence the general statement follows) by drawing a circle s' concentric with s_i and lying between it and Q or any other boundary circle s_j. On s', $g(Q, M)$ neither vanishes nor becomes infinite. The function $g(Q, M)$ will then be contained between two particular harmonic functions $\underline{u}(M)$ and $\overline{u}(M)$ both of which vanish on s_i and are given constant values on s', so that they are both linear functions of $\log r_i$,

$$\underline{u}(M) = k \frac{\log r_i - \log a_i}{\log r' - \log a_i}, \qquad \overline{u}(M) = K \frac{\log r_i - \log a_i}{\log r' - \log a_i},$$

where k is > 0 but $< g(Q, M)$ on s', and K is finite but $> g(Q, M)$ on s'. But the normal derivative of any function which vanishes on s_i is merely the limit of its value on s' divided by $r_i - a_i$ as r_i tends to a_i; hence direct calculation of the derivatives of the three functions $\underline{u}$, $\overline{u}$ and g establishes the result.

Consider now one of the internal boundaries s_i. From what we have just proved, for a sufficiently small value of x the completé curve c' given by $g(Q, M) = x$ contains, as a part, a separate closed branch c'_i which encloses a single s_i in its interior.. Let us consider in particular this interior boundary and the region S' between s_i and c'_i. Analogous considerations will apply of course to s_0.

* For the existence of the Green's function for S see e. g., Lichtenstein, *Neuere Entwicklung der Potentialtheorie*, Encyklopädie der mathematischen Wissenschaften, Band II, Teil 3, Leipzig (1921) § 17, § 20. See also Chap. VII below.

The points on s_i and on c'_i correspond to the same values of h, say the range of values $h' \leq h < h''$, where $h = h(Q, M)$ is the (non-uniform) harmonic function conjugate to $g(Q, M)$. If we map conformally the region S' on an annular region S'' by the transformation

$$w = e^{-(g+hi)\frac{2\pi}{h''-h'}},$$

the curves $g = $ const. become concentric circles in S'' between $r = e^{-x}$, corresponding to c'_i, and $r = 1$, corresponding to s_i, the map being conformal within and on the boundaries.

Assume now that in S' the integral $\int_{h'}^{h''} |u(M)|\, dh(Q, M)$ extended along a curve $g(Q, M) = y$ is bounded as y approaches zero. That implies that in S'' the integral $\int_0^{2\pi} |u|\, d\varphi$ is bounded $e^{-x} \leq r < 1$, and hence, as a property of the integral formulae of Theorem 3, that $\lim_{r=1} \int_0^\theta u(r, \theta)\, d\theta$ exists for all θ. In S' it follows therefore that the integral $\int_{h_1}^{h_2} u(M)\, dh(Q, M)$, extended along $g = y$, has a limit $U(h_1, h_2)$ as y approaches zero for all h_1, h_2 in the given interval $h' \leq h_1 \leq h_2 \leq h''$.

For the purpose of discussing the function U it is perhaps simpler to avoid using the Riemann surface for the non-uniform function h, and make it single valued, by means of cuts, from 0 to 2π. It may be then that for a given boundary circle the corresponding values of h will not form a complete interval, but rather a finite number of distinct intervals $(h'_1, h''_1), \cdots, (h'_p, h''_p)$. We shall speak of the end points of these intervals as cut-points of the boundary. In this case we may use the mapping function

$$w = e^{-2\pi \frac{g+i\left\{h - h'_i + \sum_1^{i-1}(h''_r - h'_r)\right\}}{\sum_1^p (h''_r - h'_r)}}, \qquad h'_i \leq h < h''_i.$$

By thus defining $U(h_1, h_2)$ on successive portions of the complete boundary of S, we see that if $\int_0^{2\pi} |u|\, dh$ is bounded the quantity

(6) $$\lim_{y=0} \int_H^{h} u(M)\,dh(Q,M) = U(P)$$

exists, and may be used to define a function $U(P)$ on the complete boundary which is single valued as a function of h, $0 \leq h \leq 2\pi$; it is also single valued as a function of position P on the boundary except for the one point $P_0 = P_{2\pi}$. It is a function of limited variation as a function of h or of any continuous monotonic function of h, such as the arc length, properly defined, along the complete boundary. Moreover its continuities are regular except at the cut-points h_i', h_i'', and the direction of increase of h along the boundary is, except at the cut-points, the same as for the h-functions of the corresponding simply connected regions S_i; the latter point is an immediate consequence of Lemma 1. In fact, as follows from Lemma 1, the branch points of the analytic function $g + hi$ cannot lie on the boundary of S.

The derivatives of h, considered as a function on the Riemann surface for $g + hi$, with respect to r_i and θ_i are single valued and continuous on and in the neighborhood of the boundary of S; for the analytic function $g + hi$, whose various branches differ merely by constant values, may be extended analytically across any boundary circle.* Moreover the quantity

$$\frac{\partial h}{\partial \theta_i} = -\frac{\partial g}{\partial n}$$

does not vanish on the boundary. We wish to consider the quantity

$$\frac{dh(y|A,M)}{dh(y|Q,M)} = \frac{\partial h(y|A,M)}{\partial s_M} \bigg/ \frac{\partial h(y|Q,M)}{\partial s_M},$$

which represents the rate of change of the h-function with pole A for the curve $g(Q,M) = y$ with respect to the h-function with pole Q for the same curve, along that curve. In particular, $g(y|Q,M) = g(0|Q,M) - y$ and $h(y|Q,M) = h(0|Q,M)$, whence the denominator fraction of the right hand member of

* Osgood, *Funktionentheorie*, Leipzig (1912), p. 672.

96   CHAPTER VI

the above equation is continuous and different from zero, as y approaches zero and M approaches a point P of the boundary of S.

Consider now the numerator. We can take $y > 0$ near enough to 0 so that the function $g(y'|A, M)$ for $y' \leq y$ may be extended harmonically over a region which contains in its interior the whole region defined by $g(Q, M) = -Y$, where Y is some positive number, sufficiently small. For this purpose consider the region S'', which we have already used, associated say with s_0. If x_0 is sufficiently small the image of the point A will not lie in S''. Since the curves $g(Q, M) = y$ are transformed into the circles $r = e^{-y}$ and the transform $\overline{g}(M)$ of the function $g(y|A, M)$ vanishes when M is on the curve of index y, the function $\overline{g}(M)$ may be extended harmonically through a region which includes at least the circle inverse to $r = e^{-x_0}$ in the circle $r = e^{-y}$; that is, the circle of radius e^{x_0-2y}.* Hence if $Y_0 < x_0$, we may choose $y = y_0$ sufficiently small so that $x - 2y_0 > Y_0$; moreover the transform of the function $g(Q, M)$ is itself harmonically extensible through a region which contains the same circle of radius e^{Y_0}. In other words, if $y' \leq y_0$ the region of harmonic extension of $g(y'|A, M)$ and $g(Q, M)$ in the neighborhood of s_0 includes the points for which $g(Q, M) \geq -Y_0$.

If we proceed similarly with the other portions of the boundary $s_1, \ldots, s_n$, and define similarly other pairs of numbers (y_i, Y_i), we may choose y as the smallest of the numbers $y_0, \ldots, y_n$ and Y as the smallest of the numbers $Y_0, \ldots, Y_n$. But $g(Q, M) = 0$ consists of $n+1$ circles; and therefore $g(Q, M) \geq -Y$ defines a region which includes all these circles in its interior, for $g(Q, M)$ is harmonically extensible across these circles. In particular Y may be taken near enough to zero so that the curve $g(Q, M) = -Y$ consists of $n+1$ analytic closed curves.

The function $g(y'|A, M) - g(y''|A, M)$ is harmonic at A, in particular, and therefore an application of Green's theorem to the region bounded by $g(Q, M) = -Y$ yields the equation

------
* Osgood, *loc. cit.*

$$2\pi \{g(y'|A, M) - g(y''|A, M)\}$$
$$= \int_{(-Y)} \{g(y'|A, P') - g(y''|A, P')\} \frac{\partial g(-Y|P', M)}{\partial n_{P'}} ds_{P'},$$

and, by differentiation, the equation

$$2\pi \left\{ \frac{\partial g(y'|A, M)}{\partial \mu_M} - \frac{\partial g(y''|A, M)}{\partial \mu_M} \right\}$$
$$= \int_{(-Y)} \{g(y'|A, P') - g(y''|A, P')\} \frac{\partial^2 g(-Y|P', M)}{\partial n_{P'} \partial \mu_M} ds_{P'},$$

where μ_M is an arbitrary fixed direction at M, M is excluded from a given arbitrary neighborhood of A but is otherwiese arbitrary in S or on its boundary, and Y satisfies the conditions of the previous paragraph. But $g(Y|P', M)$ and all its derivatives are bounded, since P' is on $g(Q, P') = -Y$, and M is in S or on its boundary; moreover $g(y'|A, P') - g(y''|A, P')$ is of one sign. Hence we may write

$$2\pi \left| \frac{\partial g(y'|A, M)}{\partial \mu_M} - \frac{\partial g(y''|A, M)}{\partial \mu_M} \right|$$
$$\leq K \int_{(-Y)} |g(y'|A, P') - g(y''|A, P')| ds_{P'}.$$

The integrand of the right hand member is bounded for all y', y'' which are $\leq y$ and ≥ 0, and it approaches 0 as $y' - y''$ tends to 0. Hence

$$\lim_{y''=y'} \frac{\partial g(y''|A, M)}{\partial \mu_M} = \frac{\partial g(y'|A, M)}{\partial \mu_M},$$

and uniformly, in M and in μ_M.

Let now M' be a second point of the same sort as M and $\nu_{M'}$ an arbitrary direction at M'. We have, holding y' fast,

$$\lim_{\substack{M=M' \\ \mu_M=\nu_{M'}}} \frac{\partial g(y'|A, M)}{\partial \mu_M} = \frac{\partial g(y'|A, M')}{\partial \nu_M},$$

where the limit is uniform for all M' and all $\nu_{M'}$. Hence

$$\lim_{\substack{y''=y'\\M=M'\\\mu_M=\nu_{M'}}} \frac{\partial g(y''|A, M)}{\partial \mu_M} = \frac{\partial g(y'|A, M')}{\partial \nu_{M'}},$$

uniformly in M' and $\nu_{M'}$.

In particular

$$\lim_{\substack{y''=0\\M=P}} \frac{\partial g(y''|A, M)}{\partial n_M} = \frac{\partial g(0|A, P)}{\partial n_P},$$

uniformly in P, where M is on $g(Q, M) = y''$ and P is on $g(Q, P) = 0$, and therefore

(7) $$\lim_{\substack{y''=0\\M=P}} \frac{\partial h(y''|A, M)}{\partial s_M} = \frac{\partial h(0|A, P)}{\partial s_P},$$

uniformly in P. We have, therefore, returning to the ratio

$$\frac{dh(0|A, P)}{dh(0|Q, P)} = \frac{dh(A, P)}{dh(Q, P)},$$

the following proposition.

LEMMA 2. Let Q and A be two points of S and let $h(y|A, M)$ be the h-function with pole A for the curve $g(Q, M) = y > 0$. If the $h(y|A, M)$ and $h(y|Q, M) = h(0|Q, M) = h(Q, M)$ are considered on the Riemann surfaces of their respective functions $g + hi$, the quantity

$$\frac{dh(y|A, M)}{dh(y|Q, M)} = \frac{dh(y|A, M)}{dh(Q, M)},$$

which represents the rate of change along the curve $g(Q, M) = y$, is nevertheless a single-valued, continuous function of M in a neighborhood of the boundary of S. Moreover

$$\lim_{\substack{y=0\\M=P}} \frac{dh(y|A, M)}{dh(Q, M)} = \frac{dh(A, P)}{dh(Q, P)}$$

uniformly in P, where P is on the boundary of S, and this latter expression is positive for all P.

We are now in position to prove a fundamental theorem.

THEOREM 5. *Let*

$$I(y, h) = \int_0^h u(M)\,dh(Q, M), \quad T(y, h) = \int_0^h |u(M)|\,dh(Q, M)$$

represent integrals along the curve $g(Q, M) = y$ *with respect to* $h(Q, M)$ *made single-valued by means of cuts. A necessary and sufficient condition that* $u(r, \theta)$ *be the difference of two not negative functions, each harmonic in S is that* $u(r, \theta)$ *be harmonic in S and that* $T(y, 2\pi)$ *remain bounded as y approaches 0; that is, that* $I(y, h)$ *be of uniformly limited variation as a function of h, as y approaches 0.*

An equivalent theorem may easily be stated in terms of the Riemann surface for $g+hi$, but the present form is perhaps more closely connected with the physical concept of the curves $h = $ const. as lines of flow.

Suppose then that u is the difference $u' - u''$ of two not negative functions, harmonic in S. We have for u' the equation*

$$2\pi u'(Q) = \int_0^{2\pi} u'(M)\,dh(Q, M) = \int_0^{2\pi} |u'(M)|\,dh(Q, M),$$

when the integration is extended over a complete curve $g(Q, M) = y$, with a similar equation for u''. Hence

$$\int_0^{2\pi} |u(M)|\,dh(Q, M) \leq 2\pi\{u'(Q) + u''(Q)\},$$

and the condition is satisfied.

Suppose now that the condition is satisfied. We shall show first that for any point A of S the function $u(A)$ is given by a formula

(8) $$u(A) = \frac{1}{2\pi} \int_0^{2\pi} \frac{dh(A, P)}{dh(Q, P)}\,dU(P),$$

where $U(P)$ is a function of limited variation on the boundary, $0 \leq h(Q, P) \leq 2\pi$.

* This is merely the familiar formula

$$2\pi u'(Q) = \int_{[g=y]} u'(M)\frac{\partial g(Q, M)}{\partial n}\,ds.$$

In order to verify this equation we notice that the familiar equation

$$u(A) = \frac{1}{2\pi} \int_{(g=y)} u(M) \frac{\partial g(y \mid A, M)}{\partial n_M} ds_M$$
$$= \frac{1}{2\pi} \int_{(g=y)} u(M) \frac{\partial h(y \mid A, M)}{\partial s_M} ds_M$$

may, by means of (10), Chap. I, be written in the form

(9) $$u(A) = \frac{1}{2\pi} \int_{(g=y)} \frac{dh(y \mid A, M)}{dh(Q, M)} d\int_0^{h_M} u(M') dh(Q, M')$$
$$= \frac{1}{2\pi} \int_{(g=y)} \frac{dh(y \mid A, M)}{dh(Q, M)} dI(y, h).$$

By hypothesis $I(y, h)$ is of limited variation uniformly in y, and, as we have shown, has a limit $U(P)$ for every value of h, $U(P)$ being a function of limited variation. Moreover, by Lemma 2, the continuous function $dh(y \mid A, M)/dh(Q, M)$ approaches $dh(A, P)/dh(Q, P)$ uniformly, as a function of $h(Q, P)$, $0 \leq h(Q, P) \leq 2\pi$. Hence we may apply the Helly-Bray theorem [Theorem 3, Chap. I], which supplies at once the desired equation (8).

But by Lemma 2, $dh(A, P)/dh(Q, P) > 0$, and $U(P)$ being of limited variation, as a function of $h(Q, P)$, is the difference of two non-decreasing functions. Hence $U(A)$ is the difference of the two corresponding not negative functions, harmonic in S. Thus the theorem is proved.

COROLLARY 1. A necessary and sufficient condition that $u(A)$ be given by (8), where $U(P)$ is of limited variation, $0 \leq h(Q, P) \leq 2\pi$, and Q is an arbitrary fixed point of S is that $u(A)$ belong to the class of functions considered in Theorems 3, 4 and 5. This equation furnishes the solution of the generalized Dirichlet problem for S.

COROLLARY 2. If the condition that $T(y, h)$ be bounded holds with respect to a given pole Q, it holds for any other pole Q'.

EXERCISE 1. State Theorem 5 in terms of the Riemann surfaces, and describe the discontinuities of the limiting function $U(P)$ on the surface belonging with $h(Q, P)$.

FINITELY CONNECTED PLANE REGIONS 101

EXERCISE 2. Obtain a formula for the boundary values of $U(A)$ as A approaches P along a curve $h(Q, P) = $ const., and rewrite (8) for the case that $U(P)$ is absolutely continuous as a function of $h(Q, P)$.

39. Boundary integrals and Stieltjes integral equations. So far we have not established any relation between the boundary function $U(P)$, of limited variation, and the $n+1$ boundary functions of limited variation $U_i(\theta_i)$

(10) $$U_i(\theta_i) = \lim_{r_i = a_i} \int_0^{\theta_i} u(r_i, \theta_i) d\theta_i, \quad i = 0, 1, \cdots, n,$$

and although we know that the former function is arbitrary, we have not established the same fact for the latter functions, nor do we know that for a given harmonic function of the general class that we have been describing the functions $F_i(\theta_i)$ of Art. 37 are essentially determined. This subject will be considered in the present section. We first solve a simple type of linear Stieltjes integral equation.

THEOREM 6. *Let $U(x)$ be of limited variation, $0 \leq x \leq X$, and $K(x)$ continuous in the same closed interval, and positive. The equation*

(11) $$U(x) - U(0) = \int_0^x K(y) dV(y)$$

has the solution

(12) $$V(y) - V(0) = \int_0^y \frac{1}{K(x)} dU(x)$$

in which $V(0)$ is arbitrary and $V(y)$ a function of limited variation; and this is the only solution in which $V(y)$ is a function of limited variation.

In fact, if $V(y)$ is of limited variation it is the difference of two non-decreasing functions, hence by means of (11) the function $U(x)$ is also the difference of two non-decreasing functions, and is of limited variation. Consequently, if $V(y)$ is of limited variation, we may form the integral

$$I(y) = \int_0^y \frac{1}{K(x)} dU(x) = \int_0^y \frac{1}{K(x)} d\int_0^x K(\alpha) dV(\alpha)$$

and we shall have

$$\left| I(y) - \sum_{0}^{n-1} \frac{1}{K(x_i)} \int_{x_i}^{x_{i+1}} K(\alpha)\, dV(\alpha) \right| \leq \omega_\delta \left(\frac{1}{K}\right) C_1,$$

where $x_0 = 0$, $x_1, \cdots, x_n = y$ are points which divide $(0, y)$ into n subintervals each of length $\leq \delta$, $\omega_\delta\left(\dfrac{1}{K}\right)$ is the maximum oscillation of the reciprocal of $K(x)$ in the interval $(0, X)$ and C_1 is a constant, independent of y. But

$$\left| \int_{x_i}^{x_{i+1}} K(\alpha)\, dV(\alpha) - K(x_i)\{V(x_{i+1}) - V(x_i)\} \right|$$
$$\leq \omega_\delta(K)\{t_V(x_{i+1}) - t_V(x_i)\}$$

where $t_V(x)$ is the total variation function of $V(x)$. Therefore

$$\left| \sum_{0}^{n-1} \frac{1}{K(x_i)} \int_{x_i}^{x_{i+1}} K(\alpha)\, dV(\alpha) - \sum_{i=0}^{n-1} \{V(x_{i+1}) - V(x_i)\} \right| \leq \omega_\delta(K) C_2,$$

where C_2 is a constant, $\leq t(X)/\min. K(x)$. Accordingly,

$$|I(y) - \{V(y) - V(0)\}| \leq \omega_\delta(K) C_2 + \omega_\delta\left(\frac{1}{K}\right) C_1,$$

and, since δ is arbitrary,

$$I(y) = V(y) - V(0).$$

Similarly if $U(x)$ is of limited variation, and $V(y)$ is given by (12) it follows that $U(x)$ is given by (11). Hence the theorem is proved.

COROLLARY. If the relation between $U(x)$ and $V(y)$ is given by (11), and φ_U, φ_V, ψ_U, ψ_V are used to denote the positive and negative variation functions of U and V, then

(13) $\quad \varphi_U(x) = \displaystyle\int_0^x K(y)\, d\varphi_V(y), \quad \psi_U(x) = \int_0^x K(y)\, d\psi_V(y)$

and these integrals may be inverted by Theorem 6.

The direct proof of (13) is more tedious than difficult. It follows however almost immediately from Theorem 6. In fact, if we write

$$\overline{\varphi}_V(y) = \int_0^y \frac{1}{K(x)} d\varphi_U(x), \quad \overline{\psi}_V(y) = \int_0^x \frac{1}{K(x)} d\psi_U(x)$$

we shall have $\overline{\varphi}_V(y)$ a non-decreasing and $\overline{\psi}_V(y)$ a non-increasing function, and

Hence
$$V(y) - V(0) = \overline{\varphi}_V(y) + \overline{\psi}_V(y).$$

$$\overline{\varphi}_V(y) \geq \varphi_V(y), \quad \overline{\psi}_V(y) \leq \psi_V(y),$$

since $\varphi_V(y)$ is the least upper bound and $\psi_V(y)$ is the greatest lower bound for their respective generating sums. Also the functions

$$\overline{\varphi}_U(x) = \int_0^x K(y) d\varphi_V(y), \quad \overline{\psi}_U(x) = \int_0^x K(y) d\psi_V(y)$$

satisfy the inequalities

$$\overline{\varphi}_U(x) \geq \varphi_U(x), \quad \overline{\psi}_U(x) \leq \psi_U(x).$$

But since $\overline{\varphi}_V \geq \varphi_V$, $\overline{\psi}_V \leq \psi_V$, we have

$$\int_0^x K(y) d\overline{\varphi}_V(y) \geq \int_0^x K(y) d\varphi_V(y)$$
$$\int_0^x K(y) d\overline{\psi}_V(y) \leq \int_0^x K(y) d\psi_V(y),$$

so that, by Theorem 6,

$$\varphi_U(x) = \int_0^x K(y) d\overline{\varphi}_V(y) \geq \overline{\varphi}_U(x), \quad \psi_U(x) \leq \overline{\psi}_U(x)$$

Hence the equality sign alone is possible and the Corollary is proved.

We may now, utilizing these results, find the desired relations between the functions $U_i(\theta_i)$ given by (10) and the function $U(P) = U(h)$ given by (6). We consider

$$\bar{I}_{0\theta_i} = \int_{0 \atop (r)}^{\theta_i} u(M) d\theta = \frac{1}{2\pi} \int_{0 \atop (r)}^{\theta_i} d\theta_M \int_{(g=0)} \frac{dh(M,P)}{dh(Q,P)} dU(P)$$

where (r) indicates integration over the circle of radius $r_i = r$, concentric with s_i, and $(g = 0)$ indicates integration over the complete boundary of S. Since M is not on this boundary we may invert the order of integration and write*

$$I_{0\theta_i} = \frac{1}{2\pi} \int_{(g=0)} dU(P) \int_0^{\theta_i} \left[\frac{dh(M,P)}{dh(Q,P)}\right]_{g=0} d\theta_M$$
$$= \frac{1}{2\pi} \int_{(g=0)} \frac{dU(P)}{\frac{\partial g(Q,P)}{\partial n_P}} \int_0^{\theta_i} \frac{\partial g(M,P)}{\partial n_P} d\theta_M$$
$$= \frac{2\pi r_M}{1} \int_{(g=0)} \frac{dU(P)}{\frac{\partial g(Q,P)}{\partial n_P}} \int_0^s \frac{\partial g(M,P)}{\partial n_P} ds_M.$$

where s is the absolute arc length corresponding to the value θ_i.

Now $g(M,P)$ is harmonic, $M \neq P$, even if M and P are points on the boundary, and can therefore be extended across the boundary s_i as a function of M or P when M and P are isolated from each other. Hence $\partial g(M,P)/\partial n_P$ approaches zero uniformly if $|\theta_P - \theta_M|$ remains different from zero, $\geq \varepsilon$. But $g(M,P)$ minus the Green's function for the circle s_i is harmonic in the neighborhood of the boundary s_i if $M \neq P$, and remains bounded in the neighborhood of $M = P$, even when M and P are points of s_i. Hence

$$\left|\frac{\partial g(M,P)}{\partial n_P} - \frac{1}{a_i} \frac{r_M^2 - a_i^2}{a_i^2 + r_M^2 - 2a_i r_M \cos(\varphi_P - \theta_M)}\right|$$

remains bounded, say $< K$, since by Lebesgue's theorem the above difference $g - g_i$ is harmonic as a function of either point, except for unnecessary discontinuities. The following equations may therefore be verified, $i = 1, \cdots, n$, and analogous ones if $i = 0$.

(14) $$\lim_{r=a_i} \int_0^s \frac{\partial g(M,P)}{\partial n_P} ds_M = 2\pi, \quad 0 < \varphi_P < \theta_i,$$
$$= 0, \quad \theta_i < \varphi_P < 2\pi,$$
$$= \pi, \quad \varphi_P = 0 \text{ or } \varphi_P = \theta_i.$$

* On integration under the integral sign in a Stieltjes integral, see for instance, Evans, *Problems of potential theory*, Rice Institute Pamphlets, vol. 7 (1920). p. 258.

In fact, to take the case where $\varphi_P = \theta_i$, we have

$$\lim_{r=a_i}\int_0^s \frac{\partial g(M,P)}{\partial n_P} ds_M = \lim_{r=a_i}\int_{s-r\varepsilon}^s \frac{\partial g(M,P)}{\partial n_P} ds_M$$

$$= \lim_{r=a_i}\int_{\theta_i-\varepsilon}^{\theta_i} \frac{(r_M^2-a_i^2)\,d\varphi_M}{a_i^2+r_M^2-2a_i r_M \cos(\varphi_M-\theta_i)} + \lim_{r=a_i} \eta(r)$$

where $|\eta(r)| \leq rK\varepsilon$, ε being arbitrarily small. From this the third of equations (14) is evident. A similar demonstration applies to the other formulae.

If then we define the function $\Psi(P)$,

$$\Psi(P) = 1, \quad P \text{ an interior point of } (0, \theta_i),$$
$$= 0, \quad P \text{ outside } (0, \theta_i),$$
$$= \frac{1}{2}, \quad P \text{ at } 0 \text{ or } \theta_i,$$

we may express $\lim_{r=a_i} I_{0\theta_i}$ as a Daniell integral, in terms of $\Psi(P)$,*

$$\lim_{r=a_i} I_{0\theta_i} = \frac{1}{a_i}\int_{s_i} \frac{1}{\frac{\partial g(Q,P)}{\partial n_P}} \Psi(P)\,dU(P),$$

whence

$$U_i(Q_i) = -\int_{(s_i)} \frac{\Psi(P)\,dU(P)}{\frac{dh(Q,P)}{d\theta_P}}$$

since $dh(Q,P)/d\theta_P = a_i dh(Q,P)/ds_P = -a_i \partial g(Q,P)/\partial n_P$.

The above integral may be further simplified by remembering that $\Psi(P)$ may be considered as the limit of the sequence of continuous functions which are defined by the equations

$$\Psi_n(P) = 1, \quad \varepsilon \leq \varphi_P \leq \theta_i - \varepsilon,$$
$$= 0, \quad \theta_i+\varepsilon \leq \varphi_P \leq 2\pi-\varepsilon,$$

and are given linearly in the remaining intervals. In fact, these functions are bounded in their set. Hence, since the

---

* See Evans, *loc. cit.*, p. 257, or for a systematic study, P. J. Daniell, *A general form of integral*, Annals of Mathematics, vol. 19 (1918), and later papers.

order of integration with respect to $h(Q, P)$ is the reverse of that with respect to θ, we have

$$U_i(\theta_i) = -\int_{h(\theta_i-0)}^{h(0+)} \frac{d\theta_P}{dh(Q,P)} dU(P)$$

$$- \frac{1}{2} \left[\frac{d\theta}{dh(Q,P)}\right]_{\theta_i} \{U(P_{\theta-0}) - U(P_{\theta+0})\}$$

$$- \frac{1}{2} \left[\frac{d\theta_P}{dh(Q,P)}\right]_0 \{U(P_{0-}) - U(P_{0+})\}$$

or, finally,

(15) $$U_i(\theta_i) = \int_{h(0)}^{h(\theta_i)} \frac{d\theta_P}{dh(Q,P)} dU(P),$$

if $U(P)$ is considered on the Riemann surface corresponding to $h(Q, P)$, where the discontinuities are regular, i. e., where

$$U(P_\theta) - U(P_{\theta-0}) = \frac{1}{2} \{U(P_{\theta+0}) - U(P_{\theta-0})\}, \text{ etc.}$$

By Theorem 6, the unique solution of (15), of limited variation, is given by the equation

(16) $$U(P_{\theta_i}) - U(P_0) = \int_0^{\theta_i} \frac{dh(Q,P)}{d\theta_P} dU_i(\theta_P),$$

the integrations in (15) and (16) being carried out on the boundary s_i. Consequently we are able to state the following theorem.

THEOREM 7. *The functions $U(P)$, given by (6) when the integration is carried out on the Riemann surface for $h(Q, P)$, and the discontinuities of $U(P)$ are consequently regular, and $U_i(\theta_i)$, given by (10), satisfy the equations (15), (16); so that $U(P)$, except for additive constants, is determined in terms of the $U_i(\theta_i)$, and vice-versa.*

COROLLARY. The equations (15), (16) remain valid if $U(P)$ or the $U_i(\theta_i)$ are absolutely continuous; and if $U(P)$ is absolutely continuous as a function of $h(Q, P)$ the $U_i(\theta_i)$ are all absolutely continuous as functions of their respective arguments, and *vice-versa*.

EXERCISE. Show that a necessary and sufficient condition that $u(M)$ belong to the class of functions considered in Theorems 3, 4 and 5 is that

$$\int_{0 \atop (g_i=y_i)}^{2\pi} |u(M)|\, dh_i(Q_i, M) < K_i, \qquad i = 0, 1, \cdots, n,$$

in the neighborhood of $y_i = 0$, Q_i being an arbitrary fixed point in the simply connected region of boundary s_i which includes S, and h_i the conjugate to its Green's function.
Show that if

$$W_i(P) = \lim_{y_i=0} \int_{0 \atop (g_i=y_i)}^{h_i} u(M)\, dh_i(Q_i, M),$$

then

$$U_i(\theta_i) = \int_{h_i(0)}^{h_i(\theta_i)} \frac{d\theta_P}{dh_i(Q_i, P)}\, dW_i(P),$$

$$W_i(P) - W_i(P_0) = \int_0^{\theta_i} \frac{dh_i(Q, P)}{d\theta_P}\, dU_i(\theta_P),$$

the integration being carried out on the boundary s_i.

40. General regions of finite connectivity. Consider a connected open region T in a plane, without isolated point boundaries, and of finite connectivity, say with $n+1$ boundaries. By means of a conformal transformation the region T may be transformed into a particular region S of the kind already considered, bounded by $n+1$ distinct circles, properly situated.* Therefore, whatever results are stated in terms invariant of conformal transformations for the general S with circular boundaries will also hold for T. In particular, Theorem 5 and equation (8) describe such results, interpretable in terms of the accessible points of the boundary of T, as in Chap. V. A special case is that where $U(P)$ is absolutely continuous as a function of $h(Q, P)$ and we may write

* The properties of the conformal transformation are developed in Hurwitz-Courant, *loc. cit.*, p. 322 ff. The region may lie on a Riemann surface provided it is sufficiently like a plane so that any simple closed curve on it divides it into two parts.

$$U(P) - U(P_0) = \int_{h_0}^{h} f(P)\, dh(QP)$$

with $f(P)$ independent of the position of the pole Q. In fact, $f(P)$ represents the boundary values of $u(M)$, which are taken on at almost all accessible points.

The generalized Dirichlet problem is uniquely solvable in the class of functions which are the differences of functions harmonic and not negative in T. In the more specialized problem, the harmonic function of class such that

$$\int_{(g=y)} u(M)\, dh(Q, M)$$

is absolutely continuous, uniformly with respect to y as y approaches 0, is uniquely determined by giving boundary values $f(P)$ almost everywhere on the accessible points of the boundary. The function $f(P)$ is arbitrary provided that it is summable in the Lebesgue sense.

The generalized Neumann problem, which is the important problem of that kind, is uniquely solvable in the class of functions of bounded total flux: i. e., such that

$$\int_{(g=y)} \left|\frac{\partial u}{\partial n}\right| ds < K,$$

as y approaches zero. In this case the function is uniquely determined by the limiting values of

$$\int_{(A)}^{(B)} \frac{\partial u}{\partial n}\, ds$$

as the points A, B, lying on the same curve $g = y$, approach the boundary along curves $h = \text{const}$. The values of the above integral are of course independent of the path in T joining A and B, and its boundary values represent a function of limited variation of one of the end points on the boundary, say P_B. It is sufficient to know this boundary function $F(P_B)$, of limited variation, of total change zero around the boundary, but otherwise arbitrary, merely on the accessible points of continuity for $F(P)$.

By means of the results of Art. 39, we may evidently express the classes of functions, the boundary conditions and the order of accessible boundary points in terms of the $n+1$ associated simply connected regions and their Green's functions. In particular, the kind of resolution given in Theorem 3 may be generalized so as to apply to the general region T. The invariant expressions are given in terms of the Exercise in Art. 39.

A word or two in regard to *isolated* boundary points of regions of finite connectivity is desirable. Here we shall consider merely uniform functions. If a function, single valued and harmonic in the region T, remains bounded in the neighborhood of such a point, say the point 0, Lebesgue's theorem implies that it be harmonic (except for an unnecessary discontinuity) at the point; and therefore for bounded functions the point 0 will not be effectively a boundary point.

If the harmonic function $u(M)$ is not bounded in the neighborhood of 0 it is evident, by means of equation (3), that $u(M)$ may be written as a function which is harmonic in the neighborhood of 0 and at 0, plus a function of the form

(17) $\quad u_0(M) = k \log r + \sum_1^\infty {}' \frac{1}{r^m} \{a_m \cos m\varphi + b_m \sin m\varphi\},$

where the part of this expression involved in the summation is the real part of a function which is holomorphic in the entire plane except at 0, and therefore where

(18) $\quad \lim_{m=\infty} \sqrt[m]{|a_m|} = \lim_{m=\infty} \sqrt[m]{|b_m|} = 0.$*

THEOREM 8. *A necessary and sufficient condition that a single valued function $u(M)$ be harmonic in the neighborhood of 0, except at 0, is that it be the sum of a function harmonic in*

---

* A direct application of the theorem of Cauchy-Hadamard, that the radius of convergence of the power series $\Sigma a_m \zeta^m$ is given by

$$\frac{1}{R} = \overline{\lim_{m=\infty}} |a_m|^{\frac{1}{m}}.$$

the neighborhood of 0 and at 0, plus a function of form (17), where the a_m, b_m are subject to (18).

Consider now functions which are the differences of two functions harmonic and not negative in T, and, in particular, in the neighborhood of the isolated boundary point 0. We shall show first that such a function, if it remains not negative in the neighborhood of 0, can have merely a logarithmic singularity at 0, i. e., will be of the form

(19) $$u(M) = k \log r + u_1(M),$$

with k a negative constant and $u_1(M)$ harmonic at 0 and in its neighborhood.

In fact, we may, in accordance with Theorem 8, find constants C, k so that the function

$$u'(M) = C + k \log r + \sum_1^\infty \frac{1}{r^p}(a_p \cos p\theta + b_p \sin p\theta),$$

which differs from $u(M)$ by a function harmonic at 0 and in its neighborhood, will be not negative in the neighborhood of 0. Hence for r sufficiently small, we shall have

(20) $$\int_0^{2\pi} u'(M)\, d\theta = \int_0^{2\pi} |u'(M)|\, d\theta = 2\pi C + 2\pi k \log r,$$

so that, in particular,

(21) $$\int_0^{2\pi} u'(M)\, d\theta = \int_0^{2\pi} |u'(M)|\, d\theta < \frac{\varphi_n(r)}{r^n}, \quad n = 1, 2, \cdots,$$

with $\lim_{r=0} \varphi_n(r) = 0$.

We have need of the following lemma, which we prove by means of a device described by Poincaré, slightly generalized.*

LEMMA. Consider a function single valued and harmonic in the neighborhood of 0, except at 0, and therefore of the form described in Theorem 8. If

(22) $$\int_0^{2\pi} |u(r, \theta)|\, d\theta \leq \frac{\varphi(r)}{r^n},$$

---

* Poincaré, *Théorie du potentiel newtonien*, Paris (1899), p. 208.

where n is a positive integer and $\lim_{r=0} \varphi(r) = 0$, then $a_n = b_n = 0$.

We write
$$X_p = a_p \cos p\theta + b_p \sin p\theta$$
and form, in accordance with (17), the equation
$$\int_0^{2\pi} X_n u(r, \theta) r^n d\theta = \int_0^{2\pi} u_1(r, \theta) X_n r^n d\theta + k \int_0^{2\pi} \log r X_n r^n d\theta$$
$$+ \int_0^{2\pi} \frac{X_n^2 r^n}{r^n} d\theta + \sum{}' \int_0^{2\pi} \frac{X_p}{r^p} X_n r^n d\theta.$$

On account of (22),
$$\left| r^n \int_0^{2\pi} X_n u(r, \theta) d\theta \right| \leq r^n (|a_n| + |b_n|) \int_0^{2\pi} |u(r, \theta)| d\theta$$
$$\leq (|a_n| + |b_n|) \varphi(r),$$

and this expression has the limit 0. Moreover
$$\int_0^{2\pi} X_n X_p d\theta = 0, \qquad p \neq n,$$
$$\lim_{r=0} r^n \int_0^{2\pi} X_n u_1(r, \theta) d\theta = 0,$$
$$k r^n \log r \int_0^{2\pi} X_n d\theta = 0,$$
whence
$$\int_0^{2\pi} X_n^2 d\theta = 0,$$

and a_n and b_n must vanish. The lemma is proved.

The function $u'(M)$ satisfies the conditions of this lemma for all positive integral values of n, from (21). Hence
$$u'(M) = C + k \log r,$$

where, of course, $k < 0$. If therefore we return to functions which are the differences of two such not negative functions in the neighborhood of 0, only the same sort of singularity is admitted, and we have the following corollary.*

* Bôcher [Bulletin of the American Mathematical Society, vol. 9 (1903), p. 455] showed that if $u(r, \theta)$ is harmonic $0 < r < R_1$, and becomes $+\infty$, or $-\infty$, for $r = 0$, its only possible singularity at 0 is of the form $k \log r$. See also recent papers by Raynor, ibid. vol. 32 (1926), p. 537, and Kellogg, ibid. vol. 32 (1926), p. 664, and Picone, Lincei (1926).

COROLLARY 1. If $u(M)$ is the difference of two functions harmonic and not negative in the plane region T, of which 0 is an isolated boundary point, it can have a singularity at 0 merely of the form $k \log r$, and when this term is subtracted the remainder will have merely an unnecessary discontinuity at 0.

By means of the same lemma we have also the following proposition.

COROLLARY 2. If $\int |u| \, ds$, extended over a circumference of radius r and center 0, remains bounded as r approaches zero, $u(M)$ can have a singularity at 0 merely of the form

$$k \log r + \frac{1}{r}(a_1 \cos \theta + b_1 \sin \theta).$$

This proposition, by further specialization, yields another corollary.

COROLLARY 3. If $\int |u| \, d\theta$, extended over a circumference of radius r and center 0, remains bounded as r approaches zero, $u(M)$ has merely an unnecessary discontinuity at 0, and 0 is not an effective boundary point.

These corollaries to Theorem 8 show incidentally that the classes of functions, which in S were identical, are no longer co-extensive when one or more of the circles s_i is allowed to shrink down to a point.

41. Annular regions. Determination of the functions $F_0(\theta)$ and $F_1(\theta)$. Let us return now to the region S and assume it to be the annular region between two concentric circles of radii a_0 and a_1, $a_0 > a_1$. We shall take up the problem of determining the functions $F_0(\theta)$, $F_1(\theta)$ of Theorem 3 in terms of the boundary functions $U_0(\theta)$, $U_1(\theta)$.

We write, in accordance with Theorem 3,

$$(23) \quad u(r, \theta) = k \log r + \beta + \frac{1}{2\pi} \int_0^{2\pi} \frac{(a_0^2 - r^2) \, dF_0(\varphi)}{a_0^2 + r^2 - 2 a_0 r \cos(\varphi - \theta)} + \frac{1}{2\pi} \int_0^{2\pi} \frac{(r^2 - a_1^2) \, dF_1(\varphi)}{a_1^2 + r^2 - 2 a_1 r \cos(\varphi - \theta)}$$

FINITELY CONNECTED PLANE REGIONS   113

where, by introducing the constant β, we are able to assume that the $F_0(\varphi)$, as well as the $F_1(\varphi)$, is periodic. We may also assume that these functions vanish for $\varphi = 0$, since they contain an additive arbitrary constant.

THEOREM 9. *In the representation* (23) *the constants* k, β *and the functions* $F_0(\varphi)$, $F_1(\varphi)$, *assumed to be periodic and of limited variation, vanishing at* $\varphi = 0$ *and with regular discontinuities, are uniquely determined by the boundary functions* $U_0(\theta)$, $U_1(\theta)$.

In fact, by taking account of the periodicity of the functions $F_0(\varphi)$, $F_1(\varphi)$ and integrating (23) by parts, we have

$$\int_0^\theta u(r,\theta)\,d\theta = \beta\theta + k\theta \log r$$

$$+ \frac{1}{2\pi} \int_0^{2\pi} \left[\frac{a_0^2 - r^2}{a_0^2 + r^2 - 2a_0 r \cos(\varphi-\theta)}\right]_{\theta=0}^{\theta=\theta} F_0(\varphi)\,d\varphi$$

$$+ \frac{1}{2\pi} \int_0^{2\pi} \left[\frac{r^2 - a_1^2}{a_1^2 + r^2 - 2a_1 r \cos(\varphi-\theta)}\right]_0^\theta F_1(\varphi)\,d\varphi,$$

or

$$U_0(\theta) = \beta\theta + k\theta \log a_0 + F_0(\theta)$$

(24)
$$+ \frac{1}{2\pi} \int_0^{2\pi} \left[\frac{a_0^2 - a_1^2}{a_0^2 + a_1^2 - 2a_0 a_1 \cos(\varphi-\theta)}\right]_0^\theta F_1(\varphi)\,d\varphi$$

$$U_1(\theta) = \beta\theta + k\theta \log a_1 + F_1(\theta)$$

$$+ \frac{1}{2\pi} \int_0^{2\pi} \left[\frac{a_0^2 - a_1^2}{a_0^2 + a_1^2 - 2a_0 a_1 \cos(\varphi-\theta)}\right]_0^\theta F_0(\varphi)\,d\varphi.$$

The equations (24) are mixed integral equations for the determination of the unknowns, the Stieltjes integrals having been eliminated by the integration by parts.

These equations are however of simple type. In fact, if we take $\theta = 2\pi$ we have

$$U_0(2\pi) = 2\pi\beta + 2\pi k \log a_0$$
$$U_1(2\pi) = 2\pi\beta + 2\pi k \log a_1,$$

so that

(25)
$$k = \frac{1}{2\pi} \frac{U_0(2\pi) - U_1(2\pi)}{\log(a_0/a_1)},$$
$$\beta = \frac{1}{2\pi} \frac{U_1(2\pi)\log a_0 - U_0(2\pi)\log a_1}{\log(a_0/a_1)},$$

and may therefore be regarded as known in (24).

But now we have merely two simultaneous Fredholm equations for the determination of $F_0(\varphi)$, $F_1(\varphi)$ of the form

(26)
$$A_0(\theta) = F_0(\theta) + \int_0^{2\pi} K(\theta, \varphi) F_1(\varphi) d\varphi,$$
$$A_1(\theta) = F_1(\theta) + \int_0^{2\pi} K(\theta, \varphi) F_0(\varphi) d\varphi,$$

in which

(26′)
$$A_0(\theta) = U_0(\theta) - \beta\theta - k\theta \log a_0,$$
$$A_1(\theta) = U_1(\theta) - \beta\theta - k\theta \log a_1,$$

and are periodic, vanish for $\varphi = 0$, are of limited variation and have regular discontinuities.

It is evident therefore that if there are any bounded integrable solutions of (26) they will be periodic and will vanish for $\varphi = 0$. Moreover the total variation of

$$\int_0^{2\pi} K(\theta, \varphi) F_1(\varphi) d\varphi$$

will be

$$\leq N \int_0^{2\pi} d\varphi \int_0^{2\pi} \left| \frac{\partial K(\theta, \varphi)}{\partial \theta} \right| d\theta$$

if $|F_1(\varphi)| < N$. Hence from the first of equations (26) we see that $F_0(\theta)$ is of limited variation if $F_1(\varphi)$ is bounded. Similarly $F_1(\varphi)$ is of limited variation if $F_0(\varphi)$ is bounded. And if these functions are of limited variation we see from (26) that their discontinuities must be regular. Consequently if (26) has bounded solutions they will be of the type specified in the theorem. The question of the existence and uniquenes of bounded soulutions depends upon the possible continuous solutions of the corresponding homogeneous equations

(27)
$$0 = G_0(\theta) + \int_0^{2\pi} K(\theta, \varphi) G_1(\varphi) d\varphi,$$
$$0 = G_1(\theta) + \int_0^{2\pi} K(\theta, \varphi) G_0(\varphi) d\varphi,$$

and the associated homogeneous equations

(28)
$$0 = H_0(\theta) + \int_0^{2\pi} K(\varphi, \theta) H_1(\varphi) d\varphi,$$
$$0 = H_1(\theta) + \int_0^{2\pi} K(\varphi, \theta) H_0(\varphi) d\varphi.$$

If there are no such solutions of (27) other than zero, there will be none of (28), and *vice versa*; and there will be unique bounded integrable solutions of (26).*

Consider then the equations (28), which may be written explicitly in the form

$$H_0(\theta) = -\frac{1}{2\pi} \int_0^{2\pi} \frac{a_0^2 - a_1^2}{a_0^2 + a_1^2 - 2 a_0 a_1 \cos(\theta - \psi)} \Big]_{\psi=0}^{\psi=\varphi} H_1(\varphi) d\varphi,$$

$$H_1(\theta) = -\frac{1}{2\pi} \int_0^{2\pi} \frac{a_0^2 - a_1^2}{a_0^2 + a_1^2 - 2 a_0 a_1 \cos(\theta - \psi)} \Big]_{\psi=0}^{\psi=\varphi} H_0(\varphi) d\varphi.$$

Accordingly the $H_i(\theta)$ must be periodic with period 2π. Also

$$H_0(\theta) = -\frac{1}{2\pi} \int_0^{2\pi} \frac{a_0^2 - a_1^2}{a_0^2 + a_1^2 - 2 a_0 a_1 \cos(\theta - \varphi)} H_1(\varphi) d\varphi$$
$$+ \frac{1}{2\pi} \frac{a_0^2 - a_1^2}{a_0^2 + a_1^2 - 2 a_0 a_1 \cos \theta} \int_0^{2\pi} H_1(\varphi) d\varphi$$

whence

$$\int_0^{2\pi} H_0(\theta) d\theta = 0.$$

---

* The familiar device in the treatment of simultaneous Fredholm equations is to handle them as a single one, by defining a compound kernel over a compound interval [Vivanti, *Elementi della teoria delle equazioni integrali lineari*, Milan (1916), p. 278].

A similar relation holds for $H_1(\theta)$, and we have therefore the equations

(29)
$$H_0(\theta) = -\frac{1}{2\pi}\int_0^{2\pi}\frac{a_0{}^2-a_1{}^2}{a_0{}^2+a_1{}^2-2a_0 a_1\cos(\theta-\varphi)}H_1(\varphi)\,d\varphi,$$
$$H_1(\theta) = -\frac{1}{2\pi}\int_0^{2\pi}\frac{a_0{}^2-a_1{}^2}{a_0{}^2+a_1{}^2-2a_0 a_1\cos(\theta-\varphi)}H_0(\varphi)\,d\varphi.$$

Now define the functions

$$v_0(r,\theta) = \frac{1}{2\pi}\int_0^{2\pi}\frac{a_0{}^2-r^2}{a_0{}^2+r^2-2a_0 r\cos(\varphi-\theta)}H_0(\varphi)\,d\varphi$$

$$v_1(r,\theta) = -\frac{1}{2\pi}\int_0^{2\pi}\frac{r^2-a_1{}^2}{a_1{}^2+r^2-2a_1 r\cos(\varphi-\theta)}H_1(\varphi)\,d\varphi,$$

the first harmonic inside s_0, the second harmonic outside s_1, and regulat at ∞. But if the equations (29) have continuous solutions, we know from Chap. II that both of the harmonic functions, just defined, take on the values $H_0(\theta)$ on s_0 and $-H_1(\theta)$ on s_1, and, being bounded in S, are therefore identical within S. Each is therefore a unique harmonic extension of the other and the function

$$v(r,\theta) = v_0(r,\theta) = v_1(r,\theta)$$

is harmonic in the entire plane and regular at infinity, and therefore must reduce to a constant. But since

$$\int_0^{2\pi}H_0(\varphi)\,d\varphi = 0,$$

this constant must be zero. Hence both $H_0(\theta)$ and $H_1(\theta)$ are identically zero, and the theorem is proved.

The exposition which has just been completed is seen to provide another proof, independent of that in Theorem 5 and its first corollary, that the generalized Dirichlet problem is uniquely solvable for the annular region S in the class of functions of Theorems 3 and 4. For the general region S bounded by $n+1$ circles the consideration of integral equations does not give such simple results. But knowing as we do that the class of functions described in Theorems 3 and 4 is the same as that of Theorem 5 and equation (8), we know that given the boundary functions $U_i(\theta_i)$ there always exist functions $F_i(\varphi)$ so that the function $u(M)$ will be given by the representation of Theorem 3. In the next section we shall use the integral equations to show that the $F_i(\varphi)$ are essentially uniquely determined.

EXERCISE 1. The reader may show directly that there are no non-zero solutions of (27) by showing that they may be differentiated.

EXERCISE 2. Consider the class of functions $v(r, \theta)$ harmonic within the annular region S such that

$$\int_0^{2\pi} \left|\frac{\partial v}{\partial r}\right| d\theta$$

is bounded for all circles in S concentric with s_0 and s_1. By writing

$$v(r, \theta) = \alpha + k \log r$$
$$- \frac{a_0}{2\pi} \int_0^{2\pi} \log(a_0^2 + r^2 - 2 a_0 r \cos(\varphi - \theta)) \, dF_0(\varphi)$$
$$+ \frac{a_1}{2\pi} \int_0^{2\pi} \log(a_1^2 + r^2 - 2 a_1 r \cos(\varphi - \theta)) \, dF_1(\varphi)$$

show, by finding the equations to determine $F_0(\varphi)$, $F_1(\varphi)$, that except for an arbitrary additive constant there is a unique solution in the above class of the generalized Neumann problem, for which

$$\lim_{r=a_0} \int_0^\theta \frac{\partial v}{\partial r} d\theta = V_0(\theta), \quad \lim_{r=a_1} \int_0^\theta \frac{\partial v}{\partial r} d\theta = V_1(\theta),$$

provided that $V_0(\theta)$, $V_1(\theta)$ are of limited variation, with regular discontinuities, vanish for $\varphi = 0$ and are such that

$$a_0 V_0(2\pi) = a_1 V_1(2\pi).$$

The integral equations for $F_0(\varphi)$, $F_1(\varphi)$ reduce to those already treated.

42. Uniqueness of the representation of Theorem 3 for S. We shall prove the following Theorem.

THEOREM 10. *Given $u(M)$ as in Theorems 3 and 4, we may write*

$$(30) \begin{cases} u(r, \theta) = \beta + \sum_1^n k_i \log r_i \\ \qquad + \dfrac{1}{2\pi} \displaystyle\int_0^{2\pi} \dfrac{(a_0{}^2 - r_0{}^2)\, dF_0(\varphi)}{a_0{}^2 + r_0{}^2 - 2 a_0 r_0 \cos(\varphi - \theta_0)} \\ \qquad + \sum_1^n \dfrac{1}{2\pi} \displaystyle\int_0^{2\pi} \dfrac{(r_i{}^2 - a_i{}^2)\, dF_i(\varphi)}{a_i{}^2 + r_i{}^2 - 2 a_i r_i \cos(\varphi - \theta_i)} \end{cases}$$

with the functions $F_0(\varphi)$, $F_i(\varphi)$ as in Theorem 3, but vanishing at $\varphi = 0$, with regular discontinuities, and all of them periodic with period 2π. With these conditions the functions $F_0(\varphi)$, $F_1(\varphi), \cdots, F_n(\varphi)$ and the constants β, $k_1, \cdots, k_n$ are all uniquely determined.

Corresponding to equations (24) of Art. 41 we have the equations

$$(31) \begin{cases} U_0(\theta_0) = \beta\theta_0 + \sum_1^n k_i \displaystyle\int_0^{\theta_0} \log r_{i0}\, d\theta_0 + F(\theta_0) \\ \qquad + \dfrac{1}{2\pi} \sum_1^n \displaystyle\int_0^{2\pi} \left\{ \int_0^{\theta_0} \dfrac{(r_{i0}{}^2 - a_i{}^2)\, d\theta_{i0}}{a_i{}^2 + r_{i0}{}^2 - 2 a_i r_{i0} \cos(\varphi - \theta_{i0})} \right\} dF_i(\varphi) \\ U_j(\theta_j) = \beta\theta_j + k_j\theta_j \log r_j + \sum_1^{n'} k_i \displaystyle\int_0^{\theta_j} \log r_{ij}\, d\theta_j + F_j(\theta_j) \\ \qquad + \dfrac{1}{2\varphi} \displaystyle\int_0^{2\pi} \left\{ \int_0^{\theta_j} \dfrac{(a_0{}^2 - r_{0j}{}^2)\, d\theta_{0j}}{a_0{}^2 + r_{0j}{}^2 - 2 a_0 r_{0j} \cos(\varphi - \theta_{0j})} \right\} dF_0(\varphi) \\ \qquad + \dfrac{1}{2\pi} \sum{}' \displaystyle\int_0^{2\pi} \left\{ \int_0^{\theta_j} \dfrac{(r_{ij}{}^2 - a_i{}^2)\, d\theta_{ij}}{a_i{}^2 + r_{ij}{}^2 - 2 a_i r_{ij} \cos(\varphi - \theta_{ij})} \right\} dF_i(\varphi), \\ \qquad\qquad\qquad j = 1, 2, \cdots, n, \end{cases}$$

in which θ_{ij}, r_{ij} are the polar coördinates of a point on s_j with respect to the center of the circle s_i.

Since the equations (30) are necessary and sufficient that $u(M)$ be of the given class, and the member of the class is uniquely determined by the boundary functions $U_i(\theta_i)$, $i = 0, 1, \cdots, n$, the equations (31) are necessary and

FINITELY CONNECTED PLANE REGIONS 119

sufficient for (30), where the $U_i(\theta_i)$ are of limited variation, with regular discontinuities and vanish when their arguments vanish. Suppose there where two sets of solutions of the kind specified in Theorem 10; the set of differences β, k_i, F_0, F_i between the corresponding elements of the two sets of solutions would be solutions of the homogeneous Stieltjes integrals of the same kind.

From the form of these equations, since the boundary circles are all distinct, we have a right to differentiate under the integral signs, and we see that the functions $F_0, F_1, \cdots, F_n$ are all absolutely continuous. Hence we may differentiate the equations and write

(32)
$$\begin{cases} 0 = \beta + \sum_1^n k_i \log r_{i0} + f_0(\theta_0) \\ \qquad + \dfrac{1}{2\pi} \sum_1^n \int_0^{2\pi} \dfrac{(r_{i0}^2 - a_i^2) f_i(\varphi)\, d\varphi}{a_i^2 + r_{i0}^2 - 2 a_i r_{i0} \cos(\varphi - \theta_{i0})} \\ 0 = \beta + \sum_1^n k_i \log r_{ij} + f_j(\theta_j) \\ \qquad + \dfrac{1}{2\pi} \int_0^{2\pi} \dfrac{(a_0^2 - r_{0j}^2) f_0(\varphi)\, d\varphi}{a_0^2 + r_{0j}^2 - 2 a_0 r_{0j} \cos(\varphi - \theta_{0j})} \\ \qquad + \dfrac{1}{2\pi} {\sum_1^n}' \int_0^{2\pi} \dfrac{(r_{ij}^2 - a_i^2) f_i(\varphi)\, d\varphi}{a_i^2 + r_{ij}^2 - 2 a_i r_{ij} \cos(\varphi - \theta_{ij})}, \\ \qquad\qquad\qquad\qquad j = 1, 2, \cdots, n, \end{cases}$$

where

$$F_0(\varphi) = \int_0^\varphi f_0(\varphi)\, d\varphi, \qquad F_j(\varphi) = \int_0^\varphi f_j(\varphi)\, d\varphi,$$

and the functions $f_0(\varphi)$, $f_j(\varphi)$ are evidently continuous.

Define now the functions

$$v_0(M) = \beta + \dfrac{1}{2\pi} \int_0^{2\pi} \dfrac{(a_0^2 - r_0^2) f_0(\varphi)\, d\varphi}{a_0^2 + r_0^2 - 2 a_0 r_0 \cos(\varphi - \theta_0)}$$

$$v_j(M) = k_j \log r_j + \dfrac{1}{2\pi} \int_0^{2\pi} \dfrac{(r_j^2 - a_j^2) f_j(\varphi)\, d\varphi}{a_j^2 + r_j^2 - 2 a_j r_j \cos(\varphi - \theta_j)},$$
$$j = 1, 2, \cdots, n.$$

The function
$$v(M) = v_0(M) + \sum_1^n v_j(M)$$
is harmonic in S, and bounded, and by (32) vanishes everywhere on the boundary of S; hence $v(M) = 0$ identically in S.

We have

(33) $$v_0(M) = -\sum_1^n v_j(M)$$

in S, and accordingly the left hand member provides the unique extension of the right hand member everywhere inside s_0, and the right hand member provides the unique extension of the left hand member outside $s_1, \cdots, s_n$. We thus define the function $V_0(M)$ in terms of (32) and these extensions. Hence

$$\int_0^{2\pi} V_0(M)\, d\theta = 2\pi V_0(0) = 2\pi\beta.$$

But this integral may also be obtained by integrating over a large circle of radius R, and yields therefore

$$2\pi\beta = \int_0^{2\pi} \sum_1^n k_i \log R\, d\theta_0 + \int_0^{2\pi} \sum_1^n k_i \log \frac{r_i}{R}\, d\theta_0.$$

The second integral comes as near to the value 0 as we please, by taking R large enough, while the first is $(2\pi \sum k_i) \log R$. Hence we must have $\sum k_i = 0$, $\beta = 0$.

Now however the function $V_0(M)$ is harmonic all over the plane and regular at ∞. It is therefore a constant, and since its integral from 0 to 2π is zero, it must vanish identically; i. e.,

$$v_0(M) \equiv 0, \quad f_0(\varphi) \equiv 0.$$

An analogous method of procedure shows in turn that k_1, $f_1(\varphi)$, k_2, $f_2(\varphi)$, $\cdots$, k_n, $f_n(\varphi)$ all vanish. Hence the $F_0(\varphi)$, $F_i(\varphi)$ also vanish identically, and the theorem is proved.

CHAPTER VII

RELATED PROBLEMS

43. A simple discontinuous boundary value problem. Consider a region T bounded by a simple rectifiable curve s, with a unique tangent at each point; and consider a function $f(s)$, defined on the boundary in terms of the arc length, continuous except at a finite number of points $A_1, \cdots, A_n$, and at each of these points such that both $f(s_i+0) = b_i$ and $f(s_i-0) = a_i$ exist. It is easy to show that there is one and only one function $u(M)$ bounded and harmonic in S which takes on the values $f(s)$ continuously as M approaches any point of the boundary not an A_i.[*]

In fact, the function

$$\alpha(M) = -\sum_1^n \frac{b_i - a_i}{\pi} \arctan \frac{y - y_i}{x - x_i}$$

is bounded and harmonic in S and takes on, on the boundary, values $\varphi(s)$ which have precisely the same discontinuities as $f(s)$. If then we denote by $\beta(M)$ the solution of the continuous boundary value problem $f(s) - \varphi(s)$, the function

$$u(M) = \alpha(M) + \beta(M)$$

will be a solution of our problem.

That the solution is unique follows of course from the theory of Chap. V. But the uniqueness is established also by elementary means on the basis of a theorem of Zaremba.[†] If $U(M)$, harmonic in T, takes on continuously the boundary values 0 except at a finite number of points A_i, where

---

[*] Goursat, *Cours d'analyse*, vol. III, Paris (1915), p. 204.

[†] Bulletin of the Academy of Sciences of Cracow (1909), p. 561. See Goursat loc. cit. The theorem may be regarded as containing as a special case the introductory theorem of Chapter IV, there ascribed to Lebesgue, as a special case again of other results of his.

$$\lim_{r_i=0} \left| \frac{U(M)}{\log r_i} \right| = 0, \qquad r_i = \overline{A_i\, M},$$

then $U(M)$ vanishes identically in T. The proof of this proposition may be omitted since a slightly more general theorem will be discussed, in Art. 47, which depends on the same method of proof.

In the simple problem, just discussed, it was feasible to lead the question back to one where the boundary values were continuous. This remark, in fact, applies to the usual treatment of a discontinuous boundary value problem, which thus becomes a limiting case, as in the theorems on sequences cited in Chap. V, or a mere offshoot, as in the example just given. This point of view is hardly due to the physical applications of the problem, for it cannot be said that modern physics has any special affinity for the continuous, if indeed a science can be called modern which is entirely Greek in its devotion to whole numbers.* It is therefore a special strength of the Stieltjes integral that it applies the fundamental processes of integration in the representation of discontinuous quantity.

In the treatment given in these chapters the emphasis has been put directly on functions of limited variation, rather than continuous functions, on account of the relation of the former to general distributions of matter. But this statement implies more than can rightfully be asserted, since except for the simplest kinds of boundaries it was found desirable to introduce the Green's function and the theory of conformal representation. The latter is a theory which requires modification for more than two dimensions. And the Green's function brings in to some extent the continuous boundary value problem—the determination of a harmonic function which takes on the frontier values $\log r$. On the other hand, with this much admitted, there is something gained in reducing the problem to central ideas, where so much is known in detail.

---

* See the first two pages of Born, *Problems of atomic dynamics*, Cambridge (1926).

44. Continuous boundary value problems. Given a function $f(M)$ continuous in an open region T and on its frontier, the central problem is to find a function $u(M)$, harmonic in T, which takes on the value $f(P)$ continuously, as M, in T, approaches the point P, on the boundary. It is only necessary to mention the alternating method of Schwarz, Poincaré's "méthode de balayage", the minimal processes of Riemann, Hilbert, Zaremba and Lebesgue, the successive approximation method of Le Roux, and Phillips and Wiener, reducing the approximate solution of the problem to the solution of a set of linear algebraic equations, and the method of C. Neumann which culminates in Fredholm's solution of the linear integral equation with constant limits.* These provide solutions of the problem, in part, or as a whole.

The method of Poincaré, for example, applies in particular to any region bounded by a finite number of distinct simple closed curves, with continuously turning tangent at every point. Now if the region T is a bounded open plane region whose frontier consists of a finite number of distinct connected sets, these boundary sets, being closed, will be at a finite distance from each other. Hence T may be regarded as the limit of regions T_n of the sort to which the Poincaré method applies, all of the same connectivity as T, and such that T_{n+1} includes T_n with its boundary. The Green's function $g_n(Q, M)$ of T_n is for each M an increasing function of n which does not exceed $G(Q, M)$, where $G(Q, M)$ is the Green's function for a circle of center Q and radius sufficiently large; hence g_n approaches a limit function $g(Q, M)$ which is evidently independent of the mode of formation of the T_n. This function, defined as 0 for all points on the boundary of T, may be called the Green's function for T.

---

*Summaries of these matters are given in G. Bouligand, *Fonctions harmoniques. Principes de Picard et de Dirichlet* [Mémorial des sciences mathématiques, fasc. XI] Paris (1926), and in the forthcoming report of O. D. Kellogg on the continuous boundary value problem, delivered as a Symposium before the American Mathematical Society, Dec. 1925. See also Goursat, *Cours d'Analyse*, vol. III.

By means of Harnack's theorem we know that $g(Q, M)$ is the sum of $\log 1/r$ and a function harmonic at all points of T. That it takes on continuously the boundary values 0 at all boundary points P is a theorem of Osgood.* The theorem is established by means of an elementary conformal transformation which carries T into a portion of the interior of a circle and, at the same time, the given boundary point P into a point of the circumference. In this form the demonstration applies directly to a simply connected region, but the theorem is seen to be valid for our region T since the Green's function for T with pole Q cannot exceed the Green's function with pole Q for any simply connected region containing T which has a part of the boundary of T for its boundary. That the method of proof is not accidental is evident from the fact that the theorem is not necessarily true for three-dimensional regions, or for plane regions of infinite connectivity.

Some of the methods devised for the continuous boundary value problem apply also to some extent to discontinuous boundary values. The method of integral equations may be so used for simple types of boundaries. More suggestive is the method of Zaremba.† He considers the quantity

(1) $$B(f, h) = \int_T \left(\frac{\partial f}{\partial x} \frac{\partial h}{\partial x} + \frac{\partial f}{\partial y} \frac{\partial h}{\partial y} \right) d\sigma,$$

in which $f(M)$ is the given continuous function and $h(M)$ is an arbitrary function, harmonic in T, subject merely to the conditions that $A(f)$ and $A(h)$ both exist, where

(2) $$A(f) = \int_T \left\{ \left(\frac{\partial f}{\partial x}\right)^2 + \left(\frac{\partial f}{\partial y}\right)^2 \right\} d\sigma.$$

* Osgood, *Funktionentheorie*, Leipzig (1912), p. 702.

† Fourth International Mathematical Congress, vol. 2, p. 194, Bulletin of the Academy of Sciences of Cracow (1909), p. 197. We have at hand merely the summaries given in the Jahrbuch der Fortschritte der Mathematik (1909) and in Lichtenstein's article in the Encyklopädie.

Take now functions $v(M)$, not necessarily harmonic in T, but such that $A(v)$ is defined, which satisfy a boundary condition which can be expressed in the form

$$B(v, h) = B(f, h)$$

for all functions $h(M)$ of the class specified. That this is a sort of average value boundary condition is seen by applying Green's theorem. The solution of the problem depends upon finding, by means of sequences of functions $v(M)$, a particular function $v(M)$ which gives to the integral $A(v)$ its minimum value.

This method of solution does not yield a result in some cases even with simple boundaries, such as circles. Thus with values given continuously on the circumference of a circle, a corresponding function $f(M)$ can be found, continuous within the circle and on the circumference, for such a function is given by Poisson's integral. But it may not be possible to give such a function which will satisfy the condition that $A(f)$ exists.* On the other hand the method does yield a result in many cases where the boundary values are discontinuous, and is thus a direct attack on the discontinuous boundary value problem.

Some of the methods cited are special cases of a general treatment of the solution of the Dirichlet problem as a functional of the boundary curve of the region and of the values on that curve. For boundaries which are not simple the solution appears as a continuous prolongation of the functional.†

45. Regions with continuous boundaries.

A special case of the continuous boundary value problem is that of the conformal transformation of a circle into a region bounded by a simple continuous curve. It is a problem which is all the more special on account of the fact that two conjugate harmonic

---

* Hadamard, Bulletin de la Societé Mathématique de France, vol. 34 (1906).

† The most systematic discussion of this point of view will be found in three articles by G. Bouligand, Bulletin des Sciences mathématiques, vol. 48 (1924), pages 183, 205, 246.

functions are at the same time taking on continuous boundary values. We have already had an instance of such specialization (Ex. 5, Art. 27). More remarkable is the theorem of Fejér: *if $w(z)$ maps a circle into the interior of a simply covered plane region bounded by a Jordan curve, the Taylor's series for $w(z)$, with respect to the center of the circle, converges uniformly, not only in the interior, but also on the circumference of the circle.*[*] In other words, the conjugate Fourier series for the boundary values $f(\theta)$, $g(\theta)$ are uniformly convergent, $0 \leq \theta \leq 2\pi$.

The theorem rests upon the following lemma. Let

$$(3) \qquad \sum_0^\infty u_i = u_0 + u_1 + \cdots + u_n + \cdots,$$

the terms being real or complex, be summable by the method of the arithmetic mean; if, further, the series

$$(4) \qquad \sum_1^\infty i|u_i|^2 = |u_1|^2 + 2|u_2|^2 + \cdots + n|u_n|^2 + \cdots,$$

is convergent, the series (3) is also convergent. If (3) is summable uniformly by the method of the arithmetic mean, and (4) is convergent, the series (3) is uniformly convergent.

We know also, as a classical theorem, that if $f(\theta)$ is a summable function in the Lebesgue sense its Fourier series is summable by the method of the arithmetic mean for any θ where $f(\theta)$ is the derivative of its indefinite integral, and uniformly through any closed interval where $f(\theta)$ is continuous.

Now we have noticed that the integral

$$I = \int \left\{ \left(\frac{\partial u}{\partial x}\right)^2 + \left(\frac{\partial u}{\partial y}\right)^2 \right\} d\sigma = \int \left\{ \left(\frac{\partial u}{\partial r}\right)^2 + \frac{1}{r^2}\left(\frac{\partial u}{\partial \theta}\right)^2 \right\} d\sigma,$$

if it exists, extended over the interior of the circle, is the area of the region into which the circle is transformed by

[*] Comptes Rendus, t. 156 (1913), p. 46, also in *Mathematische Abhandlungen H. A. Schwarz zu seinem fünfzigjährigen Doktorjubiläum gewidmet von Freunden und Schülern*, Berlin (1914), p. 42.

means of the conformal transformation defined by the harmonic function u and its conjugate. If we write
$$u(r, \theta) = \frac{1}{2} a_0 + \sum_1^\infty r^m (a_m \cos m\theta + b_m \sin m\theta)$$
this integral for the unit circle has the value
$$I = \lim_{r=1} \pi \sum_1^\infty m r^{2m} (a_m^2 + b_m^2);$$
and if I remains bounded in the limit, the series of positive terms
(5) $$\pi \sum_1^\infty m (a_m^2 + b_m^2)$$
is convergent; and conversely, if the series is convergent, the limit I exists, and represents this area. Moreover the convergence of (5) is also sufficient that the Fourier series
$$\frac{1}{2} a_0 + \sum_1^\infty (a_m \cos m\theta + b_m \sin m\theta)$$
shall be summable by the method of the arithmetic mean, almost everywhere, to a function $f(\theta)$ for which the a_m, b_m are the Fourier coefficients. In fact the existence of
$$\sum_1^\infty (a_m^2 + b_m^2)$$
is sufficient that $f(\theta)$ be summable in the Lebesgue sense, with its square.

If therefore the area of the region into which the circle is transformed, measured on the leaves of the Riemann surface, if necessary, remains finite, the lemma tells us that the two conjugate Fourier series converge almost everywhere on the circumference to functions $f(\theta)$, $g(\theta)$ such that
(6) $$\lim_{r=1} u(r, \theta) = f(\theta), \quad \lim_{r=1} v(r, \theta) = g(\theta).$$
A special case of this situation is that where $u(r, \theta)$, $v(r, \theta)$ are bounded and the transformed region is a simply covered plane region, or is spread on a Riemann surface of a finite number of leaves.

Suppose finally that the transformation maps the circle on a plane region interior to a simple closed continuous curve. Carathéodory's theorems, already cited, tell us that the transformation is continuous within and on the boundary, and therefore that the boundary values of the u and v are two continuous functions $f(\theta)$, $g(\theta)$. Hence the summation by the method of the arithmetic mean converges uniformly, and the theorem is proved.

EXERCISE. Interpret Fejér's theorem with respect to monotonic changes of variable in a given arbitrary continuous function $f(x)$, and development in Fourier series.

46. Regions with rectifiable boundaries. If the boundary values of the conjugate functions u, v, harmonic within the unit circle, are functions of limited variation of some parameter t on the boundary of the transformed region, they will also be functions of limited variation of θ; for they will be functions of limited variation of the arc-length s, and the latter is a monotonic function of θ. Hence the conditions given in Example 5, Art. 27 that

$$\int_0^{2\pi} \left| \frac{\partial u(r, \theta)}{\partial r} \right| d\theta, \qquad \int_0^{2\pi} \left| \frac{\partial u(r, \theta)}{\partial \theta} \right| d\theta$$

remain bounded as r approaches 1, or that $u(r, \theta)$, $v(r, \theta)$ be functions of θ of limited variation, uniformly in r, $r < 1$, are necessary and sufficient that the boundary of the transformed region be a rectifiable curve.

We wish to prove that *the transformation which carries the interior of the unit circle conformally into the interior of a region with rectifiable boundary, simply covered on a plane or on a Riemann surface, not necessarily of a finite number of leaves, is conformal almost everywhere on the boundary.* Incidentally the boundary values f, g will be seen to be *absolutely continuous* functions of θ, as they must necessarily be of s.

Consider a point P' on the circumference where $f'(\theta)$ and $g'(\theta)$ both exist, and $f'(\theta) \neq 0$. For the corresponding point P

of the boundary of the transformed region T the slope of the tangent is defined and is equal to

$$\frac{dv}{du} = \frac{g'(\theta)}{f'(\theta)} = \tan \beta.$$

Let s_1' be an arc in S which leads to P'; the corresponding arc s_1 in T leads to P (which may of course be a part of a multiple point of the boundary). We suppose that the direction of the arc s_1' is defined at every point, and changes continuously as its generic point M' approaches and coincides with P', making at P' an angle α', which is not zero,

$$\tan \alpha' = -\frac{dr}{r d\theta},$$

with the tangent direction.

On the curve s_1 we have

$$\frac{dv_1}{du_1} = \frac{\frac{\partial v}{\partial r} dr + \frac{\partial v}{\partial \theta} d\theta}{\frac{\partial u}{\partial r} dr + \frac{\partial u}{\partial \theta} d\theta}$$

But since $\partial u/\partial r$, $\partial u/\partial \theta$, $\partial v/\partial r$, $\partial v/\partial \theta$ approach their values at P' when M' approaches P' in the wide sense, by Art. 23, Chap. IV, we have

$$\lim_{M'=P'} \frac{dv_1}{du_1} = \frac{\frac{g'(\theta)}{f'(\theta)} - \frac{dr}{r d\theta}}{1 + \frac{g'(\theta)}{f'(\theta)} \frac{dr}{r d\theta}}$$
$$= \tan(\beta + \alpha'),$$

when $dr/d\theta$ approaches a finite value, and

$$\lim_{M'=P'} \frac{dv_1}{du_1} = -\frac{f'(\theta)}{g'(\theta)}$$

when $dr/d\theta$ approaches $\mp \infty$. Hence in all cases

$$\alpha' = \arctan \left(\frac{dv_1}{du_1}\right)_P - \beta$$

and the transformation is seen to be conformal at P'. Similar reasoning is valid if $g'(\theta)$ is not zero at P'.

EXERCISE 1. Let M_i' be a sequence of points in S and M_i the corresponding sequence in T. Show that a necessary and sufficient condition that M_i approach P and the direction of $M_i P$ approach a limiting direction not tangent to the boundary of T is the existence of the same situation in S, provided that $f'(\theta)$ or $g'(\theta)$ does not vanish at P'.

But the points where $f'(\theta)$ or $g'(\theta)$ fails to exist form a set of measure 0 on the circumference. The points where $f'(\theta)$ and $g'(\theta)$ both exist and vanish are points at which the transforming function $w(z)$ has a derivative $w'(z)$ which takes on the value 0 in the wide sense. Here we can make use of a remarkable theorem of Lusin and Priwaloff.* *The function $w'(z)$, holomorphic within the circle, cannot vanish on the circumference, in the wide sense, at a set of more than zero measure.* Hence the transformation defined by $w(z)$ is conformal for almost all θ.

The result may be made still more precise. In the above cited memoir the authors show that in the conformal transformation of the interior of the circle into a simply covered region with rectifiable boundary s, any point set of zero measure on the circumference will correspond to a set of zero measure on s.† The same is true for the general simply connected region with rectifiable boundary. In fact, if there were a set of positive measure on s, corresponding to a set of zero measure on the circumference, there would be such a set in one of the leaves of the Riemann surface, since the number of leaves is denumerable; and there would be a portion of this set, of positive measure, contained between two h-curves of finite length, drawn from a point 0.

But on the circle, the corresponding set would be of zero

---

* Annales scientifiques de l'École Normale Supérieure, t. 42 (1925), p. 159.

† Lusin and Priwaloff, loc. cit., p. 156. Reference is there made also to earlier papers: Lusin, *Sur la représentation conforme*, Bull. de l'Inst. Pol. Ivanovo-Vosn. (1919), and Priwaloff, *Intégrale de Cauchy*, Bull. de l'Univers. à Saratov. (1918).

measure, lying on an arc contained between two circles from a point O', drawn normally to the circumference. If we map this portion of the circle on the whole circle, the set on the arc will become a set on the circumference, still of zero measure; and by means of a transformation which is the succession of the inverses of the two just indicated, taken in the proper order, we shall transform the whole circle into the simply covered region whose rectifiable boundary is made up of a portion of s and the h-curves from O. Thus we arrive at a contradiction with the theorem cited.

The coördinates of a point on the boundary are thus seen to be absolutely continuous functions of θ as well as s. If we combine this result with that of Example 5, Art. 27, we notice that *if two conjugate Fourier series represent functions of limited variation, those functions must be absolutely continuous*. Finally, we are able to say that the transformation is conformal on the boundary except possibly at a set of zero measure, measured on s. This was the theorem to be proved.

At any boundary point P for T where the corresponding values of $f'(\theta)$ and $g'(\theta)$ are defined and not both zero, we may speak about approach to the boundary *in the wide sense*. For if we consider the transformation into the circle, any curve which leads to P, eventually coming and remaining within an angle at P whose sides make positive angles with the tangent, corresponds to a curve in the circle, leading to P', which eventually comes and remains within any angle whose sides make smaller angles with the tangent at P'—and *vice versa*. Hence we can generalize at once Fatou's theorem to regions with rectifiable boundaries, considering approach in the wide sense.

Let $f(s)$ be any function summable in the Lebesgue sense on s, and $F(s)$ a function of limited variation on s; they will then have the same properties with respect to θ, since s is a monotonic absolutely continuous function of θ. Hence the Dirichlet and Neumann problems may be stated directly in terms of $f(s)$ and $F(s)$. In particular, and this is the generalization of Fatou's theorem, *if $\zeta(M)$ is harmonic in T*

and of class (i) *it takes on boundary values almost everywhere on s in the wide sense*; in the subclass (ii) *it is determined by those boundary values.*[*]

For the Neumann problem we must have $F(s)$ periodic or

$$\int_s f(s)\,ds = 0.$$

Let $\zeta(M)$ be of the class (j). Then $\partial\zeta/\partial g$, taken along curves $h = $ const., takes on its boundary values in the wide sense, almost everywhere. But if we denote by n the direction of the interior normal to the curves $g = $ const., we have

$$\frac{\partial\zeta}{\partial n} = \frac{\partial\zeta}{\partial g}\frac{\partial g}{\partial n} = \frac{\partial\zeta}{\partial g} \Big/ \left(r^2\left(\frac{\partial u}{\partial r}\right)^2 + \left(\frac{\partial u}{\partial \theta}\right)^2\right)^{\frac{1}{2}},$$

where $w = u+iv$ maps the circle on T. Hence $\partial\zeta/\partial n$ *takes on its boundary values almost everywhere in the wide sense on s*; *and if $\zeta(M)$ is of the subclass* (jj) *it is determined by those boundary values*. It is in this sense that the Neumann problem is particularly related to regions with rectifiable boundaries, as distinguished from the generalized problem treated in Chap. V.

From the considerations of Chap. VI, we see that 'the theorems on these classes of functions apply to plane regions of finite connectivity bounded by $n+1$ rectifiable distinct curves. The conformal transformations involved refer to the simply connected regions determined by the boundaries separately.

EXERCISE 2. Show that θ is also an absolutely continuous function of s.

47. Regions of infinite connectivity. Consider a plane region T, bounded externally by a simple closed rectifiable curve C_0, and internally by a denumerable infinity of curves of the same kind, $C_1, C_2, \cdots$. The complete boundary of T consists of the set of points formed by these curves and all the limiting points of them. If $u(M)$, bounded and harmonic

[*] Plancherel, Bull. des Sc. Math., vol. 34 (1910), p. 111, for more special boundaries shows that if $\lim_{g=0}\int_0^h |u|\,ds = \int_0^h |f(s)|\,ds$, there can be not more than one solution.

in T, takes on the boundary values 0 continuously on C_0, $C_1, C_2, \cdots$ does it vanish identically in T?

Let E_k be the collection of limiting points of the set consisting of the points of

$$C_k + C_{k+1} + \cdots.$$

It is a closed set which includes all the points of C_k, $C_{k+1}, \cdots$. Define also

$$E = \lim_{k=\infty} E_k = E_0 \cdot E_1 \cdot E_2 \cdots,$$

which is also a closed set. The complete boundary of T is the set

$$E + C_1 + C_2 + \cdots.$$

A partial answer to our problem may now be stated as follows: *With the given hypotheses, if E is a reducible set, $u(M)$ vanishes identically in T.* We assume that E is finitely reducible.

In order to prove the proposition, let Q be an arbitrary point of the open region T. We shall show that if $\eta > 0$ is given, $|u(Q)| < \eta$. Suppose that E has successive derived sets $E', E'', \cdots, E^{(l)}, E^{(l)}$ consisting merely of a finite number of points

$$A_1^{(l)}, A_2^{(l)}, \cdots, A_{n_l}^{(l)}.$$

With these points as centers we draw circles of radii $\varrho_i^{(l)}$, respectively, small enough so that Q is exterior to them, and so that no point of $E^{(l-1)}$ lies on any circumference; but otherwise arbitrary. Outside of these circles there will remain a finite number of points of $E^{(l-1)}$, say

$$A_1^{(l-1)}, A_2^{(l-1)}, \cdots, A_{n_{l-1}}^{(l-1)}.$$

With these points as centers we draw circles of radii $\varrho_i^{(l-1)}$, respectively, distinct from the previous ones, small enough so that Q is exterior to them, and so drawn that no point of $E^{(l-2)}$ lies on any circumference; but otherwise arbitrary. Proceeding in this way we finally isolate all the points of E.

Let T' be the connected portion of T, exterior to these circles and containing Q. There will be only a finite number of the C_k, say $C_1, \cdots, C_q$, which have points outside or on the circumferences of these circles. The region which is defined by the curves $C_0, C_1, \cdots, C_q$, as complete boundary, is finitely connected, and the constructed circles either insert additional distinct boundaries or form cross or return cuts in this region. Hence the connectivity of T' is finite, and and it has no isolated point boundaries.

Let H be the diameter of T, and write

$$\eta' = \eta'' \cdots = \eta^{(l)} < \eta/(l+1)$$

Choose now the sets of constants, defined successively for $j = l, l-1, l-2, \cdots, 1, 0$, as follows:

(7) $\begin{cases} \eta_i^{(j)} = \eta^{(j)}/n_j, \quad \varepsilon_i^{(j)} = \eta_i^{(j)}/\log \dfrac{H}{A_i^{(j)} Q} \\ \varrho_i^{(j)} \text{ small enough so that } |u(M)| < \varepsilon_i^{(j)} \log H/\varrho_i^{(j)} \\ \text{when } M \text{ is on the circle of center } A_i^{(j)} \text{ and radius } \varrho_i^{(j)}. \end{cases}$

The function

(8) $$v(M) = \sum_{j=1}^{l} \sum_{i=1}^{n_j} \varepsilon_i^{(j)} \log H/\overline{A_i^{(j)} M}$$

is harmonic in T' and continuous in the region consisting of T' and its boundary points. It is moreover everywhere positive in T'. But on the boundary of T', $|u(M)| < v(M)$. Hence throughout T',

$$-v(M) < u(M) < v(M),$$

so that

$$|u(Q)| < v(Q) = \sum_{j=1}^{l} \sum_{i=1}^{n_j} \eta_i^{(j)} < \eta. \quad \text{[Q. E. D.]}$$

For what has so far been obtained the hypothesis that the boundaries $C_0, C_1, \cdots$ be rectifiable is unnecessary; in fact, beyond the requirement that any finite number of them counting from C_0, $C_0 + C_1 + \cdots + C_q$, form the boundary of a bounded finitely connected region without isolated point boundaries, it is merely assumed that E be reducible.

A different kind of theorem is however not so easily extended to general regions.

Let us assume the C_i rectifiable again, but instead of requiring the $u(M)$ to take on the value 0 continuously, let us suppose merely that the function, remaining bounded, takes on the value 0 on each C_i almost everywhere in the narrow sense, that is, as M approaches the boundary along the curves $h_i(Q, M) = $ const., where h_i is the conjugate of the Green's function for the region bounded by C_i and containing T. Then *if E is reducible, $u(M)$ vanishes identically in T.* This is a discontinuous boundary value problem.

From Art. 40 we know that we can express $u(M)$ in T' by means of the Green's function integral, which now takes this form (see Art. 39)

$$u(M) = \frac{1}{2\pi} \int_s u(P) \frac{\partial g(M, P)}{\partial n_P} ds_P$$

in terms of its boundary values. The same applies to the function $v(M)$ given by (8). By comparing these explicit expressions we find again $|u(Q)| < v(Q) < \eta$, and thus prove the theorem.

EXERCISE. Replace the condition that $u(M)$ be bounded by a condition of uniform absolute continuity with respect to each C_i, assuming that $u(M)/\log PM$ has the limit 0 as M approaches P, P being a point of E.

We have been concerned here with a set E of *zero capacity.*[*] Capacity is measured by

$$\frac{1}{2\pi} \int_s \frac{\partial u}{\partial n} ds,$$

where s is any curve or system of curves enclosing E, and u is the lower bound of all conductor potentials which take the value 1 on curves s enclosing E. Incidentally, a function harmonic in a certain region except at a set of interior points of zero capacity has only unnecessary discontinuities, if it

[*] Wiener, Journal of Mathematics and Physics of the Massachusetts Institute of Technology, vol. 3 (1924), p. 26.

remains bounded. In fact, if we denote by B a closed set of frontier points of T which are such that $B + T$ is still an open region, a necessary and sufficient condition that

$$\lim_{M=P} G(M, P) > 0$$

is that B be of capacity 0.* The same concept is valuable for the discussion of the regularity or irregularity, as far as the continuous boundary value problem is concerned, of arbitrary boundary points.†

Let us in conclusion sketch the proof of a general theorem. Let $f_i(P)$ be assigned, bounded, in numerical value less than a fixed number N, and integrable in the Lebesgue sense, on C_i, $i = 0, 1, \cdots$. Let $u_0(M)$ be the harmonic function determined in the region T_0 with boundary C_0 by the boundary values $f_0(P)$, and $u_i(M)$ be the harmonic function determined in the region T_k with boundary $C_0 + C_1 + \cdots + C_k$, with boundary values $f_0(P), f_1(P), \cdots, f_k(P)$. Consider $\lim_{k=\infty} u_k(M)$. The functions u_k are bounded in their set.

If E is reducible, it must be reducible on each C_i. Hence it is possible to show (by taking each $\varrho_i^{(j)} < \delta_j/n_j$ where $\sum_1^\infty \delta_j$ is given with sum arbitrarily small, $< \delta$) that $|u_{k+p} - u_k|$ can be made uniformly small, n great enough, for all points on each given C_i or in T distant from E by $\geq \delta'$, where δ' is fixed but arbitrary. In particular then, the convergence of u_k to some function u is uniform in the neighborhood of any interior point of T, and $u(M)$ is harmonic in T.‡ Also $u(M)$ takes on the boundary values $f_i(P)$ at almost all points of each C_i. By means of this reasoning we can state the theorem:

* O. D. Kellogg, Proceedings of the National Academy of Sciences, vol. 12 (1926), p. 400. The set E above is not a set B. Compare also the theorem of Art. 50, below.

† For such a discussion of necessary and sufficient conditions we refer again to Bouligand's Mémorial. The reader will find references to work of Zaremba, Lebesgue, Bouligand, Wiener and Kellogg.

‡ Osgood, loc. cit., p. 652.

With the given hypotheses on T, if E is reducible, a function $u(M)$ bounded and harmonic in T takes on boundary values in the wide sense almost everywhere on each C_i. If these values are assigned arbitrarily, bounded in their set and integrable in the Lebesgue sense, there is one and only one function $u(M)$, bounded and harmonic in T, which takes on these functions as boundary values almost everywhere on each C_i in the narrow (and therefore in the wide) sense.

48. Remarks on necessary and sufficient conditions.

Our general point of view, especially in the first six chapters, has been the discussion of necessary and sufficient conditions. In other words we have dealt with the characterization of certain classes of harmonic functions. One feels somehow that it is desirable to have the characterization as simple as possible. But does it mean anything to say that a necessary and sufficient condition is simple? Let us dwell on this point merely long enough to answer the question in the negative—at least, if the reader will agree that it is not merely a question of simplicity in language, or choice of the smallest possible number of words.

We may suggest that simplicity means availability for application. On this hypothesis the term requires opposite qualities for necessary and sufficient conditions. For what makes a necessary condition simple is that it gives as much information as possible about a class of functions, and what makes a sufficient condition simple is that it requires as little information as possible. In other words, if we are dealing with a class of harmonic functions, a sufficient condition is simple if it states as little as possible in order to make the function belong to the class, beyond the fact that the function is harmonic; and thereby uses the harmonic property as much as possible in the demonstration of it; on the other hand, a necessary condition states as much as possible beyond the fact that the function is harmonic. The terms are necessarily vague, but perhaps an example will make clear the point of view.

Consider a function $u(r, \theta)$ harmonic within a unit circle.

The necessary and sufficient condition of Bray and the author (see Chap. III) that such a function be given by a Poisson integral in terms of its boundary values is that there should be a sequence of values r_i, $r_{i+1} > r_i$ with $\lim r_i = 1$, such that the absolute continuity of the integral

$$\int u(r_i, \theta) \, d\theta$$

be uniform for all i. Noaillon's condition (see also Chap. III) is that $u(r, \theta)$ should converge in the mean, of order 1, to a summable function $f(\theta)$. Consider the sufficiency condition.

It will be shown in Art. 49 that for any sequence of summable functions which converges in the mean, of order 1, the absolute continuity of the integral is uniform throughout the sequence. On the other hand, the sequence of functions $f_i(\theta) = \text{const.} = N_i$, where $\lim N_i = \infty$, is a sequence which satisfies the condition of Bray and the author without satisfying the condition of Noaillon; i. e., aside from the condition of being harmonic, less information is required for the former condition than for the latter. It is evident, for instance, that any bounded harmonic function satisfies the uniform absolute continuity condition, but not that it satisfies that of convergence in the mean. Similar remarks may be made for other special cases. Hence in spite of an apparent complexity of statement, it is clear that, as a sufficient condition, that of uniform absolute continuity is more useful. Perhaps there is a place for some other sufficient condition, still more simple than either; one which requires less of the general theory of integration of a sequence of functions.

On the other hand, as a necessary condition, Noaillon's condition states a simple and useful fact that is not a bare consequence of the other, but yields additional information which the other furnishes only through interrogating again the properties of harmonic functions. Moreover Noaillon's condition suggests an interesting classification, which we shall pursue further.

49. Convergence in the mean of positive order less than one.*

With reference to an interval (a, b) we say that $f_m(x) \geq 0$ converges in the mean, of order ν, $\nu > 0$, if $(f_m(x))^\nu$ is summable, and if

$$(9) \quad \lim_{n=\infty} \int_a^b |f_m(x) - f_k(x)|^\nu\, dx = 0, \qquad m, k \geq n.$$

In fact, since f_m and f_k are not negative we have

$$|f_m(x) - f_k(x)|^\nu \leq \{f_m(x)\}^\nu + \{f_k(x)\}^\nu,$$

and therefore the left hand member of the inequality represents a summable function of x.

We say that $f_m(x) \geq 0$ converges in the mean, of order ν, $\nu > 0$, to $f(x)$ if $\{f_m(x)\}^\nu$ and $\{f(x)\}^\nu$ are summable, and if

$$(10) \quad \lim_{n=\infty} \int_a^b |f(x) - f_n(x)|^\nu\, dx = 0.$$

On the basis of these definitions a number of simple theorems may be established. For this purpose a well known fact for the case $\nu = 2$ will be assumed, viz., if $f_m(x)$ converges in the mean, of order 2, there will be a subsequence $f_{m_i}(x)$ which converges to a function $f(x)$ as a true limit, almost everywhere.

LEMMA 1. *If $f_m(x) \geq 0$ is convergent in the mean, of order ν, $\{f_m(x)\}^{1/2}$ is convergent in the mean, of order 2ν.*

In fact,

and evidently
$$f_m{}^\nu = \{f_m{}^{1/2}\}^{2\nu}$$

$$|f_m - f_k| \geq (f_m{}^{1/2} - f_k{}^{1/2})^2,$$
$$|f_m - f_k|^\nu \geq (f_m{}^{1/2} - f_k{}^{1/2})^{2\nu},$$

from which the proposition follows.

LEMMA 2. *If $f_m(x) \geq 0$ is convergent in the mean, of order ν, it is convergent in the mean of order $\varkappa$, where $\varkappa$ is arbitrary, less than ν.*

---

* The theorems given below apply also to $r > 1$. They are to some extent proved in Hobson, *The theory of functions of a real variable*, vol. II, Cambridge (1926), p. 239 ff., but for convenience in presentation are attached here to the classical case, $\nu = 2$.

Evidently $f_m^\varkappa$ is summable, if f_m^ν is summable. We are given further that
$$\int_a^b |f_m - f_k|^\nu \, dx < \varepsilon, \qquad m, k \geq n.$$
Denote by $E_{m,k}$ the set
$$E_{m,k} = E(|f_m - f_k|^\nu \geq \varepsilon^{1/2}).$$
Then meas. $E_{m,k} \leq \varepsilon^{1/2}$, and in $E_{m,k}$ we have
$$|f_m - f_k|^\varkappa \leq 1 \text{ or } < |f_m - f_k|^\nu.$$
Hence
$$\int_{E_{m,k}} |f_m - f_k|^\varkappa \, dx < \int_{E_{m,k}} \{1 + |f_m - f_k|^\nu\} \, dx$$
$$< \varepsilon^{1/2} + \int_a^b |f_m - f_k|^\nu \, dx < \varepsilon^{1/2} + \varepsilon.$$
But in $CE_{m,k}$, $|f_m - f_k|^\varkappa < \varepsilon^{\varkappa/2\nu}$, and
$$\int_{CE_{m,k}} |f_m - f_k|^\varkappa < \varepsilon^{\varkappa/2\nu}(b-a).$$
Hence finally,
$$\int_a^b |f_m - f_k|^\varkappa \, dx < \varepsilon^{\varkappa/2\nu}(b-a) + \varepsilon^{1/2} + \varepsilon, \qquad m, k \geq n,$$
which has the limit 0 as n becomes infinite.

THEOREM 1. *If $f_m(x) \geq 0$ is convergent in the mean, of order ν, there is a subsequence $f_{m_i}(x)$ and a function $f(x)$ such that*

(11) $$\lim_{i = \infty} f_{m_i}(x) = f(x), \text{ almost everywhere.}$$

If $\nu \geq 2$, it follows from Lemma 2 that $f_m(x)$ is convergent in the mean, of order 2; and the theorem reduces to the classical one.

If $0 < \nu < 2$, there will be, by Lemma 1, an integer p great enough so that
$$(f_m)^{1/2^p} = \varphi_m$$

will be convergent in the mean of order $2^p \nu \geq 2$. Hence there will be a subsequence φ_{m_i} and a function φ such that

$$\lim_{i=\infty} \varphi_{m_i} = \varphi, \text{ almost everywhere.}$$

We write then

$$f(x) = \varphi^{2^p},$$

and the theorem is proved.

LEMMA 3. *If $f_m(x) \geq 0$ is convergent in the mean, of order ν, the absolute continuity of $\int f_m^\nu\, dx$ is uniform for all m.*

Let us assume the contrary. If the absolute continuity is not uniform, there will be an $\eta > 0$ such that no matter what δ is given, arbitrarily small, and what p_0 arbitrarily large, there will be a set E, meas. $E < \delta$, and a p, $p > p_0$, for which

$$\int_E f_p^\nu\, dx > \eta.$$

Choose now p_0 great enough so that

$$\int_a^b |f_p(x) - f_{p_0}(x)|^\nu\, dx < \eta/2^{\nu+1}, \qquad p \geq p_0,$$

and take δ small enough so that

$$\int f_{p_0}^\nu\, dx < \eta/2^{\nu+1} \text{ when meas. } e < \delta.$$

This yields a contradiction for some $p > p_0$. In fact,

$$\int_E f_p^\nu\, dx = \int_E |f_{p_0} + f_p - f_{p_0}|^\nu\, dx$$
$$\leq 2^\nu \int_E f_{p_0}^\nu\, dx + 2^\nu \int_E |f_p - f_{p_0}|^\nu\, dx < \eta,$$

which for some $p > p_0$ contradicts the first inequality.

An application of De la Vallée Poussin's theorem of Chap. I to Lemma 3 yields at once the fact that f^ν is summable, and that

$$\lim_{m=\infty} \int_{x_1}^{x_2} f_{m_i}^\nu\, dx = \int_{x_1}^{x_2} f^\nu\, dx.$$

Also
$$|f_m-f|^\nu = |f_m-f_{m_i}+f_{m_i}-f|^\nu$$
$$\leq 2^\nu \{|f_m-f_{m_i}|^\nu+|f_{m_i}-f|^\nu\}$$
and
$$\int_a^b |f_m-f|^\nu \, dx \leq 2^\nu \varepsilon + \int_a^b |f_{m_i}-f|^\nu \, dx.$$

But the absolute continuity of $\int |f_{m_i}-f|^\nu \, dx$ is uniform, since
$$|f_{m_i}-f|^\nu \leq 2^\nu (f_{m_i}^\nu+f^\nu),$$
and therefore
$$\lim_{i=\infty} \int_a^b |f_{m_i}-f|^\nu \, dx = 0.$$
Hence
$$\lim_{m=\infty} \int_a^b |f_m-f|^\nu \, dx = 0,$$

and $f_m(x)$ converges in the mean, of order ν, to $f(x)$.

There cannot be any function $F(x)$ which differs from $f(x)$ on a set of positive measure, and possesses the property (11) or (10). For
$$|F-f|^\nu \leq 2^\nu \{|F-f_m|^\nu+|f-f_m|^\nu\}.$$

Hence we have the following theorem:

THEOREM 2. *If $f_m(x) \geq 0$ converges in the mean, of order ν, then*
$$\lim_{i=\infty} \int_{x_1}^{x_2} f_{m_i}^\nu \, dx = \int_{x_1}^{x_2} f^\nu \, dx,$$

for the m_i of Theorem 1, and $f_m(x)$ converges in the mean, of order ν, to $f(x)$. The function $f(x)$, except on an arbitrary set of zero measure, is the only such limit function.

We have also
$$|f_m-f_k|^\nu \leq 2^\nu \{|f-f_m|^\nu+|f-f_k|^\nu\},$$

so that we can state immediately the converse of Theorem 2:

THEOREM 3. *If $f_m(x) \geq 0$ converges in the mean, of order ν, to $f(x)$, it converges in the mean, of order ν.*

RELATED PROBLEMS

On the basis of these theorems we may prove a proposition which separates sequences which converge in the mean from those whose integrals become infinite.

THEOREM 4. *Let $f_m(x)$ be a sequence of not negative functions which has the limit 0 for almost all x. Then if $\nu > 0$ exists so that $f_m(x)$ fails to converge in the mean, of order ν, there will be a subsequence $f_{m_i}(x)$ such that for any $\varkappa$, $\varkappa > \nu$, we shall have*

$$\lim_{i=\infty} \int_a^b f_{m_i}^{\varkappa} \, dx = \infty.$$

Assume the theorem to be false. There will then be a subsequence $f_{m_k}(x)$ such that

$$\int_a^b f_{m_k}^{\nu} \, dx \geq A > 0, \qquad k = 1, 2, \cdots.$$

Given a decreasing sequence of constants δ_i with $\lim \delta_i = 0$, there will be a corresponding sequence of sets E_i, with meas. $E_i < \delta_i$, and a subsequence of the f_{m_k}, namely f_{m_i}, such that

$$\begin{cases} \int_{E_i} f_{m_i}^{\nu} \, dx > \eta, & \eta \text{ some constant} > 0, \\ (f_{m_i}(x))^{\nu} > \delta_i^{-1/2}, & x \text{ in } E_i. \end{cases}$$

In fact, we know from De la Vallée Poussin's theorem that there is an $\eta' > 0$ and a sequence of sets E_i' such that for a certain subsequence f_{m_i} we shall have

$$\int_{E_i'} f_{m_i}^{\nu} \, dx > \eta', \qquad \text{meas. } E_i' < \delta_i.$$

But if E_i'' is the portion of E_i' for which $(f_{m_i}(x))^{\nu} \leq \delta_i^{-1/2}$, we shall have

$$\int_{E_i''} f_{m_i}^{\nu} \, dx \leq \int_{E_i'} \delta_i^{-1/2} \, dx \leq \delta_i^{1/2}.$$

If we write then $E_i = E_i' - E_i''$, take $\eta < \eta'$ and begin the sequence m_i late enough, all the conditions will be satisfied.

If now ϱ is any positive number,

$$\int_{E_i} f_{m_i}{}^{\nu+\varrho} dx > \delta_i{}^{-\varrho/2\nu} \int_{E_i} f_{m_i}{}^{\nu} dx > \eta\, \delta_i{}^{-\varrho/2\nu},$$

so that the limit of the left hand member becomes infinite with i. But all the more

$$\lim_{i=\infty} \int_a^b f_{m_i}{}^{\nu+\varrho} dx = \infty,$$

and this is what was to be proved.

We may apply this theorem directly to the Poisson-Stieltjes integral in terms of the following proposition.

THEOREM 5. *If $u(r, \theta)$ is given inside the unit circle by the Poisson-Stieltjes integral*

$$u(r, \theta) = \frac{1}{2\pi} \int_0^{2\pi} \frac{(1-r^2)\, dF(\varphi)}{1+r^2-2r\cos(\varphi-\theta)}$$

with $F(\varphi)$ of limited variation, and $F(\varphi+2\pi) = F(\varphi) + F(2\pi) - F(0)$, then, if we write $f(\theta) = F'(\theta)$ where that derivative exists, we have

$$\lim_{r=1} \int_0^{2\pi} |f(\theta) - u(r, \theta)|^{\nu}\, d\theta = 0,$$

for all $\nu < 1$.

For if this equation were not satisfied, there would be, for some $\nu < 1$, a sequence r_i of values r such that $|f(\theta) - u(r_i, \theta)| = f_i(\theta)$ did not converge in the mean, of order ν. But $\lim f_i(\theta) = 0$ almost everywhere; hence by Theorem 4 there would be a subsequence r_j, with $\lim r_j = 1$, so that

$$\lim_{j=\infty} \int_0^{2\pi} |f(\theta) - u(r_j, \theta)|\, d\theta = \infty$$

since $1 > \nu$. Hence

$$\lim_{j=\infty} \int_0^{2\pi} |u(r_j, \theta)|\, d\theta = \infty,$$

which is impossible, since $u(r, \theta)$ is of class (i).

In other words, if $u(r, \theta)$ is given by a Poisson-Stieltjes integral, but not by a Poisson integral, $u(r, \theta)$ fails to converge in the mean, of order 1, by Noaillon's theorem, but converges in the mean, of all orders < 1.

50. Integro-differential equations of Bôcher type.

In a study in which are considered the behaviour of a function in general, its values almost everywhere, the values of its integral, and so on, one may well ask if the point of view should not be broader still, and if instead of Laplace's equation itself, some integral equation, or some equation in average values should not more properly have been considered. The answer to this demand is that under very general conditions, any equation which seeks to express in more general language the physical idea behind Laplace's equation will imply Laplace's equation itself.* One may use properties that depend on the mean value theorem, even to the extent of a definition by "médiation spatiale" after Zaremba and Lebesgue one may generalize the Laplacian operator by using a second order limit, or by regarding the operator as a functional derivative, or one may, with Bôcher, replace the equation of Laplace by suitable integro-differential equations. The latter seems closest to physical concepts of the operator.

The following theorem is a generalization of that of Bôcher.

Consider a class of simple rectifiable curves in T, each with only a finite number of vertices and such that the integral

$$\int_s \frac{|\cos nr|}{r} ds$$

is bounded for each curve, where $r = PP'$, the distance between two points on the curve. Let $u(M)$ be summable on every curve of this class, and let the partial derivatives of u in the x and y directions be summable over the interior of such curves, and further, let the equations

(12)
$$\int_s u\, dy = \int_\sigma \frac{\partial u}{\partial x} d\sigma,$$
$$\int_s u\, dx = -\int_\sigma \frac{\partial u}{\partial y} d\sigma$$

be valid.

---
* See G. Bouligand, Mémorial, loc. cit., Arts. 3, 4, 5, 13.

If the equation

(13) $$\int_s \frac{\partial u}{\partial x} dy - \frac{\partial u}{\partial y} dx = 0$$

holds for all s in T of the given class, the function u has merely unnecessary discontinuities at a set of points of superficial measure 0, *and when these are removed the function* $u(M)$ *becomes harmonic in T*.*

It is sufficient to use instead of the derivatives $\partial u/\partial x$ and $\partial u/\partial y$ the derivatives in any two directions α, β and define the derivatives in other directions vectorially. One may, in fact, merely assume that the quantity

$$D_\alpha u = \lim_{\sigma=0} \frac{1}{\sigma} \int_s u \, d\alpha'$$

(α' the direction $\pi/2$ in advance of α, σ a "regular" family), is defined almost everywhere in T for two given directions α, β, and satisfies the equations corresponding to (12). Then $D_s u$ is determined for an arbitrary direction s, and is a vector; and equation (13) takes the form

(14) $$\int_s D_n u \, ds = 0.$$

These latter considerations become important for the consideration of the natural generalization of Poisson's equation

* G. C. Evans, Rice Institute Pamphlets, vol. 7 (1920), p. 286. For formula (23), p. 282 and that on the bottom of p. 279 there should be required an additional hypothesis, say that $\partial v/\partial x$ and $\partial w/\partial y$ remain bounded. This does not affect the validity of the theorem of Art. 5.33 (although it requires an obvious modification in the first part of the proof) nor of the theorem given above.

The theorem given above is true if the equation (13) is required merely for circles, as G. Bouligand remarks, on the basis of a general theorem of Zaremba. In fact, the demonstration cited above uses nothing but circles, and does not really require *all* circles in T. The equation (13) may also accordingly be required merely for rectangles. Bôcher's theorem is for circles, but assumes continuity. These considerations are not so important if one regards the subject from a physical point of view [see Evans, loc. cit., p. 261, also Cambridge Colloquium of the American Mathematical Society, Part I, New York (1918), p. 81 footnote].

(15) $$\int_s D_n u \, ds = F(s),$$

where $F(s)$ is an additive function of curves of limited variation, corresponding therefore to an arbitrary completely additive function of point sets in the plane,—i. e., to an arbitrary distribution of mass in the plane.* The solution of Laplace's equation is therefore the key to the solution of (15) and to other partial differential equations of elliptic type, their corresponding integro-differential equations, and their discontinuous boundary value problems.

On a different side, the study of harmonic functions is of course interpreted by the theory of functions of a complex variable. But the problem becomes sharpened by the presence of two conjugate harmonic functions. From this point of view the discontinuous boundary value problem is allied to the study of the singularities of a Taylor's series on the circumference of the circle of convergence, and the determination of a Taylor's series in terms of its singularities in the plane or on its Riemann surface.†

* G. C. Evans, Rice Institute Pamphlets, loc. cit.

† See Hadamard and Mandelbrojt, *La série de Taylor et son prolongement analytique*, Scientia 41, 2nd edition, Paris (1926).

INDEX

Absolute continuity, 12; uniform, 13
Accessible boundary points, 74; order of, 75
Bôcher, 91, 111, 145
Borel, 7, 12, 22
Born, 122
Bouligand, 123, 136, 145, 146
Bray, 14, 51, 138; Helly-Bray theorem, 15, 16, 60, 100
Capacity, 135
Carathéodory, 74, 75, 128
Cauchy, 9, 109; integral formula of, 65
Conformal transformation, 1, 26, 69; of finitely connected regions, 107
Courant, 74; Hurwitz - Courant, *Funktionentheorie*, 74, 107
Daniell, 79, 105
Dirichlet problem, 1, 46, 54, 73, 78, 100, 108, 116, 125, 131
Discard, 62
Fatou, 22, 43, 131
Fejér, 24, 30, 64, 126
Fischer-Riesz theorem, 68
Fredholm, 115, 123
Function, of limited variation, 1, 79; of limited variation on E, 5; negative variation, 3; positive variation, 2, 5; summable, 12; total variation, 3, 39
Goursat, *Cours d'analyse*, 121, 123
Green's function, 71, 93, 104, 122
Hadamard, 109, 125, 147
Harmonic function, 26; of physical character, 59

Harnack, 124
Helly, 14; see Bray
Hilbert, 123
Hobson, *Real variable*, 68, 139
Hurwitz, see Courant
Integral, of Lebesgue, 12; Poisson, 28, 30, 43, 87, 88; Poisson-Stieltjes, 35, 88; Riemann-Stieltjes, 8; Stieltjes, 1, 7
Kellogg, 43, 50, 111, 123
Laplace's equation, 26, 145
Law of the mean, 8, 10, 11, 17
Lebesgue, 12, 86, 104, 109, 123, 136, 145
Le Roux, 123
Lichtenstein, 85, 93, 124
Lusin, 130
Mandelbrojt, 147
Multiple boundary point, 75
Neumann, C., 123
Neumann problem, 1, 55, 58, 84, 108, 117, 131
Noaillon, 50, 138, 144
Osgood, 12, 30, 74, 82, 124; and Taylor, 74; *Funktionentheorie*, 56, 69, 71, 90, 95, 96, 124, 136
Phillips, see Wiener
Plancherel, 132
Plemelj, 55
Poincaré, 110, 123
Poisson's equation, 146
Potential, of double layer, 1, 60; of single layer, 1, 55
Primenden, 78

Priwaloff, 130
Rademacher, 22
Raynor, 111
Riemann, 123
Riemann surface, 94, 95, 99, 106, 107, 127
Riesz, see Fischer-Riesz theorem
Robinson, 21
Schwarz, 123, 126
Semi-continuous, 81
Stieltjes integral equations, 101, 113, 118

Summability of series, 22
Taylor, see Osgood
Vallée Poussin, De la, 13, 14, 34, 78, 141, 143
Vitali, 7, 61, 62
Vivanti, 115
Volterra, 59
Wide sense, 39
Wiener, 79, 135, 136; Phillips and, 123
Zaremba, 121, 123, 124, 136, 145, 146

FUNDAMENTAL EXISTENCE THEOREMS

BY

GILBERT AMES BLISS

CONTENTS

	Pages
INTRODUCTION	1

CHAPTER I
ORDINARY POINTS OF IMPLICIT FUNCTIONS

1. The fundamental theorem ... 7
2. Equations in which the functions are analytic ... 12
3. Goursat's method of approximation ... 16
4. Bolza's extension of the fundamental theorem ... 19
5. The unique sheet of solutions associated with an initial solution ... 21
6. Auxiliary theorems and definitions ... 28
7. A criterion that a sheet of solutions be single-valued ... 33
8. Transformations of n variables and a modification of a theorem of Schoenflies ... 37

CHAPTER II
SINGULAR POINTS OF IMPLICIT FUNCTIONS

Introduction ... 43
9. The preparation theorem of Weierstrass ... 49
10. The zeros of $\varphi(u, v)$, $\psi(u, v)$, or their functional determinant ... 53
11. Singular points of a real transformation of two variables ... 60
12. The case where the functional determinant vanishes identically ... 67
13. A generalization of the preparation theorem of Weierstrass ... 70
14. Applications of the preceding theory ... 78

CHAPTER III

Existence Theorems for Differential Equations

Introduction..	86
15. The convergence inequality........................	88
16. The Cauchy polygons and their convergence over a limited interval..............................	89
17. The existence of a solution extending to the boundary of the region R.............................	93
18. The continuity and differentiability of the solutions	95
19. An existence theorem for a partial differential equation of the first order which is not necessarily analytic..	98

FUNDAMENTAL EXISTENCE THEOREMS

BY

GILBERT AMES BLISS

INTRODUCTION

The existence theorems to which these lectures are devoted have been the subject of a long sequence of investigations extending from the time of Cauchy to the present day, and have found application at the basis of a variety of mathematical theories including, as perhaps of especial importance, the theory of algebraic functions and the calculus of variations. If a single solution $(a; b) = (a_1, a_2, \cdots, a_m; b_1, b_2, \cdots, b_n)$ of a set of equations

$$f_\alpha(x_1, x_2, \cdots, x_m; y_1, y_2, \cdots, y_n) = 0 \quad (\alpha = 1, 2, \cdots, n)$$

is known, then in a neighborhood of $(a\,;b)$ there is one and only one other solution corresponding to each set of values x in a properly chosen neighborhood of the values a, and in the totality of solutions $(x\,;y)$ so defined the variables y are single-valued and continuous functions of the x's. If a set of initial constants $(\xi, \eta_1, \eta_2, \cdots, \eta_n)$ is given, then in a neighborhood of these values there is one and but one continuous arc

$$y_\alpha = y_\alpha(x) \quad (\alpha = 1, 2, \cdots, n)$$

satisfying the differential equations

$$\frac{dy_\alpha}{dx} = g_\alpha(x, y_1, y_2, \cdots, y_n) \quad (\alpha = 1, 2, \cdots, n)$$

and passing through the initial values η when $x = \xi$.

The formulation and first satisfactory proofs of these theorems, at least for the case where only two variables x, y are involved, seem to be ascribed with unanimity to Cauchy. For the implicit functions his proof rested upon the assumption that the function f should be expressible by means of a power series, and the solution he sought was also so expressible, a restriction which was later removed with remarkable insight by Dini. For a differential equation, on the other hand, Cauchy assumed only the continuity of the function g and its first derivative for y, and his method of proof, with the well-known alteration due to Lipschitz, retains to-day recognized advantages over those of later writers.

In the following pages (§§ 1, 16) the two theorems stated above are proved with such alterations in the usual methods as seemed desirable or advantageous in the present connection. The proof given for the fundamental theorem of implicit functions is applicable when the independent variables x are replaced by a variable p which has a range of much more general type than a set of points in an m-dimensional x-space.* It is not necessary always to know an initial solution in order that others may be found. In the treatment of Kepler's equation, for example, which defines the eccentric anomaly of a planet moving in an elliptical orbit in terms of the observed mean anomaly, one starts with an approximate solution only and determines an exact solution by means of a convergent succession of approximations. This procedure is closely allied to a method of approximation due to Goursat (§ 3), suggested apparently by Picard's treatment of the existence theorem for differential equations.

One of the principal purposes of the paragraphs which follow, however, is to free the existence theorems as far as possible from

---

* The notion of a general range has been elucidated by Moore, The New Haven Mathematical Colloquium, page 4, the special cases which he particularly considers being enumerated on page 13. An application of the method of § 1 of these lectures when the range of p is a set of continuous curves, has been made by Fischer, "A generalization of Volterra's derivative of a function of a line," Dissertation, Chicago (1912).

the often inconvenient restriction which is implied by the words
" in a neighborhood of," or which is so aptly expressed in German
by the phrase " im Kleinen." It is evident from very simple
examples that the totality of solutions $(x; y)$ associated continuously with a given initial solution of a system of equations
$f = 0$ of the form described above, can not in general have the
property that the variables y are everywhere single-valued
functions of the variables x, and the result of attempting,
perhaps unconsciously, to preserve the single-valued character
of the solutions has been the restriction of the region to which the
existence theorems apply. In order to avoid this difficulty and
to characterize to some extent the totality of solutions associated
continuously with a given initial one in a region specified in
advance, the writer has introduced (§ 5) the notion of a particular
kind of point set called a sheet of points. In a suitably chosen
neighborhood of a point $(a; b)$ of the sheet there corresponds
to every set of values x sufficiently near to the values a exactly
one point $(x; y)$ of the sheet, and the single-valued functions
y so determined are continuous and have continuous first derivatives. This condition does not at all imply that there are
no other points of the sheet outside the specified neighborhood
of the point $(a; b)$ and having a projection x near to a. With
the help of the notion of a sheet of points it can be concluded that
with any initial solution $(a; b)$ of the equations $f = 0$ there is
associated a unique sheet S of solutions whose only boundary
points are so-called exceptional points where the functions f
either actually fail, or else are not assumed, to have the continuity
and other properties which are demanded in the proof of the
well-known theorem for the existence of solutions in a neighborhood of an initial one. It is important oftentimes to know
whether or not a sheet of solutions is actually single-valued
throughout its entire extent, and a criterion sufficient to ensure
this property has also been derived (§ 7).

On the basis of these results some important theorems concerning the transformation of plane regions into regions of

another plane by means of equations of the form

$$x_1 = \psi_1(y_1, y_2), \quad x_2 = \psi_2(y_1, y_2),$$

as in the theory of conformal transformation, have been deduced (§ 8). If the functions ψ have suitable continuity properties and a non-vanishing functional determinant in the interior of a simply closed regular curve B in the y-plane, and if B is transformed into a simply closed regular curve A of the x-plane, then the equations define a one-to-one correspondence between the interiors of A and B, and the inverse functions so defined have continuity properties similar to those of ψ_1 and ψ_2. This is but a sample of the theorems which may be stated. Others are also given (§ 8) which apply to the transformation of regions not necessarily finite, and to systems containing more than two equations.

The theory of the singularities of implicit functions is of considerable difficulty and has been but incompletely developed. For a transformation of the form above in which the functions ψ_1, ψ_2 are analytic, the singular point to be studied, at which the functional determinant $D = \partial(\psi_1, \psi_2)/\partial(y_1, y_2)$ vanishes, as well as its image in the x-plane, may both without loss of generality be supposed at the origin. The most general case under these circumstances is that for which the determinant D does not vanish identically and the equations $\psi_1 = 0$, $\psi_2 = 0$ have no real solutions in common near the origin except the values $y_1 = y_2 = 0$ themselves. It is found that the branches of the curve $D = 0$ bound off with a suitably chosen circle about the origin a number of triangular regions. Each of these regions is transformed in a one-to-one way into a sort of Riemann surface on the x-plane which winds about the origin and is bounded by the image of the boundary of the triangular region (see § 11, Fig. 6). If the signs of D in two adjacent triangular regions are opposite, then their images overlap along the common boundary; otherwise they adjoin without overlapping. At any point of one of the Riemann surfaces the inverse functions defined by the transformation are continuous and in the interior of the surface they have everywhere continuous derivatives. These results are obtained by means of applications of the theorem described above for the transformation of the interior of a simply closed curve B; and the same method of procedure would undoubtedly be of service when the curves $\psi_1 = 0$, $\psi_2 = 0$ have real branches through the origin in common, which must occur whenever they have common points in every neighborhood of the values $y_1 = y_2 = 0$. The case where the determinant D vanishes identically is also considered (§ 12).

For the singularities of implicit functions defined by a system of equations $f = 0$ there is a generalization of the preparation theorem of Weierstrass (§ 9) suggested to the writer by some remarks in the introduction of Poincaré's Thesis, and by a study of the elimination theory of Kronecker for algebraic equations. The theorem is presented here (§ 13) for two equations and two variables y_1, y_2 in the form originally given at the time of the Princeton Colloquium, but the method of proof is similar to that of a later paper* and applies with suitable modifications to a system containing more equations and independent variables. These results can not by any means be said to afford a complete characterization of the singularities of implicit functions, but it is hoped that they may be useful in paving the way for researches of a more comprehensive character.

The writer published some years ago a paper† concerning the extensibility of the solutions of a system of differential equations, of the form specified above, from boundary to boundary of a finite closed region R in which the functions g_α are supposed to have suitable continuity properties. In the last chapter of these lectures the character of the region has been generalized so that no restrictions as to its finiteness or closure are made, and it is shown that the approximations of Cauchy converge to a solution over an interval

---

* See the footnote to page 73.

† "The solutions of differential equations of the first order as functions of their initial values," *Annals of Mathematics*, 2d series, vol. 6 (1904), page 49.

in the interior of which the limiting curve is continuous and interior to R, while at the ends of the interval the only limit points of the curve are at infinity or else are on the boundary of the region. The solutions so defined are continuous and differentiable with respect to their initial values, a property which once proved is of great service in many of the applications of the existence theorems. One situation in which these results have an important bearing is related to a partial differential equation of the first order

$$F(x, y, z, \partial z/\partial x, \partial z/\partial y) = 0.$$

When this equation is analytic, any analytic curve C, which is not a so-called integral curve, defines uniquely an analytic surface containing the curve and satisfying the differential equation. The uniqueness in this case is a consequence, in the first place, of the fact that an analytic surface is completely determined when an initial series defining its values in a limited region is given, and, in the second place, of the theorem that at a given point and normal of the initial curve C satisfying the differential equation there is but one series defining an integral surface including the points of C and having the given initial normal. It is not self evident in what sense a solution of a non-analytic equation is uniquely determined by an initial curve, as may be seen by very simple examples. An initial curve which is not an integral curve will in general have associated with it, however, a strip of normals which satisfy the partial differential equation, and whose elements as initial values determine a one-parameter family of characteristic strips simply covering a region R_{xy} of the xy-plane about the projection of the initial curve C. There is one and but one integral surface of the differential equation with a continuously turning tangent plane and continuous curvature, which is defined at every point of the region R_{xy} and contains the initial curve C and its strip of normals (§ 19).

CHAPTER I

ORDINARY POINTS OF IMPLICIT FUNCTIONS

§1. The Fundamental Theorem

The fundamental theorem of the implicit function theory states the existence of a set of functions

$$y_\alpha = y_\alpha(x_1, x_2, \cdots, x_m) \quad (\alpha = 1, 2, \cdots, n)$$

which satisfy a system of equations of the form

(1) $\quad f_\alpha(x_1, x_2, \cdots, x_m; y_1, y_2, \cdots, y_n) = 0 \quad (\alpha = 1, 2, \cdots, n)$

in a neighborhood of a given initial solution $(a; b)$. Dini's method,* for the case in which the functions f are only assumed to be continuous and to have continuous first derivatives, is to show the existence of a solution of a single equation, and then to extend his result by mathematical induction to a system of the form given above, a plan which has been followed, with only slight alterations and improvements in form, by most writers on the theory of functions of a real variable. In a more recent paper† Goursat has applied a method of successive approximations which enabled him to do away with the assumption of the existence of the derivatives of the functions f with respect to the independent variables x.

One can hardly be dissatisfied with either of these methods of attack. It is true that when the theorem is stated as precisely as in the following paragraphs, the determination of the neighborhoods at the stage when the induction must be made is rather inelegant, but the difficulties encountered are not serious. The introduction of successive approximations is an interesting step,

---

*Lezioni di Analisi infinitesimale, vol. 1, chap. 13. For historical remarks, see Osgood, Encyclopädie der mathematischen Wissenschaften, II, B 1, § 44 and footnote 30.

† Bulletin de la Société mathématique de France, vol. 31 (1903), page 185.

though it does not simplify the situation and indeed does not add generality with regard to the assumptions on the functions f. The method of Dini can in fact, by only a slight modification, be made to apply to cases where the functions do not have derivatives with respect to the variables x. The proof which is given in the following paragraphs seems to have advantages in the matter of simplicity over either of the others. It applies equally well, without induction, to one or a system of equations, and requires only the initial assumptions which Goursat mentions in his paper.

Where it is possible without sacrificing clearness, the row letters f, x, y, p, a, b will be used to denote the systems

$$f = (f_1, f_2, \cdots, f_n), \quad x = (x_1, x_2, \cdots, x_m),$$
$$y = (y_1, y_2, \cdots, y_n), \quad a = (a_1, a_2, \cdots, a_m),$$
$$b = (b_1, b_2, \cdots, b_n), \quad p = (a_1, a_2, \cdots, a_m; b_1, b_2, \cdots, b_n).$$

In this notation the equations (1) have the form

$$f(x; y) = 0,$$

the interpretation being that every element of f is a function of $x_1, x_2, \cdots, x_m; y_1, y_2, \cdots, y_n$, and every f_i is to be set equal to zero. The notations p_ϵ, a_ϵ, b_ϵ represent respectively the neighborhoods

$$|x - a| < \epsilon, \quad |y - b| < \epsilon; \quad |x - a| < \epsilon; \quad |y - b| < \epsilon$$

of the points p, a, b.

With these notations in mind the fundamental theorem which is to be proved may be stated as follows:

Hypotheses:

1) *the functions $f(x; y)$ are continuous, and have first partial derivatives with respect to the variables y which are also continuous, in a neighborhood of the point $(a; b)$ which will be denoted by p;*

2) $f(a; b) = 0$;

3) *the functional determinant $D = \partial(f_1, f_2, \cdots, f_n)/\partial(y_1, y_2, \cdots, y_n)$ is different from zero at p.*

FUNDAMENTAL EXISTENCE THEOREMS.   9

Conclusions:

1) a neighborhood p_ϵ can be found in which there corresponds to a given value x at most one solution $(x; y)$ of the equations $f(x; y) = 0$;

2) for any neighborhood p_ϵ with the property just described a constant $\delta \leqq \epsilon$ can be found such that every x in a_δ has associated with it a point $(x; y)$ which satisfies the equations $f(x; y) = 0$;

3) the functions $y(x_1, x_2, \cdots, x_m)$ so found are continuous in the region a_δ.

For the neighborhood p_ϵ let one be chosen in which the continuity properties of the functions f are preserved. If $(x; y)$ and $(x; y')$ are two points in p_ϵ, it follows, by applying Taylor's formula to the differences $f(x; y') - f(x; y)$, that

$$f_1(x; y') - f_1(x; y) = \frac{\partial f_1}{\partial y_1}(y_1' - y_1) + \cdots + \frac{\partial f_1}{\partial y_n}(y_n' - y_n),$$

$$\cdot \quad \cdot \quad \cdot \quad \cdot \quad \cdot \quad \cdot \quad \cdot \quad \cdot \quad \cdot \quad \cdot$$

$$f_n(x; y') - f_n(x; y) = \frac{\partial f_n}{\partial y_1}(y_1' - y_1) + \cdots + \frac{\partial f_n}{\partial y_n}(y_n' - y_n),$$

where the arguments of the derivatives $\partial f_\alpha/\partial y_\beta$ have the form $x; y + \theta_\alpha(y' - y)$, and $0 < \theta_\alpha < 1$. The determinant of these derivatives is different from zero when $(x; y') = (x; y) = (a; b)$, and hence must remain different from zero if p_ϵ is restricted so that in it the functional determinant D remains different from zero. It is then impossible that $(x; y)$ and $(x; y')$ should both be solutions of the equations $f(x; y) = 0$, if y is distinct from y'.

In the corresponding region b_ϵ the function

$$\varphi(a; y) = f_1^2(a; y) + f_2^2(a; y) + \cdots + f_n^2(a; y)$$

has a minimum for $y = b$, since for that value it vanishes and for every other it is positive. In particular

$$\varphi(a; \eta) - \varphi(a; b) > m > 0$$

when η ranges over the *closed* set of points η forming the boundary of b_ϵ, on account of the continuity of φ, and the inequality

$$\varphi(x; \eta) - \varphi(x; b) > m$$

remains true for all values x in a suitably chosen domain a_δ. Hence for a fixed x in a_δ the minimum of $\varphi(x; y)$ is attained at a point y *interior* to b_ϵ. At such a point, however,

$$\frac{1}{2}\frac{\partial \varphi}{\partial y_1} = f_1\frac{\partial f_1}{\partial y_1} + f_2\frac{\partial f_2}{\partial y_1} + \cdots + f_n\frac{\partial f_n}{\partial y_1} = 0,$$

$$\cdots \cdots \cdots \cdots \cdots$$

$$\frac{1}{2}\frac{\partial \varphi}{\partial y_n} = f_1\frac{\partial f_1}{\partial y_n} + f_2\frac{\partial f_2}{\partial y_n} + \cdots + f_n\frac{\partial f_n}{\partial y_n} = 0,$$

and this can happen only when all the elements of f are zero, since the functional determinant D is different from zero in p_ϵ. It follows that to every point x in a_δ there corresponds in p_ϵ a solution $(x; y)$ of the equations $f(x; y) = 0$.

The functions $y(x_1, x_2, \cdots, x_m)$ defined in this way over the region a_δ are all continuous. For consider the values y and $y + \Delta y$ corresponding to two points x and $x + \Delta x$. By applyng Taylor's formula it follows from the relations

$$f(x; y + \Delta y) - f(x; y) = f(x; y + \Delta y) - f(x + \Delta x; y + \Delta y),$$

which are true because $(x; y)$ and $(x + \Delta x; y + \Delta y)$ both make $f = 0$, that

$$\frac{\partial f_1}{\partial y_1}\Delta y_1 + \frac{\partial f_1}{\partial y_2}\Delta y_2 + \cdots + \frac{\partial f_1}{\partial y_n}\Delta y_n$$
$$= f_1(x; y + \Delta y) - f_1(x + \Delta x; y + \Delta y),$$

(2) $\qquad \cdots \cdots \cdots \cdots \cdots$

$$\frac{\partial f_n}{\partial y_1}\Delta y_1 + \frac{\partial f_n}{\partial y_2}\Delta y_2 + \cdots + \frac{\partial f_n}{\partial y_n}\Delta y_n$$
$$= f_n(x; y + \Delta y) - f_n(x + \Delta x; y + \Delta y),$$

where the arguments of the derivatives $\partial f_\alpha/\partial y_\beta$ have the form $x; y + \theta_\alpha \Delta y$ $(0 < \theta_\alpha < 1)$. The determinant of these derivatives is different from zero on account of the way in which p_ϵ was chosen, and the second members of the equations approach zero with Δx. Hence the same must be true of the quantities

Δy, and thus the functions $y(x_1, x_2, \cdots, x_m)$ are seen to be continuous.

A similar application of Taylor's formula leads to the conclusion:

If the functions f have derivatives of the first order with respect to x_k which are continuous in the neighborhood of p, so have also the functions $y(x_1, x_2, \cdots, x_m)$ in the region a_δ; and if the f's have all derivatives of the nth order continuous, so have the functions $y(x_1, x_2, \cdots, x_m)$.

For suppose

$$\Delta x_1 \neq 0, \quad \Delta x_2 = \Delta x_3 = \cdots = \Delta x_m = 0.$$

Then by applying Taylor's formula to the second members of equations (2) it follows that

$$\frac{\partial f_1}{\partial y_1}\frac{\Delta y_1}{\Delta x_1} + \frac{\partial f_1}{\partial y_2}\frac{\Delta y_2}{\Delta x_1} + \cdots + \frac{\partial f_1}{\partial y_n}\frac{\Delta y_n}{\Delta x_1} + \frac{\partial f_1}{\partial x_1} = 0,$$

$$\cdots \cdots \cdots \cdots \cdots \cdots \cdots$$

$$\frac{\partial f_n}{\partial y_1}\frac{\Delta y_1}{\Delta x_1} + \frac{\partial f_n}{\partial y_2}\frac{\Delta y_2}{\Delta x_1} + \cdots + \frac{\partial f_n}{\partial y_n}\frac{\Delta y_n}{\Delta x_1} + \frac{\partial f_n}{\partial x_1} = 0,$$

where the arguments of the derivatives $\partial f_\alpha / \partial x_1$ have the form $x + \theta_\alpha' \Delta x;\, y + \Delta y$. Hence as Δx_1 approaches zero the quotients $\Delta y_\alpha / \Delta x_1$ approach limits $\partial y_\alpha / \partial x_1$ which satisfy the equations

(3)
$$\frac{\partial f_1}{\partial y_1}\frac{\partial y_1}{\partial x_1} + \frac{\partial f_1}{\partial y_2}\frac{\partial y_2}{\partial x_1} + \cdots + \frac{\partial f_1}{\partial y_n}\frac{\partial y_n}{\partial x_1} + \frac{\partial f_1}{\partial x_1} = 0,$$

$$\cdots \cdots \cdots \cdots \cdots \cdots \cdots$$

$$\frac{\partial f_n}{\partial y_1}\frac{\partial y_1}{\partial x_1} + \frac{\partial f_n}{\partial y_2}\frac{\partial y_2}{\partial x_1} + \cdots + \frac{\partial f_n}{\partial y_n}\frac{\partial y_n}{\partial x_1} + \frac{\partial f_n}{\partial x_1} = 0,$$

where the arguments of the derivatives of f are now $(x; y)$. A similar consideration shows the existence of the first derivatives with respect to the variables $x_2, x_3, \cdots, x_m$. The existence of the higher derivatives follows from the observation that the solutions of equations (3) for the quotients $\partial f_\alpha / \partial y_\beta$ are differentiable $n - 1$ times with respect to the variables x, on account of the assumption that the functions f are differentiable n times.

§ 2. Equations in which the Functions are Analytic

It seems necessary to proceed differently in order to prove that when the functions f in equations (1) are analytic with coefficients and variables permitted to assume imaginary values, the solutions $y = y(x_1, x_2, \cdots, x_m)$ are also analytic functions of the variables x. The following theorem can first be proved:

When the functions f are formal series in the variables x; y with literal coefficients and having no constant terms, then there exists one and but one set of series

$$(4) \qquad y_a = y_a(x_1, x_2, \cdots, x_m)$$

for the variables y, which vanish with the x s and satisfy identically the equations $f(x; y) = 0$. Each coefficient in the series y is rational in a finite number of those of the functions f, the only denominators occurring being powers of the determinant R of the coefficients of the linear terms in y.

To prove this let the equations $f = 0$ be written in the form

$$a_{11}y_1 + a_{12}y_2 + \cdots + a_{1n}y_n = g_1(x; y),$$
$$\cdots \cdots \cdots \cdots \cdots \cdots \cdots$$
$$a_{n1}y_1 + a_{n2}y_2 + \cdots + a_{nn}y_n = g_n(x; y),$$

where the functions g have no linear terms in y. By multiplying these equations by proper factors and adding, they may be made to take the form

$$(5) \qquad y_a = h_a(x; y) \qquad (\alpha = 1, 2, \cdots, n),$$

where the series h have still no linear terms in y and have coefficients which are rational in those of the functions f, the only denominators occurring being the determinant R. Any series for y which satisfy formally the original equations must satisfy the last equations, and vice versa.

Consider now a set of series (4) in which the coefficients are indeterminates c. If they satisfy the equations (5) identically, then by comparison of coefficients on the two sides it is seen that any coefficient c_ν of a term of degree ν must be equal to a polynomial, with positive integral coefficients, in a finite number of the coefficients of the functions h and in the coefficients $c_{\nu-k}$ of terms in the functions y of lower degree than ν. For there are at most a finite number of terms on the right of any given degree ν, and since the functions h have no linear terms in the variables y it follows that wherever the term containing c_ν occurs it is always multiplied by a y or by a power of some of the variables x, and hence c_ν can only appear in terms of degree greater than ν. Since the coefficients of the linear terms in the functions y are equal respectively to corresponding coefficients in the functions h, it follows by an easy induction that every coefficient in the functions y must be a polynomial with positive integral coefficients in a finite number of the coefficients of the functions h. There is evidently but one set of series (4) of the kind described satisfying formally the equations (5), or what is the same thing, the equations $f = 0$.

For any numerical choice of the coefficients of the functions f in the domain of real or imaginary numbers for which the series f converge and the determinant $R = |\, a_{\alpha\beta} \,|$ is different from zero, the series (4) for y will also be well-determined and convergent.

For, a set of equations

(6) $$y_\alpha = H_\alpha(x; y) \qquad (\alpha = 1, 2, \cdots, n)$$

can be constructed whose coefficients are all positive and greater numerically than the corresponding coefficients in the functions h, and for which the corresponding series $y = Y(x_1, x_2, \cdots, x_m)$ converge. The coefficients in the functions Y will be greater numerically than the corresponding coefficients of the series $y(x_1, x_2, \cdots, x_m)$, and hence the series y will also converge.

To show this suppose that ρ is a positive constant smaller than the radii of convergence of the functions $h(x; y)$. Then the series $h(\rho;\rho)$ are convergent, and each term is numerically smaller than a constant M chosen greater than the sum of the absolute values of the terms in any one of the series $h(\rho;\rho)$. The coefficient of any term in $h(x;y)$ is less than M/ρ^ν where ν is the degree of the term. The series

$$H_a(x;y) = \frac{M}{\left(1 - \dfrac{x_1 + x_2 + \cdots + x_m}{\rho}\right)\left(1 - \dfrac{y_1 + y_2 + \cdots + y_n}{\rho}\right)} - M - M\frac{y_1 + y_2 + \cdots + y_n}{\rho}$$

are similar to the series $h(x;y)$ in the matter of missing terms, and dominate them in the manner described above, since the coefficient of any term of degree ν is M/ρ^ν or greater.

The unique series satisfying equations (6) will evidently be convergent if a convergent series u in x can be determined satisfying

$$u = \frac{M}{\left(1 - \dfrac{x_1 + x_2 + \cdots + x_m}{\rho}\right)\left(1 - \dfrac{nu}{\rho}\right)} - M - M\frac{nu}{\rho},$$

for then every series y can be put equal to that series u. The latter equation is however a quadratic in u and has the solution

$$u = \frac{\rho^2}{2n(\rho + Mn)}\left\{1 - \sqrt{1 - \frac{4Mn(\rho + Mn)}{\rho^2}\cdot\frac{\dfrac{x_1 + x_2 + \cdots + x_m}{\rho}}{1 - \dfrac{x_1 + x_2 + \cdots + x_m}{\rho}}}\right\}$$

vanishing with x. This will certainly be representable by a convergent series in x provided that

$$|x_i| < \frac{\rho^3}{m(\rho + 2Mn)^2} \quad (i = 1, 2, \cdots, m),$$

since then the second term under the radical is numerically less than unity.

FUNDAMENTAL EXISTENCE THEOREMS. 15

The two theorems which have just been proved enable one to make the following statement concerning the solutions whose existence was proved in § 1:

If the functions $f(x; y)$ are analytic in the region p_ϵ, then the solutions (4) of the equations $f(x; y) = 0$ are analytic at every point of the region a_δ.

It is only necessary to transform the origin of coordinates to the particular point $(x; y)$ of the solution which it is desired to investigate.

Furthermore when the domain in which the equations $f = 0$ are to be studied is the domain of complex numbers, a theorem analogous to that of § 1 may be stated.

If in the domain of complex numbers the functions $f(x; y)$ are analytic at a point $p(a; b)$ at which

$$f(a; b) = 0, \quad D(a; b) = \left[\frac{\partial(f_1, f_2, \cdots, f_n)}{\partial(y_1, y_2, \cdots, y_n)}\right]_{\substack{x=a\\y=b}} \neq 0,$$

then there exists a neighborhood p_ϵ in which any x corresponds to at most one solution $(x; y)$, either real or complex, of the equations $f(x; y) = 0$. For any such choice of p_ϵ a neighborhood a_δ ($\delta \leq \epsilon$) can be found such that every point x in a_δ has associated with it a solution $(x; y)$ of the equations $f = 0$ in p_ϵ, and the values y for these solutions are defined by a set of functions

(7) $$y_\alpha = y_\alpha(x_1, x_2, \cdots, x_m) \quad (\alpha = 1, 2, \cdots, n)$$

which are expressible as series in the differences $x - a$ convergent in the region a_δ.

The existence of the neighborhood p_ϵ is provable by the argument used in § 1, since for any two points $(x; y)$ and $(x; y')$ in the common domain of convergence of the functions f, equations of the form

$$f_\alpha(x; y') - f_\alpha(x; y) = A_{\alpha 1}(y_1' - y_1) + \cdots + A_{\alpha n}(y_n' - y_n),$$
$$(\alpha = 1, 2, \cdots, n)$$

hold, where the coefficient $A_{\alpha\beta}$ is a convergent series in the differences $x - a$, $y - b$, $y' - b$ with constant term equal to $a_{\alpha\beta}$. The existence of the coefficients A can be established by considering two analogous terms in $f(x; y)$ and $f(x; y')$. The difference of such a pair of terms will always be linearly expressible in terms of the differences

$$(y_\alpha' - b_\alpha) - (y_\alpha - b_\alpha) = y_\alpha' - y_\alpha \quad (\alpha = 1, 2, \cdots, n).$$

Furthermore for $(x, y, y') = (a, b, b)$ the derivative of the first member with respect to y_β' reduces to $a_{\alpha\beta}$, while that of the second is the constant term in $A_{\alpha\beta}$. Hence for these values of the variables the determinant $|A_{\alpha\beta}|$ reduces to $D(a, b) \neq 0$.

By transforming the origin of coordinates to the point (a, b) and applying the first two theorems of this section, it follows that there exists a set of convergent series (7) satisfying the equations $f = 0$ identically; and for a sufficiently small region a_δ the points $(x; y)$ which they define will all lie in the neighborhood p_ϵ.

§ 3. Goursat's Method of Approximation

The method of approximation which is to be presented in the following paragraphs is of interest primarily because it affords a direct method of finding the values of implicit functions, and justifies computations sometimes used in the applications of the theory. In order to exhibit this method suppose again that the functions f have the properties described in the principal theorem of § 1, and consider the following set of equations suggested by Taylor's formula:

(8)
$$\begin{aligned}
f_1(x; y) + a_{11}(y_1' - y_1) + a_{12}(y_2' - y_2) + \cdots \\
+ a_{1n}(y_n' - y_n) = 0, \\
\cdots \cdots \cdots \cdots \cdots \cdots \cdots \cdots \cdots \\
f_n(x; y) + a_{n1}(y_1' - y_1) + a_{n2}(y_2' - y_2) + \cdots \\
+ a_{nn}(y_n' - y_n) = 0,
\end{aligned}$$

in which the coefficient $a_{\alpha\beta}$ is the value of $\partial f_\alpha/\partial y_\beta$ at the point p. When solved for the variables y', these equations take the form

(9) $$y_\alpha' = \varphi_\alpha(x; y) \quad (\alpha = 1, 2, \cdots, n),$$

and one verifies readily by substitution of these expressions in equations (8) that the functions φ and all of their first derivatives with respect to the elements of y are continuous near p; and at the point p itself φ_a has the value b_a, while all of its derivatives with respect to the y's vanish.

A sequence of systems $y^{(k)} = (y_1^{(k)}, y_2^{(k)}, \cdots, y_n^{(k)})$ beginning with the set
$$y' = [\varphi_1(x;b),\ \varphi_2(x;b),\ \cdots,\ \varphi_n(x;b)]$$
can now be defined by means of the recursion formulas (9), which are equivalent to
$$y_a^{(k)} = \varphi_a(x; y^{(k-1)}) \qquad (\alpha = 1, 2, \cdots, n).$$

Let p_ϵ be any neighborhood of p in which the continuity properties of f are retained, and in which the derivatives of φ remain numerically less than θ/n where $0 < \theta < 1$. If the values of x are restricted to a region $a_\delta (\delta \leq \epsilon)$ so small that every element of the set y' satisfies the inequality

(10) $$|y_a' - b_a| < \epsilon(1 - \theta),$$

then the points $(x; y^{(k)})$ will all lie in the neighborhhood p_ϵ and will approach uniformly a limiting point $(x; y)$ which is a solution of the equations (1).

To prove these statements one needs only to apply successively the inequality
$$|y_a^{(k)} - y_a^{(k-1)}| = |\varphi_a(x; y^{(k-1)}) - \varphi_a(x; y^{(k-2)})|$$
$$\leq \frac{\theta}{n}\{|y_1^{(k-1)} - y_1^{(k-2)}| + |y_2^{(k-1)} - y_2^{(k-2)}|$$
$$+ \cdots + |y_n^{(k-1)} - y_n^{(k-2)}|\},$$
which follows readily by an application of Taylor's formula. Since the inequalities (10) hold, the last formula successively applied shows that
$$|y_a^{(k)} - y_a^{(k-1)}| \leq \theta^{k-1}\epsilon(1 - \theta).$$
Consequently the sum $y_a^{(k)}$ of the first $k + 1$ terms of the series

(11) $$b_a + (y_a' - b_a) + (y_a'' - y_a') + \cdots + (y_a^{(k)} - y_a^{(k-1)}) + \cdots$$

differs in absolute value from b_a by a quantity which is less than

$$\epsilon(1 - \theta)(1 + \theta + \theta^2 + \cdots + \theta^{k-1}) = \epsilon(1 - \theta^k) < \epsilon.$$

Hence the points $(x; y)$ all lie in the neighborhood p_ϵ, and the series (11) is uniformly convergent in the neighborhood a_δ.

The limiting point $(x; y)$ evidently satisfies the equations $f = 0$. For at every stage the values $(x, y, y') = (x, y^{(k-1)}, y^{(k)})$ satisfy the equations (8), and the first members of these equations approach uniformly the values $f(x; y)$.

The process of determining the solutions described above is evidently one of trial and error. The values $y = b$ being first substituted, the equations (9) determine approximately the correction $y' - b$ which must be added to b in order to obtain a solution for any value of x near to a. For the values so corrected the equations (9) give again a new correction $y'' - y'$, and so on.

It is ordinarily presupposed that an initial solution $(a; b)$ is given, *but the process may also lead to the discovery of a solution in case only an initial point which approximately satisfies the equation is known.* To show this suppose that the functions f are continuous and have continuous first partial derivatives with respect to the variables y in a closed region R of points $(x; y)$ in which the functional determinant $D(x; y)$ is different from zero. The functions φ in equations (9) are to be thought of as depending upon $(x; y)$, and also upon the variables $(a; b)$ which enter in the derivatives $a_{\alpha\beta}$. Then the expressions $\varphi(x, y, a, b)$, $\varphi_y(x, y, a, b)$ are continuous when $(x; y)$, $(a; b)$ lie in R, and all of the derivatives φ_y vanish identically when $(x; y) = (a; b)$. The value of $\varphi(a, b, a, b)$ is not necessarily b, however, when $(a; b)$ is not a solution. Two positive constants, $\theta < 1$ and ϵ, can be determined so that

$$|\varphi_y(x, y, a, b)| < \theta/n$$

whenever $(a; b)$ and $(x; y)$ satisfy the inequalities

$$|x - a| < \epsilon, \quad |y - b| < \epsilon.$$

If now there exists a point $p(a; b)$ for which the neighborhood p_ϵ

is entirely within R, and such that

$$| \varphi(a, b, a, b) - b | < \epsilon(1 - \theta),$$

then the sequence $y^{(k)}$ defined converges uniformly as before in a neighborhood a_δ of the point a and determines a solution $(x; y)$.

As an example consider the equation

(12) $\qquad\qquad y - e \sin y = x \qquad\qquad (0 < e < 1),$

which in the theory of elliptic orbits determines the value of the eccentric anomaly y in terms of the mean anomaly x. The function φ is in this case

$$\varphi(x, y, a, b) = \frac{e(\sin y - y \cos b) + x}{1 - e \cos b}$$

and φ_y remains less than θ when

$$| y - b | < \theta \frac{1 - e}{e} = \epsilon.$$

For any given $x = a$, a value $y = b$ can be determined, by graphical methods for example, so that

$$| \varphi(a, b, a, b) - b | = \left| \frac{b - e \sin b - a}{1 - e \cos b} \right| < \theta \frac{1 - e}{e} (1 - \theta).$$

The process described above therefore converges in a suitably chosen neighborhood of $x = a$, and a solution of equation (12) can be found when an approximate solution only has been determined in advance.

§ 4. Bolza's Extension of the Fundamental Theorem*

The neighborhood P_ϵ of a set of points P in the space $(x; y)$ is the totality of points $(x; y)$ which satisfy inequalities of the form

$$| x - a | < \epsilon, \qquad | y - b | < \epsilon,$$

---

* Vorlesungen über Variationsrechnung, page 160: also *Mathematische Annalen*, vol. 63 (1906), page 247. The theorem was proved independently by Mason and Bliss, "Fields of extremals in space," *Transactions of the American Mathematical Society*, vol. 11 (1910), page 326.

where $(a; b)$ is some point of P. The sets of points (a) and (b) which belong to points $(a; b)$ of P are the projections of P in the x- and y-spaces, and will be denoted by A and B, respectively.

The fundamental theorem of § 1 remains true if in its statement the single point p is replaced by a set of points P which is finite and closed, and which furthermore has the property that no two distinct points $(a; b)$, $(a'; b')$ of P have the same projection $a' = a$. According to the conclusions of the theorem there exists then a neighborhood P_ϵ in which no two solutions of the equations $f(x; y) = 0$ have the same projection x, and a neighborhood A_δ in which every x surely belongs to a solution $(x; y)$ in P_ϵ. The single-valued functions $y(x_1, x_2, \cdots, x_m)$ so defined in A_δ are continuous, and if the functions $f(x; y)$ have continuous derivatives of the n-th order in a neighborhood of P, so have the functions $y(x_1, x_2, \cdots, x_m)$ in A_δ.

To prove the theorem suppose first that a sequence of positive constants ϵ_k ($k = 1, 2, \cdots$) approaching zero has been selected arbitrarily. If the first part of the theorem were not true, then in any neighborhood P_{ϵ_k} there would be two distinct solutions $(x; y)_k$ and $(x; y')_k$ of the equations $f(x; y) = 0$, which would satisfy, respectively, inequalities of the form

(13)
$$| x - \alpha | < \epsilon_k, \quad | y - \beta | < \epsilon_k;$$
$$| x - \alpha' | < \epsilon_k, \quad | y' - \beta' | < \epsilon_k$$

with two points $(\alpha; \beta)_k$ and $(\alpha'; \beta')_k$ of the set P. Since P is finite and closed, the sequence of values $(\alpha, \beta; \alpha', \beta')_k$ has a point of condensation $(a, b; a', b')$ for which $(a; b)$ and $(a'; b')$ are both in P. From the inequalities (13) it follows that $(a, b; a', b')$ is also a point of condensation for the sequence $(x, y; x, y')_k$, and therefore a and a' must be the same. The values b and b' must also be identical since P contains only one point $p(a; b)$ with the projection a. According to the original statement of the fundamental theorem in § 1, a neighborhood p_ϵ can be chosen in which no two solutions of the equations $(x; y) = 0$ have the same projection x. Hence the existence of the sequences $(x; y)_k$ and $(x; y')_k$ with the common point of condensation $(a; b)$ is contradicted, and it must always be possible to select a neighborhood P_ϵ in which distinct solutions of the equations $f = 0$ always have distinct projections x.

A similar argument shows that a neighborhood A_δ can be selected so that to any point of it there corresponds a solution of the equations $f = 0$. Otherwise to each δ_k of a sequence of constants approaching zero, there would correspond a point $(x)_k$ in the region A_{δ_k} which would belong to no solution in P_ϵ. To each $(x)_k$ there would correspond a point $(\alpha)_k$ in A satisfying the inequalities

$$|x - a| < \delta_k$$

with the values $(x)_k$, and the points $(\alpha)_k$ would have a point of condensation a in A, which would also be a point of condensation for the sequence $(x)_k$, since A is finite and closed when P is so. But by the original theorem of § 1, again, it is known that a neighborhood a_δ of a can be chosen in which every point x has associated with it a solution $(x; y)$ in p_ϵ, where $p(a; b)$ is the point of P having the projection a. Consequently the existence of the sequence $(x)_k$ is contradicted.

If now the region P_ϵ is so restricted that the functional determinant $D(x; y)$ remains different from zero throughout it, then the original theorem of § 1 can be applied to show that the functions $y(x_1, x_2, \cdots, x_n)$ are continuous at any point of the region A_δ and possess as many continuous derivatives as are possessed by the functions $f(x; y)$.

§ 5. The Unique Sheet of Solutions Associated with an Initial Solution

The points of the space $(x; y)$ may be divided into two classes, ordinary points and exceptional points, with respect to the functions f. An *ordinary point* is one at which the first and third hypotheses of the theorem of § 1 are postulated, that is, one near which the functions f and their first derivatives with respect to y are continuous and the functional determinant $D = \partial(f_1, f_2, \cdots,$

$f_n)/\partial(y_1, y_2, \cdots, y_n)$ is different from zero. An *exceptional point* is one at which some of these conditions are not fulfilled or are not presupposed.

A *sheet of points* in the $(m + n)$-dimensional space $(x; y)$ may be defined as a point set S with the property that for any point $p(a; b)$ belonging to the set a neighborhood p_ϵ can always be found such that no two points of S in p_ϵ have the same projection x. In other words, the variables y are single-valued functions $y(x_1, x_2, \cdots, x_m)$ in the neighborhood of the point p, for points of the sheet.

If for any neighborhood b_ϵ of the kind just described, a region a_δ ($\delta \leq \epsilon$) can be found in which every point x belongs to a point of S in p_ϵ, then p is said to be an *interior point* of the sheet S.

A *boundary point* is a limit point of points of the sheet, which is not itself an interior point and may not even belong to S.

A sheet is said to be *connected* if every pair $(x'; y')$, $(x''; y'')$ of its interior points can be joined by a continuous curve

$$x = x(t), \qquad y = y(t) \qquad (t' \leq t \leq t''),$$

consisting entirely of interior points of the sheet.

In the following pages it is always to be understood that the sheets considered are continuous and have continuous first derivatives, or in other words at any interior point of one of them the functions $y(x_1, x_2, \cdots, x_m)$ mentioned above have these properties. A sheet will be said to become infinite near a point x' if x' is the limit of the projections of a sequence of points $(x; y)$ of the sheet for which one at least of the variables y approaches infinity.

With the preceding agreements as to nomenclature in mind, it is possible to prove the following theorem:

If a point $p(a; b)$ is an ordinary point for the functions f and satisfies the equations $f = 0$, then there passes through p one and only one connected sheet of solutions of these equations, with the properties:

1) *all points of the sheet are ordinary points of the functions f*;
2) *all points are interior points*;

3) *the only boundary points of the sheet are exceptional points for the system f.*

The set of points

$$[x_1, x_2, \cdots, x_m; y_1(x_1, x_2, \cdots, x_m), \cdots, y_n(x_1, x_2, \cdots, x_m)]$$

defined over the region a_δ by the principal theorem of §1, is a sheet S_1 of solutions of the equations $f = 0$ which satisfies all the requirements of the theorem just stated except possibly the last. Its points are all interior points since the region a_δ is defined by inequalities only. If any boundary point $p'(a'; b')$ of S_1 is an ordinary point of the functions f it must satisfy the equations $f = 0$, since the f's are continuous and p' is a limit point of points on S_1. Consequently the theorem of §1 can be applied in the neighborhood of p', and the sheet S' so determined near p' forms with S_1 a new set S_2. This process may be repeated any number of times, and the totality of points which can be attained by a finite number of such extensions, constitutes the sheet S required in the theorem.

The set of points S so determined constitutes a sheet, since any point q of it is an ordinary point and a solution of the equations $f = 0$, and according to the theorem of §1 the solutions of these equations in the neighborhood of q have the property which is characteristic of a sheet. From the manner of its construction the sheet is evidently connected and consists entirely of interior points. If any boundary point q of S were an ordinary point of the functions f, the sheet could be extended to include q as an interior point by the process described in the preceding paragraph.

There could not be a second sheet Σ containing a point π not in S and having the properties stated in the theorem. For there would in that case be a continuous curve

$$x = x(t), \quad y = y(t) \qquad (t_1 \leq t \leq t_2)$$

in Σ joining p with π and consisting entirely of ordinary points. In a neighborhood of $t = t_1$ all of the points defined on the curve would also be points of S, since the solutions of the equations

$f = 0$ near the initial point p of the curve are all in S. The values of t defining points on the curve and in S would therefore have an upper bound $\tau \leq t_2$ such that τ would define on the curve a boundary point of S. But this is impossible since all of the points of the curve are ordinary points.

If the functions f are known to be continuous and to have continuous derivatives in a region R, then it follows readily from what precedes that through any ordinary solution of the equations $f = 0$ interior to R there passes one and only one sheet of solutions having the property that the only boundary points of the sheet are boundary points of R, or interior points of R at which the functional determinant vanishes. If R is finite and closed and consists entirely of ordinary points for the functions f, then there can not be more than a finite number of points of the sheet on any ordinate x. Otherwise the points common to the ordinate and the sheet would have a point of condensation p, also in R. Since p is an ordinary point there can be at most one solution of the equations in a properly chosen neighborhood p_ϵ.

It is interesting to determine a criterion which shall characterize a sheet which is at most single-valued on any ordinate. Such a criterion is derived in § 7 in connection with a theorem due originally to Schoenflies, and afterwards proved by Osgood. The proof of it involves the auxiliary notions described in § 6 and the following corollaries to the preceding theorem:

If the initial point of a continuous arc

$$(C_x) \qquad x_i = x_i(t) \qquad (i = 1, 2, \cdots, m; t' < t < t'')$$

in the x-space is the projection of a solution $p'(x'; y')$ of the equations $f = 0$ which is an ordinary point for the functions f, then there is associated with the arc C_x one and only one continuous curve

$$(C_{xy}) \quad x_i = x_i(t), \quad y_\alpha = y_\alpha(t) \qquad (i=1, 2, \cdots, m; \alpha=1, 2, \cdots, n)$$

passing through $(x'; y')$ for $t = t'$, with the properties:
1) *all of its points are solutions of the equations $f = 0$ and ordinary points of the functions f;*

FUNDAMENTAL EXISTENCE THEOREMS. 25

2) *it is defined either over the whole interval* $t' \leq t \leq t''$, *or else on an interval* $t' \leq t < \tau$ ($\leq t''$) *such that as t approaches* τ *the only limit points of the curve* C_{xy} *are at infinity or are exceptional points of the functions f.*

The truth of this statement is readily deduced from the considerations which precede, or by the following argument. The fundamental theorem of § 1 can be applied at the point $(x'; y')$. If the arc C_x is entirely within the region x_δ' then the existence and uniqueness of the curve C_{xy} is evident. In any case there will be some intervals $t' \leq t \leq t_1$ in which curves C_{xy} are defined having all the properties described in the theorem except possibly 2). Suppose that τ is the upper bound of the end values t_1 for such intervals. Then there is a curve C_{xy} well defined in the interval $t' \leq t < \tau$, and no limit point $(\alpha\,; \beta)$ of the curve as t approaches τ can be a finite ordinary point for the functions f. For if $(\alpha\,; \beta)$ were such a point, it would also satisfy the equations $f = 0$, on account of the continuity of the functions f, and the theorem of § 1 could again be applied at $(\alpha\,; \beta)$. A curve C_{xy} with all the properties of the theorem, except possibly 2), could then be defined over an interval including the interval $t' \leq t < \tau$ in its interior, which contradicts the assumption that τ is the upper bound of such intervals.

There could not be two curves C_{xy} associated with the projection C_x, having the properties described in the theorem, and having distinct points $(x; y')$ and $(x; y'')$ corresponding to the same value t_2. For if so, there would be an interval $t_3 < t \leq t_2$ in which the curves would be distinct while at $t = t_3$ they coincide. This is, however, impossible since in a neighborhood of the point corresponding to t_3 there can be but one solution of the equations $f = 0$ corresponding to a given set of values x.

Suppose that a continuum X of points $(x_1, x_2, \cdots, x_m)$ *contains no projection of a boundary point of a sheet S of solutions of the equations* $f = 0$, *and no point near which the sheet becomes infinite. Then if X contains the projection of a point on the sheet every other point of X will also be such a projection. On the other hand, if X* contains a point which is not a projection of any point of the sheet, then no point of X can be a projection of a point of S.

These statements follow readily with the help of the last theorem. For suppose that X contains the projection x' of an interior point $(x'; y')$ of a sheet of solutions of the equations $f = 0$, and let x'' be any other point of X. Since X is a continuum there exists a continuous arc C_x entirely interior to X joining x' and x'', and the corresponding continuation curve C_{xy} must be defined over the whole of the arc C_x. Hence x'' is also the projection of a point of the sheet of solutions through $(x'; y')$. The rest of the theorem follows at once.

If the curve C_{xy} in the last theorem but one is defined over the whole arc C_x, and has initial and end points p' and p'', respectively, then there always exists a positive constant ρ such that any curve Γ_x lying in the ρ-neighborhood of the curve C_x and joining x' to x'', has a unique continuation curve Γ_{xy} also joining p' and p''.

The curve
$$(\Gamma_x) \qquad x = \xi_i(u) \qquad (i = 1, 2, \cdots, m;\ u' \leqq u \leqq u'')$$
is said to lie in the ρ-neighborhood of C_x if there exists a continuous function
$$(14) \qquad t = t(u) \qquad (u' \leqq u \leqq u'')$$
taking the values t', t'' at the ends of the u-interval, and such that the point α on Γ_x, defined by any value of u, lies in the neighborhood a_ρ of the corresponding point a of C_x determined by the relation (14).

It is possible to choose two constants, ϵ and $\delta \leqq \epsilon$, so that the neighborhoods p_ϵ and a_δ have the properties described in the theorem of § 1 uniformly for every point $p(a, b)$ on the arc C_{xy}. If not, there would be a sequence of points p_k on C_{xy} with a limit point π, for which the largest possible constants ϵ_k have the limit zero. But for the point π there is an effective constant $\epsilon > 0$, and the constants ϵ_k could not therefore decrease indefinitely in size. A similar argument shows the existence of the constant δ.

FUNDAMENTAL EXISTENCE THEOREMS.

Suppose now that the interval $u' \leq u \leq u''$ is divided by values u_k ($k = 1, 2, \cdots, \nu$) into sub-intervals so small that the points of any arc $\alpha_{k-1}\alpha_k$, corresponding on Γ_x to the values $u_{k-1} \leq u \leq u_k$, all lie in the $\tfrac{1}{2}\delta$-neighborhood of the point α_{k-1}, and further so small that the same is true with respect to the point a_{k-1} of the arc $a_{k-1}a_k$ of C_x corresponding to $\alpha_{k-1}\alpha_k$ by means of the relation (14). The constant ρ is supposed to have

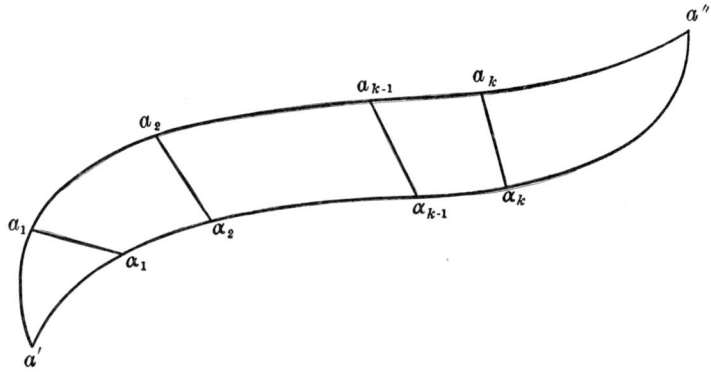

FIG. 1.

been chosen equal to $\tfrac{1}{2}\delta$, so that the curve Γ lies in the $\tfrac{1}{2}\delta$-neighborhood of C. Then the four-sided closed curve formed by the two straight lines $a_{k-1}\alpha_{k-1}$ and $a_k\alpha_k$, and the two arcs $a_{k-1}a_k$ and $\alpha_{k-1}\alpha_k$, lies entirely within the δ-neighborhood of the point a_{k-1}. The two continuation curves in the xy-space, starting with the point p_{k-1} on C_{xy} and having as projections the arcs $a_{k-1}a_k\alpha_k$ and $a_{k-1}\alpha_{k-1}\alpha_k$, respectively, lead to the same point π_k corresponding to the point α_k in the x-space.

It is possible to argue, then, that the point π_1 on the continuation curve of the arc $a'\alpha_1$ is the same as that of the continuation curve for $a'a_1\alpha_1$, since the arcs $a'\alpha_1$ and $a'a_1\alpha_1$ lie entirely within the δ-neighborhood of the point a_1. Similarly, the point π_2 for the arc $a'\alpha_2$ is the same as that for the continuation curve along $a'a_2\alpha_2$. And finally the point π'' must coincide with p'', provided always that the initial points π' and p' of the continuation curves are the same.

In particular if the curve C_{xy} is defined over the whole arc C_x, as described above, then there exists a polygon in the x-space joining a' and a'' in the ρ-neighborhood of C_x, and along which there is a continuation curve in S also joining p' and p''. The polygon can be so chosen that no two adjacent sides have more than an end point in common.

To show this, let the interval $t' \leq t \leq t''$ be divided in any way by means of points of division t', t_2, t_3, $\cdots$, t_ν, t'', and let the corresponding points on the curve C_{xy} be $(x'; y')$, $(\xi''; \eta'')$, $\cdots$, $(\xi^{(\nu)}; \eta^{(\nu)})$, $(x''; y'')$. The straight line $\xi^{(k)}\xi^{(k+1)}$ has the equations

$$x_i = \xi_i^{(k)} + \frac{t - t_k}{t_{k+1} - t_k}(\xi_i^{(k+1)} - \xi_i^{(k)}) \quad (i=1, 2, \cdots, m).$$

Since the functions defining C_x are continuous, and therefore uniformly continuous, in $t' \leq t \leq t''$, it is possible to take the points of division t', t_2, t_3, $\cdots$, t_ν, t'' so close together that the differences $x - \xi^{(k)}$, for any point x on the arc $\xi^{(k)}\xi^{(k+1)}$ of C_x, are uniformly less than an arbitrarily assigned positive constant δ; and the preceding theorem shows that the curve C_{xy} and the continuation curve along the polygon both lead from p' to p''.

If the sides $\xi^{(k)}\xi^{(k+1)}$ and $\xi^{(k+1)}\xi^{(k+2)}$ have more than the point $\xi^{(k+1)}$ in common, then one of the two would be included entirely within the other, and the continuation curve along $\xi^{(k)}\xi^{(k+2)}$ would have the same end points as that along the two successive sides. Therefore, by replacing adjacent sides by a single one whenever the two have more than one end point in common, a polygon as described in the theorem can be found.

§ 6. Auxiliary Theorems and Definitions

In this section it is proposed to record some theorems which will be of service later, especially in the proofs of the theorems of § 7. In the first place let it be agreed that a regular curve in the plane shall mean one which is continuous and has a well-defined tangent at all except possibly a finite number of points, at each of which, however, the slope of the tangent approaches definite limits as the point is approached from either side. Analytically this means that the functions

$$x = x(t), \qquad y = y(t) \qquad (t' \leqq t \leqq t'')$$

defining a regular curve are continuous in the whole interval $t' \leqq t \leqq t''$, that they are differentiable and satisfy the inequality

(15) $$(dx/dt)^2 + (dy/dt)^2 \neq 0$$

at all except possibly a finite number of values of t. At an exceptional value $t = \tau$, where the derivatives are not well defined or where the expression (15) vanishes, the angle φ defined by the equations

$$\cos \varphi = \frac{dx/dt}{\sqrt{(dx/dt)^2 + (dy/dt)^2}}, \quad \sin \varphi = \frac{dy/dt}{\sqrt{(dx/dt)^2 + (dy/dt)^2}}$$

has nevertheless a unique limit as t approaches τ on the right, and a unique limit as t approaches τ on the left. These two limits are not necessarily the same.

It is known that a simply closed regular curve C in an xy-plane divides the plane into two continua, an exterior and a finite interior.* Any two interior points can be joined by a regular curve every point of which is an interior point, and a similar statement holds for exterior points. Any continuous curve joining an interior and an exterior point must have on it at least one point of the curve C, and any point p on C can be joined with an interior point by a regular curve which has in common with C only the point p.

The interior of a simply closed regular curve

$$x = x(t), \qquad y = y(t) \qquad (t' \leqq t \leqq t'')$$

can be divided by a finite number of segments of straight lines into

---
* See for example Osgood, Lehrbuch der Funktionentheorie, Chapter V, §§ 4–6; Bliss, "A proof of the fundamental theorem of analysis situs," *Bulletin of the American Mathematical Society*, vol. 12 (1906), page 336.

regions each of which has a maximum diameter less than an arbitrarily assigned positive constant ϵ.*

Let the maximum and minimum values of y in the interval $t' \leq t \leq t''$ be y_1 and y_2, and let p_1 and p_2 be two points of C at which y has these values. It is desired to show that there is a segment $p'p''$ of the horizontal line $y = (y_1 + y_2)/2$ which forms with C two simply closed regular curves, $p'p_1p''p'$ and $p'p_2p''p'$, each containing one of the points p_1 and p_2.

The points p_1 and p_2 can be joined by a regular curve D which, except at its end points, is interior to C. Two arcs of D adjoining p_1 and p_2, can be marked off in such a way that they do not cut the line $y = (y_1 + y_2)/2$. The remaining arc D' of D is entirely interior to C and can be replaced by a continuous polygon D'' with a finite number of sides, having the same end points and consisting also of interior points of C only. Any side of D'' which has an end point in common with the line $y = (y_1 + y_2)/2$ may be rotated slightly about its other end point, and in this way it may be brought about that D'' has only interior points of its sides on the line $y = (y_1 + y_2)/2$, and actually crosses the line wherever they have a point in common.

The polygon D'' must intersect $y = (y_1 + y_2)/2$ at least once, say at a point p, since one end point of D'' is above and the other below this line. There will be a segment $p'p''$ of $y = (y_1 + y_2)/2$, containing p and such that p' and p'' are on the curve C while every other point of the segment is interior to C. There can be only a finite number of such segments $p'p''$ containing points of D'', since D'' has at most a finite number of intersections with the horizontal line. There must be at least one segment on which D'' has an odd number of intersection points, since otherwise both end points of D'' would be on the same side of $y = (y_1 + y_2)/2$. If $p'p''$ is such a segment, then it forms with C two simply closed regular curves $p'p_1p''p'$ and $p'p_2p''p'$, one of which contains p_1 and the other p_2. For after its last intersection with $p'p''$ the polygon D'' and hence p_2 is entirely exterior to the curve $p'p_1p''p'$.

---

* For a similar theorem see Osgood, loc. cit., Chapter V, § 9.

For the moment that part of a curve which does not lie in a horizontal line may be called the effective arc of the curve, in view of the fact that the altitude of the curve can not be more than one half the length of this so-called effective part. If the altitude of any curve is $\geq \epsilon$, the effective length of either of its two parts after subdivision by a horizontal line segment, as described above, will be $\leq L - \epsilon$, where L is its effective length.

If the altitude $y_1 - y_2$ of C is greater than ϵ, then the effective arc of either $p'p_1p''p'$ or $p'p_2p''p'$ will be greater in length than ϵ, and the effective length of each will also be less than $L - \epsilon$, where L is the perimeter of C. If the curve $p'p_1p''p'$, for example, has still an altitude greater than ϵ, it may be subdivided by a horizontal segment as before, and the effective arcs of the two new curves so found will be less than $L - 2\epsilon$. By a continuation of this process the interior of C will be subdivided finally by curves whose effective lengths are less than 2ϵ and whose altitudes are therefore less than ϵ.

In a similar manner the regions so formed may be subdivided by vertical segments into others whose breadths are less than ϵ, and the theorem follows at once.

A set of points in an x_1x_2-plane is *connected* if any two of its points can be joined by a continuous arc whose points all belong to the set, and it is further said to be *simply connected* if every simply closed regular curve in it has an interior which also consists only of points of the set.

It is more difficult to set down a satisfactory definition of simple connectivity for sets of points in an m-dimensional space. In the following section of these lectures, however, a special type of simple connectivity is needed which may be defined by means of some simple auxiliary conceptions.

A *normal subspace of two dimensions* in a region X of points $(x_1, x_2, \cdots, x_m)$ is a totality of points defined by equations of the form

$$x_i = \varphi_i(u_1, u_2) \qquad (i = 1, 2, \cdots, m),$$

where

1) the values (u_1, u_2) range over a simply connected region U;
2) no two distinct sets of values u define the same point x;
3) the functions φ are continuous and have continuous first derivatives in U;
4) the determinants of the second order of the matrix of derivatives $||\partial \varphi_i/\partial u_k||$ $(i = 1, 2, \cdots, m;\ k = 1, 2)$ do not all vanish simultaneously at any point of U.

A simply connected region in two dimensions is defined above, and a connected region X in a space of points $(x_1, x_2, \cdots, x_m)$ has a definition quite similar to that for two dimensions. In order to specify conveniently the properties of a region X which is simply connected, the term *elementary curve* will also be used. By an elementary curve in X is meant a simply closed continuous curve which either lies in a normal subspace of two dimensions entirely in the interior of X, or else is such that in every neighborhood of it there is a simply closed continuous curve having this property. It is thus seen that while an elementary curve may not itself be imbedded in one of the two-dimensional normal subspaces interior to X, it can nevertheless be approximated as closely as may be desired by one which does. The word neighborhood is here used in the sense described in connection with the fourth theorem of § 5 (see page 26).

If a region X is connected, then any simply closed continuous curve in its interior may be developed into two such curves by an auxiliary arc joining two of its points, and the process of development may be continued on the two arcs so formed.

If a region X is such that any simply closed continuous curve in its interior is an elementary curve, or may be developed into a number of elementary curves by means of auxiliary arcs, as just described, then X is said to be simply connected.

---

* For a discussion of the connectivity of higher spaces, see Picard and Simart, Théorie des Fonctions algébriques de deux Variables indépendantes, Chapître II, in particular §§ 11 ff. If every simply closed continuous curve interior to R lies in a normal subspace of two dimensions interior to R, one sees intuitively that a second neighboring subspace of the same kind can be passed through the curve. The closed two-dimensional subspace so formed is

§ 7. A Criterion that a Sheet of Solutions be Single-Valued

Consider in the first place a set of equations

(16) $$f_\alpha(x_1, x_2; y_1, y_2, \cdots, y_n) = 0 \quad (\alpha = 1, 2, \cdots, n)$$

in which there are but two independent variables x.

If a connected sheet S of solutions of equations (16) *consists only of ordinary points of the functions f, and furthermore has a simply connected projection X in the x_1x_2-plane such that no interior point of X is either a point where S becomes infinite or the projection of a boundary point of S, then the sheet S is single-valued over the interior of X.*

Suppose, in contradiction to the theorem, that over any interior point of X there were two points, p' and p'', of the sheet. Since S is connected there would be a continuous curve

$$(C_{xy}) \quad x_1 = x_1(t), \quad x_2 = x_2(t), \quad y_\alpha = y_\alpha(t)$$
$$(t' \leqq t \leqq t''; \alpha = 1, 2, \cdots, n)$$

consisting entirely of interior points of the sheet and joining p' with p'' in the space $(x; y)$. The projection

$$(C_x) \quad x_1 = x_1(t), \quad x_2 = x_2(t) \quad (t' \leqq t \leqq t'')$$

of this curve would necessarily be a closed curve in the x_1x_2-plane, and by the second theorem of § 5 the arc C_{xy} is the only one associated with C_x in the sheet S and having the initial point p'.

The curve C_x may be simply closed and regular; but if it is not, there will nevertheless be a curve in the region X having these properties, and for which the continuation curve analogous to C_{xy} is not closed. For, in the first place, from § 5 it is seen that the curve C_x may be supposed to be a polygon no two adjacent sides of which have more than an end point in common, provided that it is desired only to secure a continuous curve in

---

separated into two parts by the curve, and hence the number which Picard and Simart designate by p_1 is equal to unity for a simply connected region of the kind defined in the text above.

the sheet passing from p' to p''. Let the corners of this polygon in the x-plane be denoted by $\xi_1, \xi_2, \cdots, \xi_\nu$, where ξ is a symbol for a point (x_1, x_2). The side $\xi_\nu \xi_1$ touches $\xi_1 \xi_2$ at its end point ξ_1, and it can be argued therefore that there will be some first side $\xi_\lambda \xi_{\lambda+1}$ which touches some one of the preceding sides elsewhere than at its initial point ξ_λ. Let the side so touched by $\xi_\lambda \xi_{\lambda+1}$ be $\xi_\kappa \xi_{\kappa+1}$, where $\kappa + 1$ is necessarily less than λ, and let the first point of $\xi_\lambda \xi_{\lambda+1}$ which lies on $\xi_\kappa \xi_{\kappa+1}$ be ξ. If the portion of the curve C_{xy} which corresponds to the polygon

(17) $\qquad \xi, \ \xi_{\kappa+1}, \ \xi_{\kappa+2}, \ \cdots, \ \xi_\lambda, \ \xi$

is not closed, then the polygon (17) itself is a simply closed curve in X of the kind desired above, that is, one along which there exists a continuation curve in the xy-space whose end points are different.

If the portion of C_{xy} which corresponds to (17) is closed, then that part of C_{xy} which belongs to the polygon

(18) $\qquad \xi_1, \xi_2, \cdots, \xi_\kappa, \xi, \xi_{\lambda+1}, \cdots, \xi_\nu, \xi_1$

is also continuous and leads from p' to p''. Since $\kappa + 1 < \lambda$ the side $\xi_{\kappa+1} \xi_{\kappa+2}$ at least is missing in (18), and the number of sides is at least one less than that of the original polygon. By an alteration of the kind suggested in the proof of the last theorem of § 5, which also reduces the number of sides, it can be brought about, if not already true, that the polygon (18) still has no two adjacent sides with more than an end point in common.

By continuing this process one must come at some stage to a simply closed regular curve in the x-plane with a corresponding continuation curve in the xy-space which is not closed. In order not to complicate the notation too much it may be supposed that the curve C_x itself is such a curve. Every point of C_x is an interior point of the region X since the corresponding point of C_{xy} is an interior point of the sheet S. The interior of C_x is therefore also composed entirely of interior points of X, since X is simply connected. If the interior of C_x is subdivided into two parts by a segment of a straight line, as described in the preceding section, the dividing segment will also have a continuation curve on the sheet S throughout its entire length, by the second theorem of § 5. For its initial point on the curve C_x corresponds to an interior point of the sheet S and, by the hypothesis of the theorem which is to be proved, none of its points can be a point where S becomes infinite or can correspond to a boundary point of S. Hence one of the simply closed curves formed by the curve C_x and the dividing segment is a curve retaining the property that it has a continuation curve on the sheet S which is not closed. Suppose that C_x' is this curve. By continuing the process a sequence of curves $\{C_x^{(k)}\}$, with diameters approaching zero, can be found, each lying in the interior of C_x and having an unclosed continuation curve $C_{xy}^{(k)}$ on S.

If a point $p^{(k)}$ is selected arbitrarily on the curve $C_{xy}^{(k)}$, the sequence $\{p^{(k)}\}$ ($k=1, 2, \cdots, \infty$) will have a finite point of condensation $\pi(\alpha; \beta)$ in the xy-space which is an interior point of the sheet S. For the projections $x^{(k)}$ of the points $p^{(k)}$ all lie in the interior of C_x and hence must have a point of condensation α. Furthermore the points of the sequence $p^{(k)}$ whose projections are in the neighborhood of α can not become infinite or approach a boundary point of the sheet, since α is interior to X. They must therefore have at least one limit point π which is an interior point of the sheet, and with which there are associated two neighborhoods π_ϵ and α_δ by the principal theorem of § 1. Some of the points $p^{(k)}$ lie in π_ϵ, and have corresponding curves $C_x^{(k)}$ in α_δ. For such points the continuation curves $C_{xy}^{(k)}$ also lie in π_ϵ and can not be unclosed, since to any point x in α_δ there corresponds in π_ϵ at most one solution of the equations $f = 0$. The original assumption that S is multiple-valued in the interior of X is therefore contradicted.

The theorem remains true for any system of equations of the form

(19) $\quad f_\alpha(x_1, x_2, \cdots, x_m; y_1, y_2, \cdots, y_n) = 0 \quad (\alpha = 1, 2, \cdots, n).$

In this case the curves C_{xy} and C_x have equations

(C_{xy}) $$x_i = x_i(t), \quad y_a = y_a(t)$$
$$(i = 1, 2, \cdots, m;\ \alpha = 1, 2, \cdots, n;\ t' \leq t \leq t''),$$

(C_x) $$x_i = x_i(t),$$

and the question asked in the proof of the theorem just stated is whether or not the latter curve may be closed while the former has distinct end points.

It is a part of the hypothesis of the theorem that the region X is simply connected according to the definition of the preceding section; and, according to the arguments made in the paragraphs above, the curve C_x may be supposed a simply closed polygon. In any neighborhood of C_x there will be, according to § 6, on account of the simple connectivity, an elementary curve $\overline{C}_x$ lying in a normal subspace of two dimensions

(20) $$x_i = g_i(u_1, u_2) \qquad (i = 1, 2, \cdots, m)$$

entirely interior to X. If the continuation curve C_{xy} is not closed, and if $\overline{C}_x$ is taken sufficiently near to C_x, then the corresponding continuation curve $\overline{C}_{xy}$ will also not be closed.

The normal subspace (20) is defined over a simply connected domain U of points (u_1, u_2), and has no multiple points. To every point of $\overline{C}_x$ there corresponds therefore a single pair of values

(C_u) $$u_1 = u_1(t), \quad u_2 = u_2(t);$$

and the functions so defined are continuous, by the principal theorem of § 1, since at every point some pair of the equations (20) has a functional determinant for u_1, u_2 which is different from zero. The curve corresponding to the curve $\overline{C}_{xy}$ in the space $(u; y)$ may be denoted by

(C_{uy}) $u_1 = u_1(t),\quad u_2 = u_2(t),\quad y_a = y_a(t),\quad (\alpha = 1, 2, \cdots, n),$

and its initial point, corresponding to p', by $p_u'\ (u';\ y')$. Every point of C_{uy} is an ordinary solution of the equations

(21) $$\varphi_a(u_1, u_2;\ y_1, y_2, \cdots, y_n) = f_a(g_1, g_2, \cdots, g_m;\ y_1, y_2, \cdots, y_n) \parallel 0$$
$$(\alpha = 1, 2, \cdots, n).$$

FUNDAMENTAL EXISTENCE THEOREMS.   37

With a continuous curve C joining (u_1', u_2') to an arbitrarily chosen point (u_1, u_2) of U there is always associated a continuation curve of solutions of the equations (21), having the initial point p_u' and defined throughout the whole of C, since any such curve defines a curve in the x-space interior to X along the whole of which there is a corresponding continuation curve for the equations (19) in the sheet S. Hence there is a unique sheet S_u of solutions of the equations (21) whose projection in the u_1u_2-space is U; and no interior point of U is a point where the sheet becomes infinite or corresponds to a boundary point of the sheet, since the same is true of S with respect to X. The preceding argument can therefore be applied to show that the sheet S_u is single-valued over the region U, and the existence of the curve C_{uv} with the distinct end points p_u' and p_u'' is contradicted. Hence C_{xy} can not have distinct end points p' and p'', and the theorem last stated is proved.

§ 8. Transformations of n Variables and a Modification of a Theorem of Schoenflies

It is interesting to deduce by means of the preceding theorems some conclusions concerning a system of equations of the form

(22) $f_\alpha(x; y) = x_\alpha - \psi_\alpha(y_1, y_2, \cdots, y_n) = 0 \quad (\alpha = 1, 2, \cdots, n).$

The functions ψ are once for all assumed to be single-valued, continuous, and to have continuous first derivatives in a continuum Y in which the functional determinant

$$D = \partial(\psi_1, \psi_2, \cdots, \psi_n)/\partial(y_1, y_2, \cdots, y_n)$$

is different from zero. By a continuum is meant a set of points consisting only of interior points any two of which can be connected by a continuous curve lying entirely within the set. The boundary points of Y will be denoted by B, and X will represent the set of points in the x-space which corresponds to Y by means of the equations (22).

Any sequence $\{y^{(k)}\}$ of points $(y_1^{(k)}, y_2^{(k)}, \cdots, y_n^{(k)})$ $(k = 1, 2, \cdots)$ in Y, which approaches infinity or has a point of B as limit point, defines a corresponding sequence of points $\{x^{(k)}\}$ in X. The set of points of condensation for such sequences $\{x^{(k)}\}$ will be denoted by A.

The totality of solutions of the equations (22) *corresponding to points of the continuum Y form a single connected sheet S whose only boundary points have projections x and y in the sets A and B, respectively.*

For suppose that $(x'; y')$ is a first solution and $(x''; y'')$ any other. The points y' and y'' can be joined by a continuous curve interior to Y

$$y_\alpha = y_\alpha(t) \qquad (\alpha = 1, 2, \cdots, n; \ t' \leq t \leq t''),$$

and the corresponding curve

$$x_\alpha = x_\alpha(t), \qquad y_\alpha = y_\alpha(t),$$

defined by equations (22), is a curve interior to the sheet S and joining $(x'; y')$ to $(x''; y'')$, so that S is evidently connected. Any boundary point $(\alpha; \beta)$ of S must be the limit of a sequence of points $p^{(k)}$ for which the projections y are in Y. The limit β of the sequence $y^{(k)}$ can not be in Y, since then $(\alpha; \beta)$, by the theorem of § 1, would be an interior point of S. Hence β must be in B and α in A.

One may say further that if $p^{(k)}$ is a sequence of points $(x^{(k)}; y^{(k)})$ in S for which the sequence $x^{(k)}$ approaches infinity, then the only finite points of condensation possible for the sequence $y^{(k)}$ are in B. The statement is true when x and y are interchanged, on account of the definition above of the set A.

If the points of the set A are distinct from those of the image X of Y, then X is a single continuum whose only boundary points are points of A.

To prove this, consider an arbitrarily chosen point y' of Y. None of the points in a suitably chosen neighborhood of the corresponding values x' are points of A, since by the fundamental theorem of § 1 all such points correspond by means of equations (22) to points of Y, and are therefore points of X. Consider now the continuum $\overline{X}$ consisting of all points x which can be joined to x' by continuous curves containing no points of A, a continuum to which the neighborhood of x' certainly belongs, as has just been shown.

All the points of X are in the continuum $\overline{X}$, since the solutions of equations (22) corresponding to points of Y form a single connected sheet S. The curve in S joining $(x'; y')$ with any other point $(x''; y'')$ of the sheet has therefore a projection in the x-space joining x' with x'' and containing no points of the set A.

All of the points of $\overline{X}$ are points of X. For any set of values x in $\overline{X}$ can be joined to x' by a continuous curve C_x lying entirely in $\overline{X}$ and containing therefore no points of A. By the second theorem of § 5 the corresponding continuation curve C_{xy} must extend along the entire arc C_x, since otherwise the values of y for points on C_{xy} would approach infinity or else have a limit point on the boundary B of Y, and some point of C_x would in that case necessarily be a point of A. It follows that x, like x', is the image of some point y in Y.

From the initial theorem of the last section, for the case when there are more than two variables, it follows that

If A is distinct from X, and X is simply connected in the sense of §6, then the sheet S is single-valued. In other words the continuum Y is tranformed in a one-to-one way into a continuum X by means of the equations (22), and the functions

(23) $\qquad y_\alpha = y_\alpha(x_1, x_2, \cdots, x_n) \quad (\alpha = 1, 2, \cdots, n)$

so defined over X are single-valued, continuous, and have continuous first derivatives.

The character of the functions (23) near any point of X follows at once from the theorem of § 1.

Let it be supposed that the set of points A divides the x-space into exactly two continua X, Ξ such that every point of A is a bound-

ary point for each of them, and suppose furthermore that there is a particular point ξ in Ξ which does not correspond by means of the equations (22) to any point of Y. Then the image X of Y is distinct from A and coincides with X. If X is simply connected the other conclusions of the last theorem follow at once.

In the first place it can be shown that if any point ξ' of Ξ corresponds to a point of Y then every other point ξ'' of Ξ would also have this property. For ξ' and ξ'' can be joined by a continuous curve

$$x_\alpha = x_\alpha(t) \qquad (\alpha = 1, 2, \cdots, n;\ t' \leq t \leq t'')$$

entirely interior to Ξ. The corresponding continuation curve

$$x_\alpha = x_\alpha(t), \qquad y_\alpha = y_\alpha(t)$$

of solutions of equations (22) must be defined along the whole of the interval $t' \leq t \leq t''$, since otherwise as t approached any upper bound τ of the values t which could be reached by continuation, the corresponding points y of the curve would have to approach infinity or else have a point of condensation on the boundary of Y. But this is impossible, since for a sequence of points x corresponding to a sequence of points in Y approaching infinity or a boundary point of Y, the only limiting points possible are at infinity or else in the set A. It follows at once, on account of the hypothesis of the theorem, that no point of Ξ can correspond to a point of Y, and neither can any point of A, since in any neighborhood of such a point of A there are points of Ξ which in that case would also correspond to values y in Y. The image of the region Y in the x-space is a single continuum whose only boundary points are points of A. According to the preceding argument it cannot be Ξ and it must therefore be X.

A modification of a theorem of Schoenflies can be deduced readily from the results which precede. The theorem has to do with a pair of equations of the form

(24) $$x_1 = \psi_1(y_1, y_2), \qquad x_2 = \psi_2(y_1, y_2)$$

in which the functions ψ are single-valued, continuous, and have continuous derivatives on a simply closed regular curve B of the y-plane and in the interior Y of B. The functional determinant $D = \partial(\psi_1, \psi_2)/\partial(y_1, y_2)$ is supposed to be different from zero in Y. *If the curve A in the x-plane formed by transforming the simply closed regular curve B in the y-plane, by means of the equations* (24), *is distinct from the image X of the interior Y of B, then X is a simply connected continuum whose only boundary points are points of A, and the correspondence defined between X and Y is one-to-one. The single valued functions*

(25) $$y_1 = y_1(x_1, x_2), \quad y_2 = y_2(x_1, x_2),$$

*so determined in the region X, are continuous and have continuous first derivatives.**

From the preceding theorems of this section it follows that the complete image X of Y is a single finite continuum whose only boundary points are points of A. It remains to show that X is simply connected and that the correspondence between X and Y is one-to-one.

If any simply closed regular curve C_x is drawn in X, its interior must consist entirely of points of X. Otherwise there would necessarily be a boundary point of X, a point of the curve A, interior to C_x, and there would also be points of A exterior to C_x since X is finite. Hence there would necessarily be a point of the continuous curve A on C_x itself, which contradicts the assumption that A and X are distinct. It follows at once from the first paragraphs of § 7 and the simple connectivity of X just proved, that only one point y in Y corresponds to a given x in X, and by the theorem of § 1 it may be seen that the functions

---

* Schoenflies assumed only the continuity of the functions ψ_1, ψ_2, adding, however, that the correspondence defined between the regions X and Y of the two planes is to be one-to-one. In the theorem here proved ψ_1 and ψ_2 are subjected to further continuity restrictions, but the correspondence is proved to be unique. See Schoenflies, "Ueber einen Satz der Analysis Situs," *Göttinger Nachrichten* (1899), page 282. The theorem was later proved by Osgood and Bernstein in the same journal (1900), pages 94 and 98, respectively.

(25) have the continuity properties described in the theorem in the neighborhood of any particular point x.

Another theorem, slightly different in form, may be stated as follows:

If the images of the points of the simply closed regular curve B in the y-plane all lie on a simply closed regular curve A in the x-plane, then the equations (24) define a one-to-one correspondence between the interior X of A and the interior Y of B, and the functions (25) so defined have the same continuity properties as before.

In this case it can first be shown that the image x' of any point y' in Y must be distinct from A, and the rest of the proof is the same as before. For, if x' were a point of A, every point of a properly chosen neighborhood of x' would also be the image of a point of Y, since at $(x'; y')$ the functional determinant of equations (24) does not vanish. It would follow then, by continuation, that every point exterior to the curve A would also be the image of a point of Y, which is impossible since the functions ψ are finite. The continuum X is therefore identical with the interior of A, by the preceding theorems, and the correspondence between X and Y is one-to-one.

An example applying some of the theorems of §§ 5, 8 is given at the end of § 14.

CHAPTER II

SINGULAR POINTS OF IMPLICIT FUNCTIONS

The theorems which have been developed in the preceding pages of these lectures have to do with the behavior of implicit functions at ordinary points, or in regions which have no singular points in their interiors. For singular points where the functional determinant vanishes the theory is much more complicated, and no methods which can be comprehensively applied have so far been developed. There are, however, many special cases in widely different fields which have been studied with success, and it may not be out of place to glance at a few of them before proceeding to the further theorems with which these pages are primarily concerned.

Perhaps the most complete single theory which has been developed is that which has to do with the singularities of an algebraic function y of x determined by an equation of the form

$$(1) \qquad P(x, y) = 0,$$

where P is an irreducible polynomial in the two variables x and y. Suppose for convenience that the singular point to be considered is at the origin, and that the polynomial $P(0, y)$ has a lowest term of degree n in y. Then it is known that for each value of x in a sufficiently small neighborhood of $x = 0$, there exist exactly n solutions y of equation (1) in the neighborhood of $y = 0$, and the values of these solutions are given by k cycles of the form

$$(2) \qquad y = a_j x^{\mu_j/p_j} + a_j' x^{\mu_j'/p_j} + \cdots \quad (j = 1, 2, \cdots, k),$$

where the numbers μ, p are positive integers satisfying the relations

$$\mu_j < \mu_j' < \mu_j'' < \cdots, \qquad p_1 + p_2 + \cdots + p_k = n.$$

The series is one member of the cycle; the others are found by replacing x^{1/p_j} by $\omega^\nu x^{1/p_j}$ ($\nu = 1, 2, \cdots, p_j - 1$), where ω is a primitive p_j-th root of unity. The number p_j has no factor in common with the exponents $\mu_j, \mu_j', \cdots$. Otherwise the expansion would be in terms of a root of x of lower order than p_j. Thus there are in all n series in fractional powers of x which define the roots of the algebraic equation in the neighborhood of the origin. The coefficients of the series may be computed by means of the well-known Newton polygon,* or by methods due to Hamburger† and Brill.‡ If the substitution $x = t^{p_j}$ is made in the series (2), the points (x, y) which it defines may be expressed in the parametric representation

$$x = t^{p_j}, \qquad y = t^{\mu_j}\{\alpha_j + \alpha_j' t^{\mu_j' - \mu_j} + \cdots\} \qquad (j = 1, 2, \cdots, k).$$

All the solutions of the equation (1) in the neighborhood of the origin evidently belong to a finite number of such branches.

With the help of the preparation theorem of Weierstrass, which is to be studied in the following pages, results similar to those just given may be proved for the solutions of an equation $F(x, y) = 0$ in the vicinity of any point where F is analytic.

The singularities of a surface

$$F(x, y, z) = 0$$

at a point where the function F is analytic have also been extensively studied. The points of the surface in the neighborhood of a singular point are determined by means of a finite number of expansions of the form

$$x = P(u, v), \qquad y = Q(u, v),$$

where P and Q are analytic in the parameters u and v.§

---

* See Appell and Goursat, *Théorie des Fonctions algébriques*, pp. 184 ff.
† Weierstrass, *Werke*, vol. 4, Kapitel 1.
‡ *Münchener Berichte*, vol. 21 (1891), p. 207.
§ See Black, "The parametric representation of the neighborhood of a singular point of an analytic surface," *Proceedings of the American Academy of Arts and Sciences*, vol. 37 (1902), p. 281.

In the calculus of variations the construction of "fields of extremals" in the plane requires the study of the real solutions of a system of equations of the form

(3) $$x = \varphi(t, a), \quad y = \psi(t, a).$$

The extremals are the curves in the xy-plane defined by these equations for different values of a. Suppose that the parametric values

(4) $$t_0 \leq t \leq t_1, \quad a = a_0$$

define an arc E which does not intersect itself and which consists entirely of points where the functional determinant

(5) $$\Delta(t, a) = \frac{\partial(\varphi, \psi)}{\partial(t, a)}$$

is different from zero. Then to any point (x, y) in a properly chosen neighborhood of E there corresponds but one solution (t, a) of equations (3), in the neighborhood of the values (4); and the functions

$$t = t(x, y), \quad a = a(x, y)$$

so defined have continuity properties similar to those of φ and ψ themselves.* The neighborhood thus simply covered by the extremals (3) is the "field," and is perhaps the simplest example of the notion since it consists only of non-singular solutions of the equations (3).

When it is desired to find an arc C which minimizes an integral with respect to variations lying entirely on one side of C, a field of a different sort can be constructed.† The equations of the

---

The mathematical literature concerned with the singularities of a curve or surface, particularly their transformation into simpler types, is very large. The reader is referred to Pascal, Repertorium der höheren Mathematik, 2d edition, vol. 2, erste Hälfte, pp. 291 ff; and Encyclopädie der Mathematischen Wissenschaften, II B 2, p. 119, and III C 4, pp. 365 ff.

* Bolza, Vorlesungen über Variationsrechnung, pp. 249 ff.

† Bliss, "Sufficient conditions for a minimum with respect to one-sided variations," Transactions of the American Mathematical Society, vol. 5 (1904), p. 477; Bolza, "Existence proof for a field of extremals tangent to a given curve," ibid., vol. 8 (1907), p. 399.

extremals (3) can be taken so that for $t = 0$ they all intersect C and are tangent to it, and the equations

$$x = \varphi(0, a), \qquad y = \psi(0, a)$$

will then be the equations of C. If the curvatures of the two arcs at their point of contact are always different, then the extremal arcs E simply cover a portion of the plane N on one side of C and adjacent to it. In other words, the equations (3) define a one-to-one correspondence between the points of a region adjoining the axis $t = 0$ in the ta-plane, shown in the accompanying figure, and a certain neighborhood N on one side of the arc C.

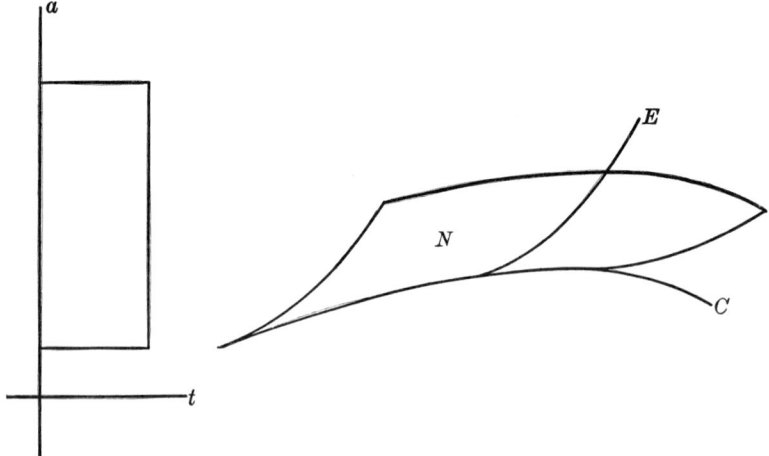

Fig. 2.

In the interior of the region N the functions $t(x, y)$, $a(x, y)$ have continuity properties similar to those of φ and ψ themselves. It is easy to see that this is a case in which the functional determinant (5) vanishes along the boundary $t = 0$ of the region to be transformed, since the curves C and E are always tangent.

In a paper published since these lectures were given, Dr. E. J. Miles[*] has considered the transformation defined by the equations

---

[*] "The absolute minimum of a definite integral in a special field," *Transactions of the American Mathematical Society*, vol. 13 (1912), pp. 37 ff.

(3) when the curve C to which the extremals E are tangent has a cusp, a situation corresponding to still another problem in the

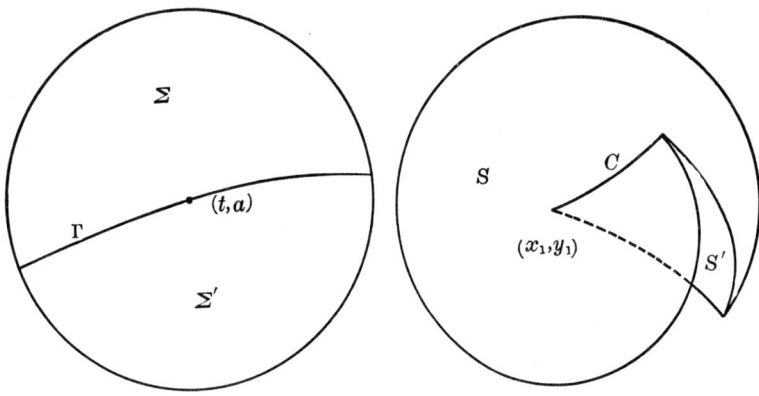

Fig. 3.

calculus of variations. In that case a point (t_1, a_1) and a curve Γ through it are transformed into a point (x_1, y_1) and a curve C as shown in the figure. One portion Σ of a neighborhood of (t_1, a_1) is then transformed in a one-to-one way into the leaf S, and the other portion Σ' into the leaf S'. At any point in the interior of one of the leaves, the variables t and a are single-valued functions of x, y having continuity properties similar to those of φ and ψ. The transformation is singular along the curve Γ.

The three examples which have been just described are only a few of the many proofs for the existence of fields involving transformations with singular points which might be cited.* Nearly all of these have to do with singularities of transformations of the form

(6) $\qquad x = \varphi(u, v), \qquad y = \psi(u, v),$

---

* Bliss, "The construction of a field of extremals about a given point," *Bulletin of the American Mathematical Society*, vol. 13 (1906), p. 47; Mason and Bliss, "Fields of extremals in space," *Transactions of the American Mathematical Society*, vol. 11 (1910), p. 325; Bill, "The construction of a space field of extremals," *Bulletin of the American Mathematical Society*, vol. 15 (1908), p. 374; Szücs, "Sur l'extrémale qui joint deux points donnés," *Mathematische Annalen*, vol. 71 (1912), p. 380. The method used by Szücs is quite closely that of Mason and Bliss in the paper mentioned above.

or
$$x = \varphi(u, v, w), \quad y = \psi(u, v, w), \quad z = \chi(u, v, w),$$

which have been studied also in a series of papers of more recent date presented as dissertations for the degree of doctor of philosophy at Harvard University.* The methods which have been used in the different cases have differed widely, and it does not seem possible at present to formulate a theory which includes them all. It is the intention of the writer, however, to show in the following pages how the transformation theorems proved above in § 7 may be applied to throw much light on the nature of real transformations of the form (6) in the neighborhoods of singular points. In the section of the lectures immediately following this introduction a simple algebraic proof of the preparation theorem of Weierstrass is given, not depending upon the theory of functions of a complex variable. A generalization of it is given in a later section which, in what might be called the general case, enables one to describe the behavior of the solutions of a system of equations of the form

$$f_i(x_1, x_2, \cdots, x_m; y_1, y_2, \cdots, y_n) = 0 \quad (i = 1, 2, \cdots, n)$$

in the neighborhood of a point where the functional determinant

$$\frac{\partial(f_1, f_2, \cdots, f_n)}{\partial(y_1, y_2, \cdots, y_n)}$$

vanishes. For these equations the variables x and y are permitted to have complex values.†

---

* Urner, "Certain singularities of point transformations in space of three dimensions," *Transactions of the American Mathematical Society*, vol. 13 (1912), p. 233; Clements, "Implicit functions defined by equations with vanishing jacobian," to appear in the same journal. Dederick, in a paper entitled "The solutions of an equation in two real variables at a point where both the partial derivatives vanish," *Bulletin of the American Mathematical Society*, vol. 16 (1909), p. 174, has discussed the singularities of a curve of the form $F(x, y) = 0$ with the help of a sort of generalization of the Weierstrass preparation theorem for a function which is not necessarily analytic.

† The proof given in these pages for the last-mentioned theorem is for the case of two variables y. For n variables see the reference in the last footnote to § 13.

§ 9. The Preparation Theorem of Weierstrass

The theorem which is to be proved may be stated in the following form:

Let $f(x_1, x_2, \cdots, x_m, y)$ be a convergent series in the variables x, y, and such that the series $f(0, 0, \cdots, 0, y)$ begins with a term of degree n. Then f is factorable in the form

$$f(x_1, x_2, \cdots, x_m, y) = (y^n + a_1 y + \cdots + a_n) \varphi(x_1, x_2, \cdots, x_m, y),$$

where $a_1, a_2, \cdots, a_n$ are convergent power series in $x_1, x_2, \cdots, x_m$ which vanish for $x_1 = x_2 = \cdots = x_m = 0$, and φ is a power series in $x_1, x_2, \cdots, x_m, y$ which has a constant term different from zero.

In the *Bulletin de la Société Mathématique de France*[*] Goursat has called attention to the fact that the proof which Weierstrass gave of this important theorem, as well as the later proofs which occur in the literature[†], make use of the notions of the function theory, while the theorem itself is essentially of an algebraic character. In the paper referred to he has given an elegant and elementary proof of the theorem which is in outline as follows:

By means of the substitution

$$y^n = -a_1 y^{n-1} - a_2 y^{n-2} - \cdots - a_n$$

the series f can be reduced to a polynomial P of degree $n-1$ in y, whose n coefficients are convergent series in $a_1, a_2, \cdots, a_n, x_1, x_2, \cdots, x_m$. By the usual theorems of implicit function theory it is shown that the n equations found by putting these coefficients equal to zero have unique solutions for $a_1, a_2, \cdots, a_n$ as power series in $x_1, x_2, \cdots, x_m$, which vanish with $x_1, x_2, \cdots, x_m$. If the values so found are substituted in the formula

$$y^n = -a_1 y^{n-1} - a_2 y^{n-2} - \cdots - a_n + \mu$$

---

[*] "Démonstration élémentaire d'un théoreme de Weierstrass," vol. 36 (1908), p. 209.

[†] See, for example, Picard, Traité d'Analyse, vol. 2, p. 243; Goursat, Cours d'Analyse, vol. 2, p. 284.

and the series f again reduced, a polynomial P_1 of degree $n-1$ in y will be found whose coefficients are series in $x_1, x_2, \cdots, x_m, \mu$. On account of the way in which the functions $a_1, a_2, \cdots, a_n$ were determined, this polynomial P_1 has a factor μ, and hence f has a factor $(y^n + a_1 y^{n-1} + \cdots + a_n)$.

Since the paper of Goursat appeared two further proofs of the theorem have been published, one by the writer[*] and the other by MacMillan,[†] each of which seems even more direct than that of Goursat. In the proof which follows use is made of the very concise and elegant method of MacMillan for determining the coefficients, while the rest of the proof is similar to that of the earlier paper of the writer cited above.

The theorem may be stated in a different form as follows:

Suppose that $f(x_1, x_2, \cdots, x_m, y)$ is a series with literal coefficients such that $f(0, 0, \cdots, 0, y)$ begins with the term $a_0 y^n$. Then there is one and but one series $b(x_1, x_2, \cdots, x_m, y)$ which satisfies formally the relation

(7) $$bf = p,$$

where p is a polynomial

$$p = a_0 y^n + a_1 y^{n-1} + \cdots + a_n$$

whose coefficients $a_k(x_1, x_2, \cdots, x_m)$ $(k = 1, 2, \cdots, n)$ are series vanishing with the x's.

Each of the coefficients in b and the a's is a rational function of a finite number of the coefficients of f with denominator a power of a_0, and the constant term in b is unity.

If the coefficients in f are chosen numerically so that f converges and $a_0 \neq 0$, then the series b and a_k $(k = 1, 2, \cdots, n)$ also converge.

The functions f, b, p may be written in the forms

(8)
$$f = a_0 y^n - y^{n+1} f_0 - f_1 - f_2 - \cdots,$$
$$b = b_0 + b_1 + b_2 + \cdots,$$
$$p = a_0 y^n - p_1 - p_2 - \cdots,$$

---

[*] *Bulletin of the American Mathematical Society*, vol. 16 (1910), p. 356.
[†] Ibid., vol. 17 (1910), p. 116.

where f_k, b_k, p_k are homogeneous expressions of degree k in $x_1, x_2, \cdots, x_m$ with coefficients which are series in y. It is desired to determine b so that the identity (7) holds, and so that the expressions p_k have coefficients which contain y only to the degree $n - 1$.

By substituting the expressions (8) in the identity (7) and equating terms of the same degree in the x's, it follows that

$$b_0(a_0 - yf_0)y^n = a_0 y^n,$$
$$b_1(a_0 - yf_0)y^n = b_0 f_1 - p_1,$$
$$b_2(a_0 - yf_0)y^n = b_0 f_2 + b_1 f_1 - p_2,$$
$$\cdots \cdots \cdots \cdots$$
$$b_k(a_0 - yf_0)y^n = b_0 f_k + b_1 f_{k-1} + \cdots + b_{k-2} f_2 + b_{k-1} f_1 - p_k,$$
$$\cdots \cdots \cdots \cdots$$

These equations are to be identities in x and y. The first one determines b_0 uniquely with constant term unity, and furthermore so that each coefficient is a quotient, in fact a polynomial with positive integral coefficients in a finite number of the coefficients of f, divided by a power of a_0. In the second equation p_1 must be chosen equal to the terms of $b_0 f_1$ which contain y to the degree $n - 1$ or less, after which b_1 is uniquely determined. Similarly in the kth equation p_k must first be chosen to cancel the terms on the right of degree $n - 1$ or less in y, and then b_k is unique.

It only remains to show that the series b and a_k are convergent in any numerical case for which f converges. There is no loss of generality in assuming that the series f converges in the domain

$$|x_i| \leq 1, \quad |y| \leq 1 \quad (i = 1, 2, \cdots, m),$$

since this can always be effected by a substitution of the form

$$x_i = \rho_i x_i', \quad y = \sigma y' \quad (i = 1, 2, \cdots, m).$$

Suppose then that K is a number greater than the absolute value of any term in the series $f(1, 1, \cdots, 1, 1)$, that is, greater than the absolute value of any coefficient in f. If A_0 is the absolute value of a_0, the series

$$F = A_0 y^n - \frac{K y^{n+1}}{1-y} - \frac{K}{1-y} X,$$

where

$$X = \frac{1}{(1-x_1)(1-x_2) \cdots (1-x_m)} - 1,$$

dominates f in the sense that every coefficient except the first has a numerical value equal to or greater than K; and the series B satisfying the relation

$$BF = A_0 y^n + A_1 y^{n-1} + \cdots + A_n$$

analogous to (7) has coefficients numerically greater than the absolute values of those of b. Hence if B converges the same will be true of b.

But it is easy to show that the series B converges. It will certainly do so if convergent series A_k, C, D can be found satisfying the relation

$$A_0 y^n (1-y) - K y^{n+1} - KX = (A_0 y^n + A_1 y^{n-1} + \cdots + A_n)(Cy + D),$$

because then B would have the value

$$B = \frac{1-y}{Cy + D}.$$

On comparing the coefficients of the two highest terms in y in the next to last equation, and for convenience denoting by α the constant value

$$\alpha = -\frac{A_0 + K}{A_0^2},$$

it is found that

$$C = \alpha A_0, \qquad D = 1 - \alpha A_1.$$

By comparing the other powers of y and substituting these values, we have

$$A_1 + \alpha A_0 A_2 = \alpha A_1^2,$$
$$A_2 + \alpha A_0 A_3 = \alpha A_1 A_2$$
$$\cdots \cdots$$
$$A_{n-1} + \alpha A_0 A_n = \alpha A_1 A_{n-1},$$
$$A_n = \alpha A_1 A_n - KX.$$

But these equations have linear terms in $A_1, A_2, \cdots, A_n$ with functional determinant different from zero, and hence have solutions, by the theorems of § 2, which are convergent series in $x_1, x_2, \cdots, x_m$ and have no constant terms.

It is evident, in any numerical case for which f is convergent, that a neighborhood of the origin may be chosen in which the series b is everywhere different from zero. In such a neighborhood all of the values $(x_1, x_2, \cdots, x_m, y)$ which make f vanish are roots of the equation $p = 0$, and vice versa.

If $f(x_1, 0, \cdots, 0, y)$ has its terms of lowest degree homogeneous and of degree n, then the polynomial $p(x_1, 0, \cdots, 0, y)$ has the same initial terms, since the first coefficient of the factor series b is unity.

§ 10. The Zeros of $\varphi(u, v), \psi(u, v)$, or their Functional Determinant

Consider a function $\varphi(u, v)$ whose values in the neighborhood of the origin in the uv-plane are given by a convergent series in u and v which vanishes for $u = v = 0$. If the series contains a factor u in every term it may be written in the form

(9) $$\varphi(u, v) = au^k \Phi(u, v),$$

where a is a constant different from zero and $\Phi(u, v)$ is a convergent series for which $\Phi(0, v)$ has a first term of the form v^m with coefficient unity. According to the results of the preceding section, all of the roots of $\Phi(u, v)$ in a neighborhood of the origin will be roots of a certain polynomial

(10) $$P = v^m + a_1 v^{m-1} + \cdots + a_m,$$

where the coefficients a_k are series in u having no constant terms.

The polynomial P may be equal to the product of two polynomials of similar form,

$$b_0 v^k + b_1 v^{k-1} + \cdots + b_k,$$
$$c_0 v^{m-k} + c_1 v^{m-k-1} + \cdots + c_{m-k},$$

where the coefficients b and c are convergent series in u. In that case the product $b_0 c_0$ must be identically unity, and by dividing the first polynomial by b_0 and multiplying the second by the same series, the two factors will have the form

$$v^k + b_1' v^{k-1} + \cdots + b_k',$$
$$v^{m-k} + c_1' v^{m-k-1} + \cdots + c_{m-k}',$$

The coefficients b' and c' are now series in u without constant terms. Otherwise the product P would have a term of lower degree than v^m, with a coefficient series whose constant term would be different from zero.

It is readily seen from this that the polynomial P is either irreducible in the sense that it can not be decomposed into a product of polynomials of the same sort, or else it is the product of a number of irreducible polynomials of lower degree.

Suppose that $Q(u, v)$ is a polynomial of the form (10) which is irreducible in the sense just described. Then its discriminant with respect to v is a series in u which does not vanish identically, since otherwise Q and Q_v would necessarily have a common factor of the form (10), and Q would not be irreducible. There is a neighborhood $0 < u \leqq u_1$ in which the discriminant is everywhere different from zero, and for any value u satisfying these inequalities the values of v making $Q = 0$ are all distinct. According to the results which have been stated above in the introduction to this chapter of the lectures, the values of v which make Q vanish for different values of u will be defined by m series of the form

(11) $$v = \alpha u^{\mu/p} + \alpha' u^{\mu'/p} + \cdots;$$

and these series must all be distinct, since for sufficiently small values $u \neq 0$, as has been seen, the roots of Q are all distinct.*

It is evident then that all the roots of $\varphi(u, v)$ in the neighborhood of the origin, including those which correspond to the factor u^k in equation (9), are given by a finite number of elements of the form

$$u = at^p, \qquad v = bt^\mu + b't^{\mu'} + \cdots,$$

where a and b do not vanish simultaneously, and p, μ, μ', $\cdots$ are positive integers having no common factor.

The product of factors of the form

(12) $$\{v - \alpha u^{\mu/p} - \alpha' u^{\mu'/p} - \cdots\},$$

corresponding to the elements of a cycle, is a polynomial $Q_1(u, v)$ of the form (10). For the product Q_1 is a series in $u^{1/p}$ and v which is unchanged when $u^{1/p}$ is replaced by $\omega^r u^{1/p}$, and Q_1 must therefore contain only powers of $u^{1/p}$ whose exponents are multiples of p, that is, positive integral powers of u.

On the other hand an irreducible polynomial Q possesses only a single cycle of elements of the form (12). Each element of a cycle belonging to Q gives rise, in fact, to a factor Q_1 of Q of the form (10). The number of elements in the cycle could not be greater than the degree of Q, and neither could it be less, since according to the argument of the paragraph just preceding, Q would then be divisible by a factor of the same form corresponding to the product of the factors (12) belonging to the cycle.

By combining these two results, it follows that *the product of the factors of the form* (12) *corresponding to the elements of a single cycle is an irreducible polynomial of the form* (10), *and conversely the elements of an irreducible polynomial of the form* (10) *form a single cycle.*

The Weierstrassian polynomial P of any function φ is a product of irreducible factors of the same form, some perhaps repeated,

---

* The method of proof for this statement in the case of a polynomial P is precisely that of the theory of algebraic functions. See the reference above (page 44) to Appell and Goursat.

to each of which there corresponds a cycle of elements. By the order of an element of φ is meant the number of times its factor (12) is repeated in the product $u^k P$. The order is evidently equal to the multiplicity in $u^k P$ of the irreducible factor to which the element belongs. If φ possesses one element of a cycle it must possess the whole cycle. For the polynomial P belonging to φ has then a common factor with the irreducible polynomial Q of the cycle, and so must be divisible by Q.

Suppose now that $\varphi(u, v)$ and $\psi(u, v)$ are two functions of the form described above, and that the functional determinant

$$(13) \qquad D(u, v) = \begin{vmatrix} \varphi_v & \varphi_v \\ \psi_u & \psi_v \end{vmatrix}$$

does not vanish identically.

If φ and ψ have an element in common, then they have in common the irreducible polynomial Q of the form (10) to which the element belongs, and Q is also factor of D.

The first part of this statement follows from the preceding paragraphs, so that φ and ψ may be supposed to have the forms

$$\varphi = QA, \qquad \psi = QB.$$

When these expressions are substituted in the functional determinant (13) the presence of the factor Q is at once evident.

A similar argument shows that *if φ has an element with corresponding factor Q of multiplicity k, and ψ has the same element and factor with multiplicity l, then D contains the element and its factor with multiplicity $k + l - 1$ at least.*

There is a sort of converse to these statements to the effect that *when φ and D have an element and its factor Q in common, then the element and Q are either multiple in φ or else are common to φ and ψ.*

To prove this let

$$\varphi = QA, \qquad D = QC,$$

FUNDAMENTAL EXISTENCE THEOREMS.

and suppose Q not a multiple factor of φ. Then

$$\begin{vmatrix} Q_u A + Q A_u & Q_v A + Q A_v \\ \psi_u & \psi_v \end{vmatrix} = QC;$$

and it follows readily that the determinant

(14)
$$\begin{vmatrix} Q_u & Q_v \\ \psi_u & \psi_v \end{vmatrix}$$

has the factor Q, since A can not have any element in common with Q. Otherwise it would contain the whole irreducible factor Q.

Since Q is irreducible, its discriminant, a series in u, can not vanish identically, and there is an interval $0 < u \leq u_1$ in which it is different from zero. For any value of u satisfying these inequalities the polynomials Q and Q_v have no common root. If

(15) $$u = at^p, \quad x = \alpha t^\mu + \alpha' t^{\mu'} + \cdots$$

is the parametric form of one of the elements of Q, then $Q(u, v)$ vanishes identically in t when these expressions are substituted, and $Q_v(u, v)$ is not identically zero in t along the element. Hence there is an interval $0 < t \leq t_1$ in which Q_v is different from zero. Since the determinant (14) has the factor Q and therefore vanishes identically along the curve (15), it follows that

$$Q_v \left(\psi_u \frac{du}{dt} + \psi_v \frac{dv}{dt} \right) \equiv \psi_v \left(Q_u \frac{du}{dt} + Q_v \frac{dv}{dt} \right) \equiv 0$$

is an identity in t. Evidently $\psi(u, v)$ must be constant along the element, and its value is everywhere zero since it vanishes for $t = 0$. Hence ψ has the element (15) in common with Q, and must have Q itself as a factor since Q is irreducible.

The real points (u, v) where one or another of the functions φ, ψ, D vanishes play an important rôle in the investigation which follows. In the discussion of them which follows it will always be understood that when u is real and positive the symbol $u^{1/p}$ stands for the real and positive pth root of u.

If the function φ has no factor u, and if each of its elements when written in the form

(16) $$v = u^{\mu/p}\{\alpha + \alpha' u^{(\mu'-\mu)/p} + \cdots\}$$

has at least one imaginary coefficient, then in a neighborhood of the origin no real point (u, v) with $u > 0$ satisfies the equation $\varphi(u, v) = 0$.

To show this, suppose for the moment that α is imaginary. Then for sufficiently small positive values of u the absolute value of $\alpha' u^{(\mu'-\mu)/p} + \cdots$ will be less than the absolute value of the imaginary part of α, and the parenthesis in the expression (16) will also be imaginary. A similar argument would show v to be complex if one of the higher coefficients were the first not real.

On the other hand, if the coefficients in the expression are all real, then for positive values of u the values of v are real, and the points (u, v) so defined lie on a real arc of the form

$$u = t^p, \qquad v = \alpha t^\mu + \alpha' t^{\mu'} + \cdots \qquad (0 \leq t \leq t_1).$$

If the elements of φ are written in the form

(17) $$v = \alpha \epsilon^\mu (-u)^{\mu/p} + \alpha' \epsilon^{\mu'}(-u)^{\mu'/p} + \cdots,$$

where ϵ is a fixed pth root of -1, then an argument similar to that just given shows that $\varphi = 0$ is satisfied by no real points in the neighborhood of the origin with negative values of u, unless at least one of the expressions (17) in $(-u)^{1/p}$ has all of its coefficients real. On the other hand any such element with real coefficients defines points (u, v) on a real arc

$$u = -t^p, \qquad v = \beta t^\mu + \beta' t^{\mu'} + \cdots \qquad (0 \leq t \leq t_1).$$

By combining these results it follows that *all of the real points, in a neighborhood of the origin, which satisfy $\varphi(u, v) = 0$, are the points of a finite number of distinct elements of the form*

(18) $$u = at^p, \qquad v = bt^\mu + b't^{\mu'} + \cdots \qquad (0 \leq t \leq t_1)$$

whose coefficients are real and such that a and b are not both zero.

It may be of interest to note in passing that if an element of φ of the form (16) has real coefficients, then the irreducible polynomial Q which belongs to that element is real. For Q is the product of

$$\{v - \alpha u^{\mu/p} - \alpha' u^{\mu'/p} - \cdots\}$$

and the other factors which arise from it by replacing $u^{1/p}$ by $\omega^\nu u^{1/p}$ ($\nu = 0, 1, 2, \cdots, p-1$). The coefficients of the product are therefore rational integral functions with real coefficients in the α's and the pth roots of unity, and symmetric in the latter. But symmetric functions of the pth roots of unity are real. A similar remark holds true for the real elements of the form (17).

Two real elements of the form (18) *are said to be distinct if there is an interval* $0 < t \leq t_1$ *on which the points* (u, v) *which they define are all distinct.* Any two elements are either distinct or else coincident throughout.

Let the two elements have the equations

$$u = at^p, \qquad v = bt^\mu + b't^{\mu'} + \cdots \qquad (0 \leq t \leq t_1),$$
$$u = ct^q, \qquad v = dt^\nu + d't^{\nu'} + \cdots \qquad (0 \leq t \leq t_2).$$

If $a = c = 0$ then the elements are distinct unless b and d have the same sign, in which case each defines the same half ray from the origin along the v-axis. If $a = 0$, $c \neq 0$ the elements are distinct. If a and c are both different from zero then the elements are distinct unless the expressions

$$v = b\left(\frac{u}{a}\right)^{\mu/p} + b'\left(\frac{u}{a}\right)^{\mu'/p} + \cdots,$$
$$v = d\left(\frac{u}{c}\right)^{\nu/p} + d'\left(\frac{u}{c}\right)^{\nu'/p} + \cdots,$$

are identical in fractional powers of u, in which case the two elements coincide.

It can readily be seen that if two functions φ and ψ have a real element in common then they must each contain the irreducible real factor which belongs to the element.

§ 11. SINGULAR POINTS OF A REAL TRANSFORMATION OF TWO VARIABLES

In this section it is proposed to study the singular points of a transformation

(19) $\qquad x = \varphi(u, v), \qquad y = \psi(u, v)$

for which φ and ψ are convergent series in u, v with real coefficients. It is presupposed that the functional determinant D of φ and ψ does not vanish identically, and that the real elements of φ and ψ described in § 10 are all distinct. There is an interval $0 \leq t \leq t_1$ for which the elements of φ, ψ, and D which are distinct have only the point $(u, v) = (0, 0)$ in common. Some of these elements may belong to both φ and D, or to ψ and D, but none are common to φ and ψ. By further restricting the interval if necessary, it can be effected that the radius

$$\rho = \sqrt{u^2 + v^2}$$

constantly increases on each element as t increases from 0 to t_1. For ρ is a series in t which does not vanish identically, and its derivative has the same character. An interval $0 < t \leq t_1$ can therefore always be selected on which both ρ and $d\rho/dt$ remain greater than zero.

It follows immediately that a constant ρ_1 can be selected so that any circle about the origin of radius ρ_1 or less is intersected once and but once by each of the elements in question. The real elements of φ, ψ, and D may therefore be represented as shown in Fig. 4.

If the value of ρ_1 is properly restricted then any one of the regions S shown in the figure is transformed in a one-to-one way by the equations (19) *into a region Σ adjoining the origin and lying entirely in one quadrant of the xy-plane. The single-valued inverse functions*

(20) $\qquad u = f(x, y), \qquad v = g(x, y)$

so defined are continuous over all of Σ and analytic in its interior.

To prove this consider the functions $r(u, v)$ and $\omega(u, v)$ defined by the equations

$$r = \sqrt{\varphi^2 + \psi^2}, \quad \cos \omega = \frac{\varphi}{r}, \quad \sin \omega = \frac{\psi}{r}.$$

If the radius ρ_1 is properly restricted, then r and ω (modulus 2π) are well defined at every point of the circle with the exception of the origin, since φ and ψ have no real roots in common aside from $(u, v) = (0, 0)$.

The value of r increases monotonically along any analytic curve

$$u = a_1 t + a_2 t^2 + \cdots, \quad v = b_1 t + b_2 t^2 + \cdots,$$

for which u and v are not identically zero, as may be seen by reasoning similar to that applied above for ρ, after noting that the series for φ and ψ can not vanish identically in t. In particular

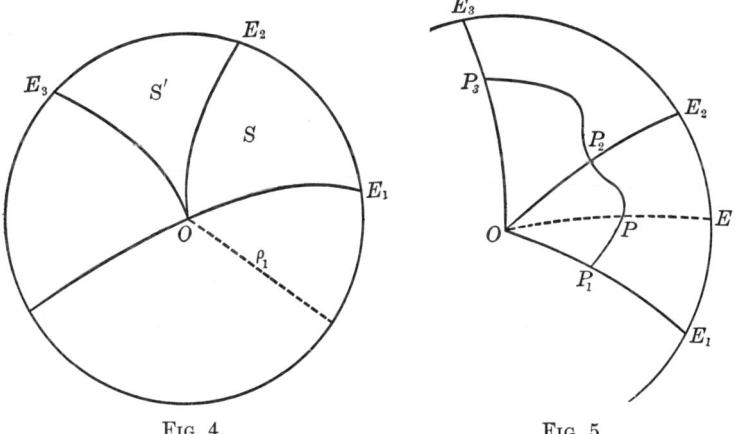

FIG. 4.   FIG. 5.

if ρ_1 is sufficiently small, then r has this property along the boundaries OE_1 and OE_2 of S, and along an auxiliary arc OE chosen arbitrarily for purposes of proof between the two elements OE_1 and OE_2.

Suppose now that k_1 is the minimum of r along the arc $E_1 E_2$, and select arbitrarily a value k between 0 and k_1. The first of

the equations
(21) $$r(u, v) = k, \qquad \omega(u, v) = z$$
is satisfied at a unique point $P(u_0, v_0)$ on the arc OE, and the corresponding value of z may be denoted by z_0. The functional determinant of r and ω has the value

$$\frac{\partial(r, \omega)}{\partial(u, v)} = \frac{D(u, v)}{r}$$

and does not vanish anywhere in the interior of S.

The domain in which the equations (21) are to be studied is that consisting of points (u, v, z) for which (u, v) is in S, and z has any real value. According to the first theorem of § 5 and the results of § 2 the equations (21) define two analytic functions

(22) $$u = u(z), \qquad v = v(z)$$

which take the initial values u_0, v_0 when $z = z_0$, and which may be continued over an interval $z_0 \leq z < \zeta''$, as described in § 5. If ζ'' is the value defining the largest such interval, the points $(u(z), v(z))$ corresponding to interior points of the interval will all be interior to S, while as z approaches ζ'' the only limit points of the values $(u(z), v(z))$ must lie on the boundary of S. Otherwise the curve (22) could be continued beyond the value ζ''.

The length of the interval $z_0 \leq z < \zeta''$ is certainly less than $\pi/2$, since in the region S neither $\sin \omega$ nor $\cos \omega$ can vanish. The curve (22) can not intersect itself, since the same values of (u, v) must define the same z by means of the second of equations (21).

As z approaches ζ'', the point $(u(z), v(z))$ approaches a unique limiting point on OE_1 or OE_2. This follows because at any limit point the value of $r(u, v)$ would have to be k, and this can happen at one point P_1 only of PE_2, and at one point P_2 only of OE_2. The curve could not have both P_1 and P_2 as limit points as z approaches ζ'', since then it would necessarily cross the arc OE at the only point P where $r(u, v) = k$, and so would intersect itself.

A similar argument shows that the equations (21) define an arc without double point over an interval $\zeta' < z \leqq z_0$, joining P with that one of the points P_1, P_2 which was not the end of the first arc. For convenience it may be assumed that ζ' is the value belonging to P_1, and ζ'' that for P_2. The preceding inequalities for z would only be reversed if the opposite were the case.

There are no other points in the region S at which $r(u, v) = k$ besides those of the arc P_1P_2 which has just been defined. If there were one not on P_1P_2, it would give rise to a second curve of the same sort joining P_1P_2. But this new curve would necessarily intersect the arc OE at P, and hence must coincide with the original arc P_1P_2 throughout.

For any value $k' < k$ there is a curve similar to P_1P_2 on which all of the points (u, v) making $r(u, v) = k'$ lie.

By means of these results it can now be shown that any two distinct points of the region OP_1P_2 are transformed into two distinct points of the xy-plane. For if (u', v') and (u'', v'') defined the same point (x', y') they would both give $r = \sqrt{x^2+y^2}$ the same value k', and hence must lie on the same curve P_1P_2. But in that case the values of ω corresponding to the two points would necessarily be different, as has been seen above, and hence (x', y') and (x'', y'') could not be the same.

From the final theorem of § 8 it follows at once that the theorem last stated above is true, provided that the circle of radius ρ_1 is altered so that the arc of it which lies between the branches OE_1 and OE_2 lies also within the region OP_1P_2. The region into which S is transformed must lie entirely in one quadrant of the xy-plane, since the values of ω which correspond to points of S are all in one quadrant. In the interior of the image of S the inverse functions (20) are analytic, since at interior points of S the determinant D is different from zero.

Some conclusions with regard to the distribution of the elements of φ, ψ, and D can be readily derived from the discussion just preceding. For example, no region S can be bounded by two elements of φ. If it were not so, then in a region bounded by two elements of φ the value of ω on the branch OE_1 would be everywhere $\pi/2$, or else everywhere $-\pi/2$, and the same is true for OE_2. But this is impossible since along the arc P_1P_2 the value of ω varies monotonically through an interval less than $\pi/2$. A similar remark holds for the elements of ψ. Hence it follows easily that

Between any elements of D the elements of φ and ψ, if there are any, must separate each other.

If the determinant D has opposite signs in two adjoining regions S and S' of the circle of radius ρ_1 in the uv-plane, shown in Fig. 5, their transforms in the xy-plane will be folded over the image of the curve OE_2 and will overlap. In order to prove this, let it first be remembered that along the element OE_2

$$\frac{dr}{dt} = r_u \frac{du}{dt} + r_v \frac{dv}{dt} \neq 0,$$

so that r_u and r_v can not vanish at any point P_2 different from the origin. Neither can they vanish at an interior point of one of the regions S, since at a point where

$$r_u = \frac{\varphi\varphi_u + \psi\psi_u}{r} = 0, \qquad r_v = \frac{\varphi\varphi_v + \psi\psi_v}{r} = 0,$$

the determinant D would necessarily have the value zero, and this does not occur in the interior of S. The equations

$$r_u \frac{du}{dz} + r_v \frac{dv}{dz} = 0, \qquad \omega_u \frac{du}{dz} + \omega_v \frac{dv}{dz} = 1$$

are satisfied everywhere between P_1 and P_2 on the arc (22). Hence

$$\frac{du}{dz} = -\frac{r}{D} r_v, \qquad \frac{dv}{dz} = \frac{r}{D} r_u.$$

As z approaches ζ'' the direction cosines of the tangent to the curve (22), for increasing z, approach the values

$$-\frac{r_v}{\sqrt{r_u^2 + r_v^2}}, \qquad \frac{r_u}{\sqrt{r_u^2 + r_v^2}}$$

on one of the arcs P_1P_2 and P_2P_3; on the other the limiting direction is exactly the opposite, since the values of D on the two arcs have opposite signs. Hence if $\omega = z$ increases along the arc P_1P_2 it must decrease along P_2P_3, and vice versa.

In the xy-plane these results mean that the images of the arcs P_1P_2 and P_2P_3 are two arcs of the circle $r = k$ which overlap near the image of P_2; the images of S and S' must therefore be superposed in the vicinity of the image of OE_2.

If the boundary OE_2 between S and S' is not one of the elements of D, the images of the two regions in the xy-plane will adjoin each other along the image of OE_2, and the inverse functions (20) will be analytic at every point of the image of OE_2 except the origin. For at such points the functional determinant D is different from zero.

By combining the results which have so far been deduced, the truth of the following theorem is established:

For a transformation

(23) $$x = \varphi(u, v), \qquad y = \psi(u, v)$$

with the characteristics described in the first paragraph of this section, a circle C can be selected in the uv-plane with center at the origin and having the following properties: The circle is intersected by each real element of the functional determinant D at some first point P. The arcs OP so determined on the different elements divide the interior of C into regions $S_1, S_2, \cdots, S_k$. The points of each region S correspond in a one-to-one way by means of equations (23) with the points of a sheet Σ of the xy-plane which winds about the origin and is bounded by the images of the boundaries of S. The single-valued functions

(24) $$u = f(x, y), \qquad v = g(x, y)$$

so determined are continuous at all points of the sheet Σ and analytic in the interior of Σ. If in two adjoining regions, say S_1 and S_2, the signs of D are opposite, then the images Σ_1 and Σ_2 overlap in the neighborhood of their common boundary $0\pi_2$; if the signs of D are the same, the regions Σ_1 and Σ_2 adjoin along $0\pi_2$ without overlapping.

The adjoining figure illustrates the case when D has four real elements and the signs of D are opposite in any two adjoining regions S. Further illustrations of the theorem are given in § 14.

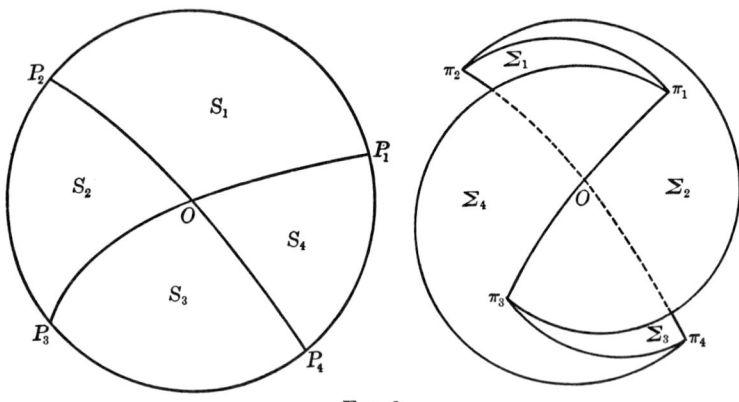

Fig. 6.

It has not been proved above that the functions (24) are continuous on a boundary 0π of one of the regions Σ. Suppose that π is a point of such a boundary, and let

(25) $\qquad \pi_1,\ \pi_2,\ \pi_3,\ \cdots$

be any sequence of points of Σ with limit π. The corresponding points

(26) $\qquad p_1,\ p_2,\ p_3,\ \cdots$

of S have condensation points in S, one of which may be denoted by p. There is then a sub-sequence

$$p_1',\ p_2',\ p_3',\ \cdots$$

among the points (26) whose limit is p; and on account of the continuity of the functions (23), the corresponding points

(27) $$\pi_1', \pi_2', \pi_3', \cdots$$

of the sequence (25) must have as limit point the image of p in Σ. But the limit of (27) is necessarily π, and π is therefore the image of p. It follows at once that the sequence (26) has a unique limit point p which is the image of π, and from this property the continuity of the functions (24) in the ordinary sense can be readily deduced.

The functions φ, ψ, and D can be expanded in the form

(28)
$$\varphi = \varphi_m + \varphi_{m+1} + \cdots,$$
$$\psi = \psi_n + \psi_{n+1} + \cdots,$$
$$D = D_{m+n-2} + D_{m+n-1} + \cdots,$$

where φ_k, ψ_k, D_k are homogeneous polynomials in u, v of degree k, and

$$D_{m+n-2} = \begin{vmatrix} \dfrac{\partial \varphi_m}{\partial u} & \dfrac{\partial \varphi_m}{\partial v} \\ \dfrac{\partial \psi_n}{\partial u} & \dfrac{\partial \psi_n}{\partial v} \end{vmatrix}.$$

If the real roots of φ_m, ψ_n, and D_{m+n-2} are all simple roots and distinct from each other, there will be an element of φ, ψ, or D in each of the corresponding directions, and a notion of the character of the transformation can be derived without difficulty. In the applications of § 14 this remark is of frequent service.

§ 12. The Case where the Functional Determinant Vanishes Identically

It is well known that when the functional determinant of two analytic functions φ and ψ vanishes identically, then near any point where not all of the derivatives φ_u, φ_v, ψ_u, ψ_v vanish the functions φ and ψ satisfy a relation of the form

$$F(\varphi, \psi) = 0$$

identically in u and v. It is possible to show that such a relation exists also near a singular point at which the four derivatives above all vanish.

If a relation can be found after a substitution of the form

(29) $$u = \alpha u_1 + \beta v_1, \qquad v = \gamma u_1 + \delta v_1,$$

for which $\alpha\delta - \beta\gamma$ does not vanish, then it will surely be satisfied when u_1 and v_1 are replaced by the original variables u, v.

Suppose then that the analytic functions φ and ψ have already been prepared by a transformation (29) in such a way that in the expansions (28) φ_m and ψ_n both contain terms in u alone. By applying the preparation theorem of Weierstrass to the functions $\varphi(u, v) - x$ and $\psi(u, v) - y$, two polynomials

$$P(u, v, x) = u^m + a_1 u^{m-1} + \cdots + a_m,$$
$$Q(u, v, y) = u^n + b_1 u^{n-1} + \cdots + b_n$$

are obtained, whose coefficients are convergent series, without constant terms, in v, x and v, y, respectively. In a certain vicinity

$$|x| < \epsilon, \quad |y| < \epsilon, \quad |u| < \epsilon, \quad |v| < \epsilon$$

the only solutions of the equations

(30) $$\varphi(u, v) - x = 0, \qquad \psi(u, v) - y = 0$$

are values (u, v, x, y) which make P and Q vanish also, and vice versa.

The resultant of P and Q is a convergent series $R(v, x, y)$ for which $R(0, x, y)$ does not vanish identically. For if all of the coefficients of $R(0, x, y)$ were zero, there would be a region

(31) $$v = 0, \quad |x| < \delta, \quad |y| < \delta \qquad (\delta \leqq \epsilon)$$

at any point of which the polynomials P and Q have a common root in absolute value less than ϵ, and the set of values $(u, 0, x, y)$ so defined satisfies also the equations (30). The existence of such a region is, however, impossible, since when y' is given

FUNDAMENTAL EXISTENCE THEOREMS. 69

satisfying (31), a value x' can always be selected which is different from the values of $\varphi(u, 0)$ at all of the n roots of $Q(u, 0, y')$. For such a set $v = 0$, x', y' in the region (31) there would be no corresponding value u' satisfying the equations (30).

The resultant $R(v, x, y)$ vanishes identically in u, v when x and y are replaced by φ and ψ. For R is expressible in the form

$$R(v, x, y) = MP + NQ,$$

where M and N are polynomials in u with coefficients which are series in v, x, y, and P and Q vanish identically when $x = \varphi$, $y = \psi$.

The series $R(0, \varphi, \psi)$ vanishes identically in u, v. If not, there would be a straight line $u = kv$ on which $R(0, \varphi, \psi)$ and φ_u are different from zero except at the origin. Let (u', v') be a point of this line near $(u, v) = (0, 0)$, at which φ and ψ have the values φ' and ψ', respectively. The series

(32) $$R(0, \varphi, \psi) + R_v(0, \varphi, \psi)v + \cdots$$

vanishes identically, in particular along the curve

(33) $$\varphi(u, v) = \varphi'$$

through the point (u', v'). Since φ_u does not vanish at (u', v'), this curve can be expressed in the form

$$u = U(v),$$

and along it

$$\frac{d}{dv}\psi(U, v) = \left[\psi_u \frac{-\varphi_v}{\varphi_u} + \psi_v\right]_{u = U(v)} = 0,$$

since the functional determinant of φ and ψ vanishes identically. On the curve (33) the function ψ has therefore the constant value ψ', and the series (32) takes the form

$$R(0, \varphi', \psi') + R_v(0, \varphi', \psi')v + \cdots$$

and vanishes identically in v. Its coefficients must therefore all vanish, since a series whose zeros have a point of condensation in the interior of its circle of convergence must have all of its coefficients equal to zero. This contradicts, however, the assumption that a point (u', v') exists at which $R(0, \varphi, \psi)$ does not vanish.

It has been shown therefore that *in case the functional determinant of the two convergent series*

$$\varphi = \varphi_m + \varphi_{m+1} + \cdots,$$
$$\psi = \psi_n + \psi_{n+1} + \cdots$$

vanishes identically, the two functions φ, ψ satisfy a relation of the form

$$F(\varphi, \psi) = 0$$

identically in u, v, where F is itself a convergent series in its two arguments. This statement is true even when φ and ψ both have singular points at the origin.

It is evident that when $D = 0$ the transformation

$$x = \varphi(u, v), \qquad y = \psi(u, v)$$

makes all of the points in the neighborhood of the origin in the uv-plane correspond to points on the various branches of the curve

$$F(x, y) = 0$$

in the xy-plane. The points (x, y) which are obtained by the transformation do not cover any region.

§ 13. A Generalization of the Preparation Theorem of Weierstrass

Consider for a moment two functions

(34) $\qquad f(u, v, x_1, x_2, \cdots, x_m), \quad g(u, v, x_1, x_2, \cdots, x_m)$

which are polynomials in the variables u, v and have for coefficients convergent series in $x_1, x_2, \cdots, x_m$. According to the usual algebraic theory of elimination, there exists a polynomial p in v

which has convergent series in the x's as coefficients, and which is linearly expressible in the form

$$p = cf + dg,$$

where c and d are polynomials of the same character as f and g. If a set of variables (u, v, x) make f and g both vanish, then v must be a root of the polynomial p; and conversely to any root of p corresponding to given values x, there exists at least one pair of values (u, v) which satisfy the two equations $f = g = 0$.

There is a generalization of the preparation theorem of Weierstrass from which similar results may be deduced with respect to two functions f and g which are not polynomials but series in the variables u and v, and with respect to the roots of such functions in a neighborhood of any set of values (u_0, v_0, x_0) making f and g vanish. As in the proof of the theorem of § 9, the point in whose neighborhood f and g are to be studied may be taken without loss of generality at the origin.

Suppose then that f and g are two convergent series in u, v, x vanishing for $(u, v, x) = (0, 0, 0)$, and such that $f(u, v, 0, 0, \cdots, 0)$ and $g(u, v, 0, 0, \cdots, 0)$ have no common factor. Then there exists a polynomial

(35) $$p = v^n + p_1 v^{n-1} + \cdots + p_n,$$

in which the coefficients p_k ($k = 1, 2, \cdots, n$) are convergent series in x having no constant terms, with the following properties: (1) *it is linearly expressible in the form*

$$p = cf + dg,$$

where c and d are convergent power series in u, v, x; (2) *in a properly chosen neighborhood*

(36) $$|u| < \epsilon, \quad |v| < \epsilon, \quad |x| < \epsilon$$

every root (u, v, x) of f and g must also make p vanish; (3) *there exists a constant $\delta \leq \epsilon$ such that for any x in the region*

(37) $$|x| < \delta$$

there is associated with each root v of p a solution (u, v, x) of the equations $f = g = 0$ satisfying the inequalities (36).*

If $f(u, v, 0, 0, \cdots, 0)$ and $g(u, v, 0, 0, \cdots, 0)$ have no common factor, then one at least of them, say f, has terms in the variable u alone, and according to the preparation theorem of Weierstrass $f(u, v, x)$ has as factor a polynomial of the form

$$(38) \qquad a_0 u^m + a_1 u^{m-1} + \cdots + a_{m-1} u + a_m = bf,$$

in which a_0 is a constant different from zero, and $a_1, a_2, \cdots, a_m$ are series in v, x without constant terms. The symmetric functions of the roots $u_1, u_2, \cdots, u_m$ of this polynomial are expressible rationally and integrally in terms of the coefficients $a_1, a_2, \cdots, a_m$, and are therefore convergent series in v, x. The product

$$(39) \qquad \prod_{k=1}^{m} g(u_k, v, x) = h(v, x)$$

is a convergent series in u_k, v, x, also symmetric in the variables u_k, and hence expressible as convergent series in v, x.

The function $h(v, 0)$ does not vanish identically, on account of the hypothesis that $f(u, v, 0, 0, \cdots, 0)$ and $g(u, v, 0, 0, \cdots, 0)$ have no common factor. If it did vanish identically, then for every sufficiently small value of v one at least of the expressions $g(u_k, v, 0)$ would vanish. But in § 10 it was seen that when $f(u, v, 0)$ and $g(u, v, 0)$ have no factor in common, there is always an interval $0 < v \leq v_1$ in which there is no value v belonging to a pair (u, v) making both of these functions vanish.

The preparation theorem of Weierstrass can therefore be applied also to the function $h(v, x)$, and the polynomial so found is the one desired in the theorem. For, in the first place, a constant ϵ can be chosen so small that every root (u, v, x) of f and g in the region (36) must be one of the sets (u_k, v, x), and must make

---

* A proof that the values of u and v belonging to the roots of a system of equations of the form (34) are roots of polynomials similar to (35) was given by Poincaré in the introduction to his Thesis, "Sur les propriétés des fonctions définies par les équations aux différences partielles," Paris (1879).

the product (39), and hence p, vanish. In the second place, a constant $\delta \leq \epsilon$ can be taken so small that every root v of p as well as the corresponding sets (u_k, v, x) lie in the domain (36). One at least of these sets must evidently satisfy $g = 0$ as well as $f = 0$. The restrictions on δ and ϵ have been stated somewhat roughly, but the reader will readily convince himself that these quantities may be selected so that the convergence of the different series and their equivalence with the corresponding polynomials are properly adjusted.

Finally, the polynomial p is linearly expressible in the form described in the theorem, in terms of f and g. To prove this, suppose that the above process has been applied to the functions $f - \alpha$ and $g - \beta$. A polynomial $P(v, x, \alpha, \beta)$ with coefficients which are series in x, α, β is then found, which may be written in the form

$$P(v, x, \alpha, \beta) = P(v, x, 0, 0) + C\alpha + D\beta,$$

where C and D are convergent series in the arguments of P. The series $P(u, x, f, g)$ vanishes identically in u, v, x since $P = 0$ must be satisfied by every set of variables (u, v, x, α, β) in a neighborhood of the origin which make $f - \alpha$ and $g - \beta$ vanish, certainly then by the set (u, v, x, f, g). Hence

$$P(v, x, 0, 0) = -Cf - Dg$$

is an identity in u, v, x, when α and β are replaced in C and D by the series f, g. But $P(v, x, 0, 0)$ is precisely the polynomial $p(v, x)$ found above, since for $\alpha = \beta = 0$ the steps in the construction of $P(v, x, 0, 0)$ are identical with those used in finding p.

If the series $f(u, v, 0, 0, \cdots, 0)$ and $g(u, v, 0, 0, \cdots, 0)$ begin with homogeneous polynomials having no common factor of degrees m and n, respectively, then the degree of the polynomial p is $\nu = mn$.

---

* In a paper of recent date the writer has developed a generalization of this theorem and the results which follow, for a system of equations of the form $f_1(x_1, x_2, \cdots, x_m; y_1, y_2, \cdots, y_n) = 0$ $(i = 1, 2, \cdots, n)$. See *Transactions of the American Mathematical Society*, vol. 13 (1912), p. 133.

Let the lowest terms of $f(u, v, 0, 0, \cdots, 0)$ and $g(u, v, 0, 0, \cdots, 0)$ be denoted by $\varphi_m(u, v)$ and $\psi_n(u, v)$, respectively. One of the two, say φ_m, has a term involving u alone with coefficient different from zero, since φ_m and ψ_n have no common factor. The terms of lowest degree in the polynomial (35) are also φ_m, since the series b has constant term unity. In the product (39) the terms may be rearranged into groups of the form $cv^\rho U$, where U is a homogeneous symmetric function of a certain degree σ in $u_1, u_2, \cdots, u_m$. The expression for such a symmetric function is isobaric and has the weight σ in the coefficients of the polynomial (35). When $x = 0$ the terms of lowest degree in U will be at least of degree σ in v, since each coefficient a_k of (35) begins with the coefficient of u^{m-k} in the polynomial $\varphi_m(u, v)$. The terms of lowest degree in v alone in the product (39) will therefore be those of the product

$$\prod_{k=1}^{m} \psi_n(u_k, v),$$

and they have the value $v^{mn} R/a_0$, in which a_0 is the coefficient of v^m in $\varphi_m(u, v)$ and R is the resultant of $\varphi_m(1, v)$ and $\psi_n(1, v)$.*
But since φ_m and ψ_n have no common factor the coefficient of v^{mn} is surely different from zero, and the theorem last stated follows at once.

If the substitution

$$v = -tu + z$$

is made, in which t is a new variable, the series

(40) $$\begin{aligned} F(u, z, x, t) &= f(u, z - tu, x), \\ G(u, z, x, t) &= g(u, z - tu, x) \end{aligned}$$

have a polynomial

(41) $$P(z; x, t) = CF + DG$$

with properties similar to those of p and of the same degree ν. In

---

* See, for example, König, Einleitung in die allgemeine Theorie der algebraischen Grössen, p. 311 and p. 271 (d).

FUNDAMENTAL EXISTENCE THEOREMS.

a properly chosen region

(42) $$|u| < \epsilon, \quad |v| < \epsilon, \quad |x| < \epsilon$$

every root (u, v, x) of f and g defines a factor $z - tu - v$ of P. If $\delta \leq \epsilon$ is sufficiently small and x a set of variables satisfying

(43) $$|x| < \delta,$$

then P has ν factors of the form $z - tu - v$, for each of which the values (u, v, x) are a solution of the equations $f = g = 0$ in the region (42).

The degree of P must be the same as that of p, since for $x = t = 0$ the series $F(u, z, 0, 0)$, $G(u, z, 0, 0)$ are identically equal to the series $f(u, v, 0)$ and $g(u, v, 0)$ when v is replaced by z. In a certain region

(44) $$|u| < \epsilon_1, \quad |z| < \epsilon_1, \quad |x| < \epsilon_1, \quad |t| < \epsilon_1,$$

where ϵ_1 is for convenience taken less than unity, every root system (u, z, x, t) of F and G makes P vanish also. If ϵ is taken less than $\epsilon_1/2$ and t is restricted to the range $|t| < \epsilon_1$, every root system (u, v, x) of f and g in the region (42) gives values u, $z = tu + v$, x, t satisfying the inequalities (44), and hence P must vanish identically in t and have $z - tu - v$ as a factor.

Suppose then that ϵ is a constant satisfying the requirements of the theorem with respect to the region (42), and that the region analogous to (37) for the polynomial P and the constant $\epsilon/2$ is

(45) $$|x| < \delta, \quad |t| < \delta;$$

and let $x = \xi$ be any set of values satisfying these inequalities. If the discriminant of P is not identically zero in t for $x = \xi$, a value $t = \tau$ can be selected also satisfying (45) and such that all the roots z of P corresponding to the values ξ, τ are distinct. There are then ν distinct root systems (u, z, ξ, τ) satisfying the inequalities (41) with ϵ_1 replaced by $\epsilon/2$. The corresponding values $(u, v = z - tu, \xi)$ are ν distinct roots of f and g lying in the region (41). According to the paragraph just preceding, P has therefore ν distinct factors $z - tu - v$.

In case the discriminant of P vanishes identically in t for $x = \xi$, the multiple factors of $P(z; \xi, t)$ can be separated out by the highest common divisor process, and the factorization of the resulting polynomial can then be discussed in a manner similar to that just explained. In either case, therefore, $P(z; \xi, t)$ has only linear factors of the form $z - tu - v$.

The number and character of the root systems (u, v, x) of the functions f and g in the neighborhood of the origin are well defined by means of the polynomial $P(z; x, t)$. To any x in the region (43) there correspond ν root systems (u, v, x) not necessarily all distinct, and the ν-valued functions $u(x), v(x)$ so defined are continuous. This is evidently true for the function $v(x)$, since its values are the roots of the polynomial $P(v; x, 0)$ whose coefficients are analytic in x. Similarly z is continuous in x, t, since its values are the roots of $P(z; x, t)$, and it follows that $u = (z - v)/t$, for a fixed value $t \neq 0$, must be continuous in x.

If P is not irreducible, that is, not decomposable into similar factors of lower degrees, its discriminant $\Delta(x, t)$ can not vanish identically in x, t. At any value $x = \xi$ where $\Delta(\xi, t)$ is not identically zero in t, the ν factors $z - tu - v$ of P are all distinct. If $t = \tau$ is selected so that $\Delta(\xi, \tau) \neq 0$, the roots of P are distinct analytic functions of x and t in the neighborhood of ξ, τ, and the corresponding values of u and v are analytic functions of x in the neighborhood of ξ.

The values $x = \xi$ near which the ν-valued functions u, v do not surely have ν distinct analytic branches, are those for which $\Delta(\xi, t)$ vanishes identically in t. At such a point some of the values of the root-systems (u, v) coincide, and only those which are distinct belong necessarily to analytic branches of the functions u, v. The values ξ which make $\Delta(\xi, t)$ identically zero must belong to one of the totalities of points defined by equating to zero the coefficients of the finite number of powers of t in the discriminant $\Delta(x, t)$.*

---

* For the characterization of these totalities after the method of Kronecker for algebraic equations, see Kistler, "Ueber Funktionen von mehreren komplexen Veränderlichen," Dissertation, Göttingen, 1905.

The only transformations of the form

$$x_1 = \Phi(x, y), \qquad y_1 = \Psi(x, y), \qquad dt_1 = \mu(x, y)dt$$

which convert every set of differential equations

(1) $$\frac{d^2x}{dt^2} = \varphi(x, y), \qquad \frac{d^2y}{dt^2} = \psi(x,y),$$

into one of the same form are those defined by (2), (2').

77. By eliminating the time from (1), giving the differential equation of the trajectories in the form (page 7)

(7) $$(\psi - y'\varphi)y''' = \{\psi_x + (\psi_y - \varphi_x)y' - \varphi_y y'^2\}y'' - 3\varphi y''^2,$$

the author proved that the only point transformations which convert every trajectory system (of a positional field) into a trajectory system are the collineations. This remains valid even in the domain of all contact transformations, as we now proceed to show.

We first consider the class of differential equations (cf. page 11)

(8) $$y''' = G(x, y, y')y'' + H(x, y, y')y''^2$$

including (7) as a special case, and characterized geometrically by the possession of property I (that is, the focal locus for each element is a circle through the given point). We prove this theorem:

The only contact transformations which convert every equation of type (8) (*that is, every system of curves with property I*) *into one of the same type are collineations and correlations.*

That no other transformations are possible is seen as follows. If a contact transformation is to convert type (8) into itself, it must convert the part common to all systems of that type into itself. The curves defined by $y'' = 0$, that is, straight lines, obviously satisfy (8) for every form of G and H. It is obvious that no other (proper) curves satisfy all such equations. But since we are dealing with contact transformations and not merely point transformations, we must replace the concept *curve* by the concept *union*. In the plane the only unions which are not (proper) curves are points. A point is regarded as made up of ∞^1 lineal elements; so x is constant, y is constant, y' is arbitrary, and therefore y'' and y''' are infinite. Point unions are to be regarded then as solutions of all equations (8). The common part thus consists of the ∞^2 straight lines and the ∞^2 points of the plane. If this is to go into itself, either points go into points and lines into lines, or else points go into lines and lines into points. We thus obtain only collineations and correlations.

That the collineations actually leave type (8) unchanged is easily verified analytically.* The work for correlations is simplified by observing that every correlation may be reduced, by means of collineations, to the form of Legendre's transformation

(9) $\qquad x_1 = -y', \qquad y_1 = xy' - y, \qquad y_1' = -x,$

(which is simply polarity with respect to the conic $x^2 + 2y - 1 = 0$). Extending (9), we find

(9') $\qquad\qquad y_1'' = \dfrac{1}{y'''}, \qquad y_1''' = \dfrac{y'''}{y''^2}.$

This converts equation (8) into one of the same form

(10) $\qquad y_1''' = G_1(x_1, y_1, y_1')y_1'' + H_1(x_1, y_1, y_1')y_1''^2,$

the new coefficient functions being related to the old as follows:

(10') $\qquad \begin{aligned} G_1 &= H(-y_1', x_1y_1' - y_1, -x_1), \\ H_1 &= G(-y_1', x_1y_1' - y_1, -x_1). \end{aligned}$

This completes the proof of the theorem stated on the previous page.

78. If we impose property II on the system (8), that is, if we consider the subclass in which

(11) $\qquad\qquad H = \dfrac{3}{y' - \omega(x, y)},$

---

* *Trans. Amer. Math. Soc.*, vol. 7 (1906), p. 420.

FUNDAMENTAL EXISTENCE THEOREMS.

has a singular point at the origin when

(51) $$\begin{vmatrix} a_{10} & a_{01} \\ b_{10} & b_{01} \end{vmatrix} = 0.$$

If one of the elements of the determinant is different from zero, it may be assumed without loss of generality to be a_{10}; then after two transformations

$$u' = a_{10}u + a_{01}v, \qquad v' = v,$$
$$x' = x, \qquad y' = -\frac{b_{10}}{a_{10}}x + y$$

the equations (50) take the form

(52) $$\begin{aligned} x &= u + a_{20}u^2 + a_{11}uv + a_{02}v^2 + \cdots, \\ y &= b_{20}u^2 + b_{11}uv + b_{02}v^2 + \cdots. \end{aligned}$$

For convenience the primes have been dropped, and the notation for coefficients of terms of higher degree than the first is the same as that in the original equation. It may further be supposed that the polynomials

$$\varphi_1 = u, \qquad \psi_2 = b_{20}u^2 + b_{11}uv + b_{02}v^2$$

have no common factor, in other words that $b_{02} \neq 0$. The origin is then a singular point for the transformation (50) of a very general type, since aside from the assumption (51) only inequalities on the coefficients of the series have been exacted.

The functional determinant has the expansion

$$D(u, v) = b_{11}u + 2b_{02}v + \cdots,$$

and hence has a single branch

$$v = -\frac{b_{11}}{b_{02}}u + \cdots,$$

along which D vanishes and on opposite sides of which D has different signs. The image Δ of this curve in the xy-plane has an ordinary point at the origin, as shown by its equations

$$x = u + \cdots, \qquad y = \frac{4b_{02}b_{20} - b_{11}^2}{b_{02}} u^2 + \cdots.$$

The region S in the figure has in it one real element of φ and at most two of ψ, since the solutions of $\varphi = 0$ lie on a single real

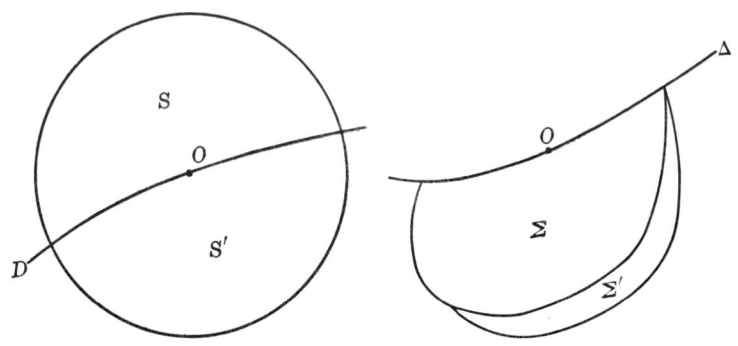

Fig. 7.

curve through the origin, and those of $\psi = 0$ are either imaginary or else lie on two real branches. Hence the region Σ which is the image of S lies on one side only of the curve Δ and overlaps the image Σ' of S'.

Since φ_1 and ψ_2 have no common factor, the theorems of § 13 show that there exist two constants, δ and ϵ, such that the equations (52) have two and only two solutions $[u_1(x, y), v_1(x, y), x, y]$, $[u_2(x, y), v_2(x, y), x, y]$ in the region

$$|u| < \epsilon, \qquad |v| < \epsilon, \qquad |x| < \epsilon, \qquad |y| < \epsilon$$

corresponding to any (x, y) in the region

$$|x| < \delta, \qquad |y| < \delta.$$

The functions u_1, v_1, u_2, v_2 so defined are everywhere continuous and the two solutions above are analytic and distinct except along the curve Δ. On one side of Δ they are imaginary, on the other real.

FUNDAMENTAL EXISTENCE THEOREMS. 81

Another interesting case is that of a transformation (50) for which again the coefficients are real, and

$$\frac{\partial \varphi}{\partial u} \equiv \frac{\partial \psi}{\partial v}, \qquad \frac{\partial \varphi}{\partial v} \equiv -\frac{\partial \psi}{\partial u}.$$

Such a transformation might be called a monogenic transformation. It follows at once that φ and ψ must begin with two homogeneous polynomials, φ_m and ψ_m, of the same degree m, which also satisfy the last equations. Consequently

$$\varphi_m + i\psi_m = (a + ib)(u + iv)^m = \rho^m(a + ib)(\cos\theta + i\sin\theta)^m$$

and

$$\varphi_m = \rho^m(a\cos m\theta - b\sin m\theta), \quad \psi_m = \rho^m(a\sin m\theta + b\cos m\theta),$$

where a and b are not both zero. These equations show that $\varphi_m(u, v)$ and $\psi_m(u, v)$ have each m real linear factors in u, v, and that no factor of φ_m is also in ψ_m.

The determinant $D(u, v)$ has an expansion

$$D(u, v) = D_{2m-1} + D_{2m} + \cdots,$$

where

$$D_{2m-1} = \begin{vmatrix} \dfrac{\partial \varphi_m}{\partial u} & \dfrac{\partial \varphi_m}{\partial v} \\ \dfrac{\partial \psi_m}{\partial u} & \dfrac{\partial \psi_m}{\partial v} \end{vmatrix} = \left(\frac{\partial \varphi_m}{\partial u}\right)^2 + \left(\frac{\partial \varphi_m}{\partial v}\right)^2.$$

The homogeneous polynomial D_{2m-1} has no real root, since such a root would necessarily belong to both $\partial \varphi_m/\partial u$ and $\partial \varphi_m/\partial v$, and from the equations

$$m\varphi_m = u\frac{\partial \varphi_m}{\partial u} + v\frac{\partial \varphi_m}{\partial v}, \quad m\psi_m = u\frac{\partial \psi_m}{\partial u} + v\frac{\partial \psi_m}{\partial v} = -u\frac{\partial \varphi_m}{\partial v} + v\frac{\partial \varphi_m}{\partial u}$$

it follows that φ_m and ψ_m would then have a common factor. Hence there are no real points at which D vanishes near the origin in the uv-plane.

The argument of § 11 shows that the elements of φ_m and ψ_m separate each other and that a neighborhood of the origin in the uv-plane is transformed into a sheet winding m times around the origin in the xy-plane, as shown in the figure. This is the

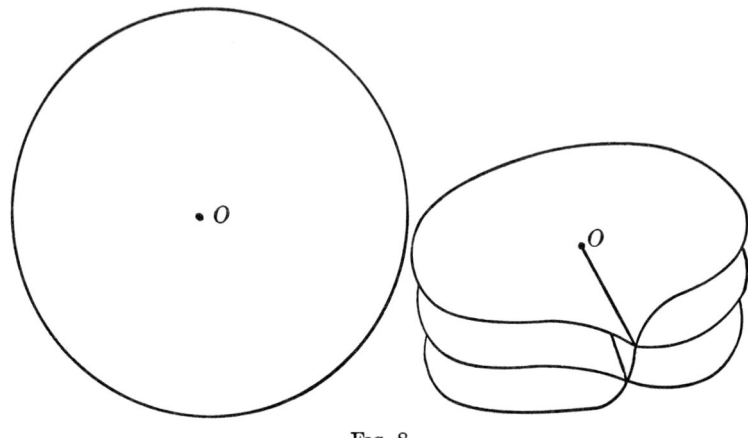

Fig. 8.

well-known transformation of the neighborhood of the origin in a complex w-plane by means of a relation of the form

$$z = Aw^m + A'w^{m+1} + \cdots,$$

where $z = x + iy$ and $w = u + iv$. The figure is drawn for $m = 3$.

There are many other special cases similar to those just given which might be elucidated by means of the theorems of the preceding sections, but for which the methods in the two examples just given are typical. It may be of interest, however, to exhibit an example which illustrates the use of the theorems of § 8, as well as the behavior of a transformation at singular points.

Suppose that the real uv-plane is transformed by means of the equations

$$(53) \qquad x = \frac{u^2}{2} - uv + \frac{v^2}{2} + \frac{u^3}{3}, \quad y = \frac{u^2}{2} + uv + \frac{v^2}{2}.$$

FUNDAMENTAL EXISTENCE THEOREMS.

The functional determinant has the value

$$D(u, v) = (u + v)(u^2 + 2u - 2v)$$

and it vanishes along the curves

$$v = -u, \qquad v = u + \frac{u^2}{2},$$

which have, respectively, the images

(54)
$$x = 2u^2 + \frac{u^3}{3}, \qquad y = 0,$$
$$x = \frac{u^3}{3} + \frac{u^4}{8}, \qquad y = \frac{u^2}{8}(u+4)^2$$

in the xy-plane. These curves are shown in the accompanying

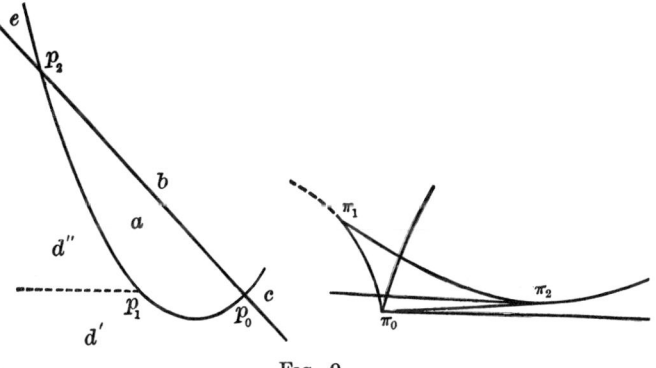

FIG. 9.

figures, the x-axis being drawn triply between $x = 0$ and $x = 32/3$ since this segment is described three times by the point (54) with varying u. To the auxiliary arc $-\infty < u \leq -2$, $v = 0$ there corresponds the curve

$$x = \frac{u^2}{2} + \frac{u^3}{3}, \qquad y = \frac{u^2}{2} \qquad (-\infty < u \leq -2)$$

shown dotted in the figure.

Consider now, for example, the region a in the uv-plane.

Its boundary is transformed into the boundary of the region α in Fig. 10. According to the generalization of the theorem of

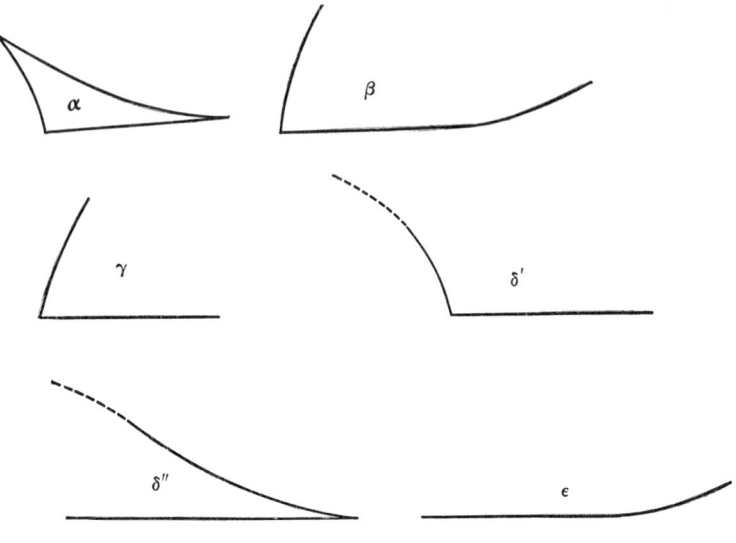

Fig. 10.

Schoenflies in § 8, the transformation defines a one-to-one correspondence between the regions a and α; and the inverse functions $u(x, y)$, $v(x, y)$ so defined are continuous over α and analytic in its interior.

Consider now the region of points (u, v, x, y) defined by the conditions that (u, v) shall lie in the region b or on its boundary, while (x, y) is unrestricted. There is but one sheet of solutions of equations (53) in this region, since any two particular solutions (u', v', x', y'), (u'', v'', x'', y'') interior to the sheet can be joined by a continuous curve lying entirely within the sheet, as may be seen by joining (u', v'), (u'', v'') by a continuous curve in b. No one of the solutions in question has a projection (x, y) outside of β, since otherwise every point exterior to β would be such a projection, according to the third theorem of § 5 or the fourth of § 8; and from the second of equations (53) it is evident that no solution (u, v, x, y) has a negative value for y. On the other hand every point of β is the projection of a solution. Since β is simply connected, it follows from the fourth theorem of § 8 that the sheet of solutions is single-valued and that the equations (53) define a one-to-one correspondence between b and β similar to that for a and α.

A similar argument can be made for each of the regions shown in the figure and its corresponding image in the xy-plane.

CHAPTER III

EXISTENCE THEOREMS FOR DIFFERENTIAL EQUATIONS

It is not within the limited scope of these lectures to give a complete account of the various methods for proving the existence of a system of solutions of a set of ordinary differential equations, nor would it be advisable, in view of the many able presentations of these fundamental theorems already well known in mathematical literature. It is rather the intention of the writer to insist on conclusions which can be derived from known methods with regard to the behavior of solutions in any region of size and shape compatible with the continuity properties of the functions by means of which the equations are defined, as over against the usual restriction of the problem to a rectangular or circular neighborhood of a particular point. It has been remarked by Picard* and Painlevé† that if a continuous solution of the differential equation

$$\frac{dy}{dx} = f(x, y) \tag{1}$$

exists over an interval $\alpha \leq x \leq \beta$, then the Cauchy polygons of approximation are defined and converge uniformly to the solution for all values of x in the interval. In § 17 below it is shown that in a region R in which the function f is continuous and satisfies the so-called Lipschitz condition, the polygons of Cauchy passing through a given initial point (ξ, η) interior to R define a priori a continuous solution of the differential equation extending to infinity or else to the boundary of the region. It follows then that there is a function

$$y = \varphi(x, \xi, \eta) \tag{2}$$

---
* *Comptes Rendus*, vol. 128 (1899), page 1363.
† *Bulletin de la Société Mathématique de France*, vol. 27 (1899), p. 151.

satisfying the differential equation (1) and defined over a region of points (x, ξ, η) of the form

$$(\xi, \eta) \text{ interior to } R, \quad \alpha(\xi, \eta) < x < \beta(\xi, \eta),$$

and as x approaches α or β the only limiting points which the points (x, y) defined by the function (2) can have are at infinity or else on the boundary of the region R.

In § 18 attention is called to the theorems of Bendixon by means of which it can be shown that the function φ is continuous, and in certain circumstances differentiable with respect to the arguments ξ, η as well as with respect to x. The "imbedding theorem" of Bolza* which asserts that any given solution, near which the function f has suitable continuity properties, can be imbedded in a one-parameter family of neighboring solutions of the differential equation, is an immediate consequence of these results, an analogue for differential equations of the fundamental theorem for implicit functions proved in § 1.

The methods mentioned above are applicable almost without change of wording to a system of equations

$$\frac{dy_\beta}{dx} = f_\beta(x, y_1, y_2, \cdots, y_n) \quad (\beta = 1, 2, \cdots, n)$$

when the symbols y and f in equations (1) are interpreted as row letters in the way apparently first introduced for differential equations by Peano.†

An interesting deduction from the theorems for a system of equations is the proof of the existence of a solution of a partial differential equation

$$F\left(x, y, z, \frac{\partial z}{\partial x}, \frac{\partial z}{\partial y}\right) = 0$$

which is not necessarily analytic in its five arguments, by means of the well-known theory of characteristic curves, as described in § 19.

---

* Vorlesungen über Variationsrechnung, page 179.

† "Intégration par séries des équations différentielles linéaires," *Mathematische Annalen*, vol. 32 (1888), p. 450.

§ 15. THE CONVERGENCE INEQUALITY

There is an inequality which is of frequent service in the existence proof of the following sections and which can be readily deduced from a simple preliminary theorem.

If u is a single-valued function of t with a well-defined forward derivative u' at each point of the interval $0 \leq t \leq t_1$, and if

$$|u'| < k|u| + l,$$

k and l being two positive constants, then u also satisfies the inequality

$$|u| \leq |u_0|e^{kt} + \frac{l}{k}(e^{kt} - 1),$$

where u_0 is the initial value of u at $t = 0$.

Consider the function

$$v = |u_0|e^{kt} + \frac{l}{k}(e^{kt} - 1)$$

satisfying the differential equation

$$v' = kv + l$$

and having $|u_0|$ as its initial value. The value of u is never greater than that of v, since otherwise the difference $u - v$ would vanish and have a positive or vanishing forward derivative at some point. At a point where u and v are equal, however,

$$|u'| < k|u| + l = kv + l = v',$$

which is a contradiction. A similar argument shows that $-u$ is always less than v.

If u is a single-valued function of x with well-defined forward and backward derivatives at each point of an interval $x_0 \leq x \leq x_1$, and such that

$$|u'| < k|u| + l,$$

then, for any ξ and x in the interval, u also satisfies the inequality

(3) $$|u| \leq |u(\xi)|e^{k|x-\xi|} + \frac{l}{k}(e^{k|x-\xi|} - 1).$$

FUNDAMENTAL EXISTENCE THEOREMS.

This may be proved from the preceding paragraphs by putting $t = x - \xi$ for values of x greater than ξ, and $t = -x + \xi$ for values less than ξ.

§ 16. THE CAUCHY POLYGONS AND THEIR CONVERGENCE OVER A LIMITED INTERVAL

It is proposed to consider a differential equation (1) for which the function $f(x, y)$ is continuous in the interior of a certain region R of the xy-plane, and such that the quotient

$$(4) \qquad \frac{f(x, y') - f(x, y)}{y' - y}$$

is finite when (x, y) and (x, y') lie in any closed region whose points are all interior to R.

A so-called Cauchy polygon for the equation (1) through a point (ξ, η) interior to R is defined by means of equations of the form

$$y_1 = \eta + f(\xi, \eta)(x_1 - \xi),$$
$$y_2 = y_1 + f(x_1, y_1)(x_2 - x_1),$$
$$\cdot \quad \cdot \quad \cdot \quad \cdot \quad \cdot \quad \cdot$$
$$y = y_{n-1} + f(x_{n-1}, y_{n-1})(x - x_{n-1}).$$

The division points

$$\xi < x_1 < x_2 < \cdots$$

may be taken for convenience at equal distances δ from each other. Any value $x > \xi$ will lie on one of the intervals $x_{n-1}x_n$, and the polygon will either be well-defined for all such values, or else there will be a constant β such that for every x in the interval $\xi \leq x < \beta$ the points of the polygon are interior to R, while for $x = \beta$ the corresponding point (x, y) will be a point of the boundary of R. The polygon defined by the equations above may be denoted by $P_1(x)$, and the analogous one when the division points are distant $\delta/2^{n-1}$ from each other by $P_n(x)$.

A common interval $\xi \leq x \leq a$ for two functions $P(x)$, $Q(x)$ with respect to any region R may be defined as one over which both are interior to R, and one such that on any ordinate of the interval all the points between $(x, P(x))$ and $(x, Q(x))$ are also interior points of R.

Consider now a closed region R_1 interior to R and containing the point (ξ, η), and let m and k be two constants greater respectively than the absolute values of $f(x, y)$ and the quotient (4) in the region R_1. *If $l > 0$ is given in advance, the partitions for any two polygons $P(x)$, $Q(x)$ through (ξ, η) can be taken so small that*

(5) $$|P(x) - Q(x)| \leq \frac{l}{k}(e^{k|x-\xi|} - 1)$$

for all values of x in any common interval of $P(x)$ and $Q(x)$ with respect to R_1. For at the point (x, y), where $y = P(x)$, the equation

$$P' = f(x, P) + \{f(x_{n-1}, y_{n-1}) - f(x, P)\} = f(x, P) + \rho$$

is satisfied by the forward and backward derivatives of the polygon P. On account of the continuity of $f(x, y)$ there exists for any l a constant μ such that

imply
$$|x - x'| < \mu, \qquad |y - y'| < \mu$$
$$|f(x, y) - f(x', y')| < l/2$$

whenever the points (x, y) and (x, y') are in R_1. If the subdivisions for $P(x)$ are taken less than μ and μ/m in length, it follows that on the polygon $P(x)$

$$|x - x_{n-1}| < \mu, \qquad |P(x) - y_{n-1}| < m|x - x_{n-1}| < \mu,$$

and hence the absolute value of ρ is less than $l/2$. Similarly $Q(x)$ satisfies an equation

$$Q' = f(x, Q) + \sigma,$$

where $|\sigma| < l/2$, provided that its intervals are less in length than μ and μ/m. The difference $P - Q$ has forward and backward derivatives which satisfy the relations

$$|P' - Q'| \leq |f(x, P) - f(x, Q)| + |\rho| + |\sigma|$$
$$< k|P - Q| + l,$$

and with the help of the lemma of § 15 the desired inequality follows at once, since P and Q have the same initial value η at $x = \xi$.

If $P(x)$ is a polygon and $Q(x)$ a solution of the differential equation, or if both are solutions, the same theorem evidently holds true, because then the function σ is identically zero, or else both ρ and σ vanish.

The polygons $P_n(x)$ all have a common interval. For take positive constants a and b such that the rectangle

(6) $$0 \leq x - \xi \leq a, \quad |y - \eta| \leq b$$

is entirely within R, and consequently has two constants m and k analogous to those above for R_1. The portions of the polygons in the rectangle (6) all lie between the straight lines

$$y - \eta = \pm m(x - \xi),$$

since the slope of any side of any one of them is numerically less than m. It follows that each is certainly well defined and within the rectangle over an interval $\xi \leq x \leq a_1$, where a_1 is the smaller of a and b/m.

The sequence of polynomials $P_n(x)$ converges uniformly, on the interval $\xi \leq x \leq \xi + a_1$, to a function $y(x)$ which has a continuous derivative and satisfies the differential equation (1). *The curve $y = y(x)$ so defined is entirely within the region R.*

For take $\epsilon > 0$ arbitrarily, and l so small that

$$\frac{l}{k}\{e^{ka_1} - 1\} < \epsilon.$$

Then

$$|P_{n'}(x) - P_n(x)| < \frac{l}{k}\{e^{ka_1} - 1\} < \epsilon,$$

provided that the intervals $\delta/2^{n'-1}$ and $\delta/2^{n-1}$ are each less than the constant μ corresponding to l. Hence the sequence $P_n(x)$ converges uniformly to a continuous function $y(x)$ on the interval $\xi \leq x \leq \xi + a_1$.

The equations

$$P_n(x) = \eta + \int_\xi^x P_n'(x)dx = \eta + \int_\xi^x \{f(x, P_n) + \rho_n\}dx$$

hold for every n, and the sequences $\{f(x, P_n)\}$ and $\{\rho_n\}$ approach uniformly the limits $f(x, y(x))$ and zero, respectively. Hence

$$y(x) = \eta + \int_\xi^x f(x, y(x))dx;$$

from which it follows by differentiation that $y(x)$ is a solution of the differential equation.

It is easy to show by means of the convergence inequality that there is only one continuous solution $y = y(x)$ of the differential equation (1) in the region R and passing through (ξ, η). For suppose there were another, $Y(x)$, distinct from $y(x)$ at a value $x' > \xi$. There would then be a value $\xi_1 < x'$ at which $y(\xi_1) = Y(\xi_1)$, and such that the two solutions would be distinct throughout the interval $\xi_1 < x \leq x'$. In a neighborhood of the point of intersection (ξ_1, η_1) interior to R a relation

$$\left|\frac{d(Y-y)}{dx}\right| = |f(x, Y) - f(x, y)| < k|Y - y|$$

would be satisfied, and hence, from the convergence inequality (3),

$$|Y - y| \leq 0.$$

This contradicts the hypothesis that $y(x)$ and $Y(x)$ are distinct throughout the interval $\xi_1 < x \leq x'$.

§ 17. The Existence of a Solution Extending to the Boundary of the Region R

It has been proved in the preceding section that, on a certain interval $\xi \leq x \leq \xi + a_1$, the polygonal curves $y = P_n(x)$ converge uniformly to a continuous solution $y = y(x)$ of the differential equation (1) lying entirely within the region R. The interval for which the proof has been given may not be the largest one on which the sequence of polygons has this property. There will, however, be a number $\beta \geq \xi + a_1$, possibly infinity, with the property that on any interval $\xi \leq x \leq \beta_1$, where $\beta_1 < \beta$, the sequence of polygons converges uniformly to a continuous solution interior to R. A continuous curve $y = y(x)$ is thus defined which has a derivative and satisfies the differential equation for all values of x in the interval $\xi \leq x < \beta$.

As x approaches β the points $(x, y(x))$ of the solution can have no limit point (β, γ) interior to the region R.

If they did, there would be for any given ϵ a value $x' < \beta$ such that

$$|x' - \beta| < \epsilon, \qquad |y(x') - \gamma| < \frac{\epsilon}{2},$$

and an integer N such that, whenever $n \geq N$, the inequality

$$|P_n(x) - y(x)| < \frac{\epsilon}{2}$$

would hold for all values of x in the interval $\xi \leq x \leq x'$. At the value x' in particular

$$|P_n(x') - \gamma| \leq |P_n(x') - y(x')| + |y(x') - \gamma| < \epsilon;$$

so that for $n \geq N$ the points $(x', P_n(x'))$ would all lie in the ϵ-neighborhood of the point (β, γ). About the point (β, γ) as center a rectangle

$$|x - \beta| \leq A, \qquad |y - \gamma| \leq B$$

could be described entirely within the region R, and in the portion

R_1 of R which lay within the rectangle or within the region

$$\xi \leq x \leq x', \qquad y(x) - \epsilon \leq y \leq y(x) + \epsilon$$

the absolute values of $f(x, y)$ and the quotient (4) would be less than two constants m and k, respectively. It can be shown without great difficulty that every polygon $P_n(x)$ for $n \geq N$ would be defined and lie within the region R for an interval extending beyond β at least a distance A_1, where A_1 is the smaller of the numbers A and $(B - \epsilon - m\epsilon)/m$. A proof similar to that of § 16 would then show that the polygons $P_n(x)$ converge uniformly to a continuous solution of equation (1) interior to R_1 over an interval $\xi \leq x \leq \beta + A_1$; and consequently β could not be the upper bound described above.

As x approaches β, therefore, the only limiting points of the solution $y = y(x)$ are at infinity or else are boundary points of the region R. If R is further a closed region, that is, one containing all of its limit points, then there is but one limit point for the curve $y = y(x)$ as x approaches β. For suppose (β, γ) to be a finite point in any neighborhood of which there are points on the curve. About (β, γ) a rectangle

(7) $$|x - \beta| \leq A, \qquad |y - \gamma| \leq B$$

can be chosen arbitrarily, and the points of R lying in it form a finite closed set in which $|f(x, y)|$ remains always less than a constant M. On the interval $\beta - A_1 < x < \beta$, where A_1 is the smaller of the numbers A and B/M, all the points of the curve $y = y(x)$ satisfy the inequality

(8) $$|y - \gamma| \leq M(\beta - x).$$

For if (x', y') is any point of the curve in the rectangle (7) and also in an ϵ-neighborhood of the point (β, γ), then the inequality

$$|y - \gamma| \leq |y' - y| + |y' - \gamma|$$
$$< M(x' - x) + \epsilon$$
$$< M(\beta - x) + \epsilon$$

must be satisfied by any preceding point $P(x, y)$ of the curve $y = y(x)$ for which the arc PP' is interior to the rectangle. It follows that the solution must lie interior to the rectangle and satisfy the last inequality, at least on an interval $x' - A_\epsilon < x < x'$, where A_ϵ is the smaller of $A - \epsilon$ and $(B - \epsilon)/M$. Hence the inequality (8) is also true on a properly chosen interval preceding $x = \beta$. It follows that as x approaches β there can be but one limit point for the curve $y = y(x)$, and this limit point is either at infinity or else is a boundary point of the region R

When the function $f(x, y)$ in the differential equation

$$\frac{dy}{dx} = f(x, y)$$

satisfies in a region R the conditions stated at the beginning of § 16, there exists through any interior point (ξ, η) of the region R one and but one continuous solution

(9) $\qquad y = \varphi(x, \xi, \eta)$

of the differential equation. This solution is defined and interior to R for all values of x interior to an interval

(10) $\qquad \alpha(\xi, \eta) < x < \beta(\xi, \eta)$,

while as x approaches one of the end values α or β, the only limiting points of the solution are either at infinity or else on the boundary of R. If the region R is closed, then the solution has a unique finite or infinite limit point as x approaches α or β.

§ 18. The Continuity and Differentiability of the Solutions

It can be shown by methods due to Bendixon* that the function $\varphi(x, \xi, \eta)$ and its derivative $\varphi_x(x, \xi, \eta)$, whose existence has been proved in the preceding sections, are continuous in all three of their arguments, and if the function $f(x, y)$ has continuous first derivatives with respect to x and y in the interior of the region R,

---

* *Bulletin de la Société Mathématique de France*, vol. 24 (1896), p. 220.

then φ and φ_x have also continuous first derivatives with respect to all of their arguments.

The continuity at any set of values (x, ξ, η) for which (ξ, η) is in R and x satisfies the inequality (10) is provable with the help of the convergence inequality of § 15. For there will always be a region R_δ about the arc S of the solution (9) over the

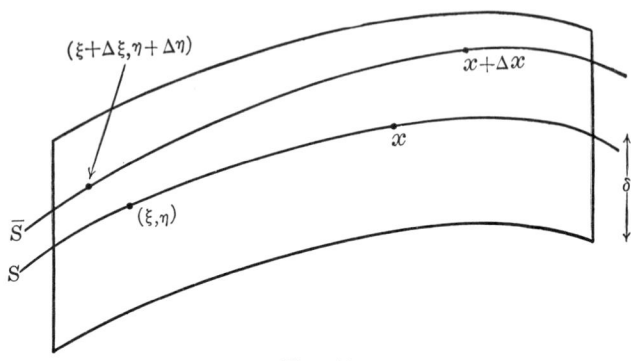

Fig. 11.

interval from ξ to x, of the kind symbolized in the figure, and so small that it lies entirely within the region R. If $(\xi+\Delta\xi, \eta+\Delta\eta)$ is any point in R, then the solution

$$(\bar{S}) \qquad y = \varphi(x, \xi + \Delta\xi, \eta + \Delta\eta)$$

satisfies the inequality

$$(11) \quad |\varphi(\xi + \Delta\xi, \xi + \Delta\xi, \eta + \Delta\eta) - \varphi(\xi + \Delta\xi, \xi, \eta)|$$
$$= |\Delta\eta + \eta - \varphi(\xi + \Delta\xi, \xi, \eta)|$$
$$\leqq |\Delta\eta| + m|\Delta\xi|,$$

where m is the maximum of the absolute value of $f(x, y)$ in R_δ, on account of the relation

$$(12) \quad |\eta - \varphi(\xi + \Delta\xi, \xi, \eta)| = \left| \int_{\xi + \Delta\xi}^{\xi} \varphi(x, \xi, \eta) dx \right| \leqq m|\Delta\xi|.$$

Hence as long as $\bar{S}$ remains within the region R_δ, it satisfies the

FUNDAMENTAL EXISTENCE THEOREMS.

convergence inequality

$$|\varphi(x, \xi+\Delta\xi, \eta+\Delta\eta) - \varphi(x, \xi, \eta)| \leq \{|\Delta\eta| + m|\Delta\xi|\}e^{|x-\xi-\Delta\xi|},$$

the initial values of the two solutions being taken at $x = \xi + \Delta\xi$. If $\Delta\xi$ and $\Delta\eta$ are sufficiently small the expression on the right is less than δ for all values of x belonging to the region R_δ, and hence $\bar{S}$ must be defined and interior to R_δ for all such values. Otherwise, for some interior value of x, it would attain one of the values $\varphi(x, \xi, \eta) \pm \delta$, which is seen to be impossible on account of the choice just made of $\Delta\xi$ and $\Delta\eta$.

Consider now the difference

$$|\varphi(x+\Delta x, \xi+\Delta\xi, \eta+\Delta\eta) - \varphi(x, \xi, \eta)|$$
$$\leq |\varphi(x+\Delta x, \xi+\Delta\xi, \eta+\Delta\eta) - \varphi(x, \xi+\Delta\xi, \eta+\Delta\eta)|$$
$$+ |\varphi(x, \xi+\Delta\xi, \eta+\Delta\eta) - \varphi(x, \xi, \eta)|.$$

By a step similar to (12), and the inequality (11), it is seen to be less than

$$m|\Delta x| + \{|\Delta\eta| + m|\Delta\xi|\}e^{k|x-\xi-\Delta\xi|}$$

whenever $\Delta\xi$ and $\Delta\eta$ have been so chosen that S lies entirely in the region R_δ. Hence the continuity of $\varphi(x, \xi, \eta)$ is proved.

To prove the differentiability of φ with respect to ξ and η, assume that $f(x, y)$ has a continuous derivative f_y in the region R, and consider the same solutions S and $\bar{S}$ in the region R_δ. The difference of their ordinates satisfies the equation

$$\frac{d\Delta\varphi}{dx} = f(x, \varphi + \Delta\varphi) - f(x, \varphi) = A\Delta\varphi,$$

where, by Taylor's formula with the integral form of remainder,

$$A = \int_0^1 f_y(x, \varphi + u\Delta\varphi)du$$

is a continuous function of x, $\Delta\xi$, $\Delta\eta$, the values ξ, η being considered as constant for the moment. Hence

$$\Delta\varphi = ce^{\int_\xi^x A\,dx}.$$

When $\Delta\xi = 0$ or $\Delta\eta = 0$, the constant c has respectively the values

$$c = \Delta\varphi|_{x=\xi} = \varphi(\xi,\,\xi,\,\eta+\Delta\eta) - \varphi(\xi,\,\xi,\,\eta) = \Delta\eta,$$

$$c = \varphi(\xi,\,\xi+\Delta\xi,\,\eta) - \varphi(\xi,\,\xi,\,\eta) = \int_{\xi+\Delta\xi}^{\xi} f(x,\,\varphi+\Delta\varphi)\,dx$$
$$= -\Delta\xi f(\xi + \theta\Delta\xi,\,\varphi(\xi + \theta\Delta\xi,\,\xi+\Delta\xi,\,\eta)),$$

where $0 < \theta < 1$. Hence the quotients $\Delta\varphi/\Delta\xi$, $\Delta\varphi/\Delta\eta$ have well-defined limiting values

$$\frac{\partial\varphi}{\partial\xi} = -f(\xi,\,\eta)e^{\int_\xi^x f_y(x,\,\phi)\,dx},\qquad \frac{\partial\varphi}{\partial\eta} = e^{\int^x f_y(x,\,\phi)\,dx}.$$

It may be remarked in conclusion that the theorems which have been proved in §§ 16–18 are true for systems of equations as well as for a single one.

§ 19. An Existence Theorem for a Partial Differential Equation of the First Order which is not Necessarily Analytic

Proofs have been given by Cauchy, Kowalewski, Darboux, and others for the theorem that in general there exists one and but one analytic surface

$$z = z(x,\,y)$$

which passes through an arbitrarily selected analytic curve C in the xy-space and, with the derivatives

$$p = \frac{\partial z}{\partial x},\qquad q = \frac{\partial z}{\partial y},$$

satisfies a differential equation of the form

$$F(x,\,y,\,z,\,p,\,q) = 0,$$

FUNDAMENTAL EXISTENCE THEOREMS. 99

where F is an analytic function of its five arguments. These proofs, however, say nothing about the solutions which may exist through a curve C whose defining functions are not expressible by means of power series; and they are not applicable when F itself has not this property. An existence proof is to be given below which is based upon much less restrictive assumptions on the functions F and the curve C. It involves the well-known theory of characteristic strips, which are solutions of a set of ordinary differential equations. If a one-parameter family of characteristic strips intersecting a given curve C is properly selected, it will generate a surface S which is a solution of the differential equation. The existence of the family and the differentiability of the surface depend, however, upon the existence and differentiability of the equations of the characteristic strips with respect to the initial values of the variables which they involve, that is, upon theorems similar to those which have been developed in the preceding sections.

Suppose that the function F is continuous and has continuous first and second derivatives in a certain region R of points (x, y, z, p, q). The differential equations satisfied by the characteristic strips have the form

(13)
$$\frac{dx}{du} = F_p, \quad \frac{dy}{du} = F_q, \quad \frac{dz}{du} = pF_p + qF_q,$$
$$\frac{dp}{du} = -F_x - pF_z, \quad \frac{dq}{dx} = -F_y - qF_z.$$

Through any initial values $(\xi, \eta, \zeta, \pi, \kappa)$ interior to R these equations have a solution with equations and initial conditions of the form

(14)
$$x = x(u, \xi, \eta, \zeta, \pi, \kappa), \quad \xi = x(0, \xi, \eta, \zeta, \pi, \kappa),$$
$$y = y(u, \xi, \eta, \zeta, \pi, \kappa), \quad \eta = y(0, \xi, \eta, \zeta, \pi, \kappa),$$
$$z = z(u, \xi, \eta, \zeta, \pi, \kappa), \quad \zeta = z(0, \xi, \eta, \zeta, \pi, \kappa),$$
$$p = p(u, \xi, \eta, \zeta, \pi, \kappa), \quad \pi = p(0, \xi, \eta, \zeta, \pi, \kappa),$$
$$q = q(u, \xi, \eta, \zeta, \pi, \kappa), \quad \kappa = q(0, \xi, \eta, \zeta, \pi, \kappa),$$

and such that each of the functions on the left and its derivative for u are continuous and have continuous first derivatives in a region of values $(u, \xi, \eta, \zeta, \pi, \kappa)$ for which $(\xi, \eta, \zeta, \pi, \kappa)$ is a point interior to R and u lies in an interval, containing the value $u = 0$, of the form,

$$\alpha(\xi, \eta, \zeta, \pi, \kappa) < u < \beta(\xi, \eta, \zeta, \pi, \kappa).$$

The points (x, y, z, p, q) so defined are all interior to the region R.

Along the solution (14) the equations

(15) $$px_u + qy_u - z_u = 0$$

(16) $$\frac{dF}{du} = F_x x_u + F_y y_u + F_z z_u + F_p p_u + F_q q_u = 0$$

are satisfied identically, so that the direction $p : q : -1$ is always normal to the curve defined by the first three equations. Evidently if F vanishes at a single point of the strip, it will also vanish at every other point. The solutions (14) along which F vanishes are called characteristic strips, and any one of the strips (14) will surely be of this type if the initial condition

$$F(\xi, \eta, \zeta, \pi, \kappa) = 0$$

is satisfied.

Consider now a continuous and differentiable strip of elements

(17) $$x = \xi(v), \quad y = \eta(v), \quad z = \zeta(v), \quad p = \pi(v), \quad q = \kappa(v)$$
$$(v_1 \leqq v \leqq v_2)$$

which lies in the interior of the region R and satisfies the conditions

(18) $$\pi \xi_v + \kappa \eta_v - \zeta_v = 0, \quad \begin{vmatrix} F_\pi & \xi_v \\ F_\kappa & \eta_v \end{vmatrix} \neq 0,$$
$$F(\xi, \eta, \zeta, \pi, \kappa) = 0,$$

where the arguments in the derivatives of F are the same as those in the second equation. The first two of these conditions imply that the direction $\pi : \kappa : -1$ is normal to the curve

(19) $$x = \xi(v), \quad y = \eta(v), \quad z = \zeta(v),$$

FUNDAMENTAL EXISTENCE THEOREMS. 101

and that the curve and its strip of normals satisfy the differential equation. The third prevents the strip from being a so-called integral strip of the differential equation, through which there does not in general pass a unique integral surface without singularities. To make the situation simpler it will be supposed that the projection of the strip (17) in the xy-plane does not intersect itself.

When the functions (17) are substituted in the equations (14), a new system

(20)
$$x = X(u, v), \quad y = Y(u, v), \quad z = Z(u, v),$$
$$p = P(u, v), \quad q = Q(u, v)$$

with the initial conditions

(21)
$$\xi(v) = X(0, v), \quad \eta(v) = Y(0, v), \quad \zeta(v) = Z(0, v),$$
$$\pi(v) = P(0, v), \quad \kappa(v) = Q(0, v)$$

is determined. There is a region

(R_{uv}) $\qquad A \leq u \leq B, \quad v_1 \leq v \leq v_2,$

where A is a negative and B a positive constant, in which the functions (20) are continuous, have continuous first derivatives, and satisfy the relation

(22) $\qquad \begin{vmatrix} X_u & X_v \\ Y_u & Y_v \end{vmatrix} \neq 0.$

For if M is the maximum of the absolute values of the functions on the right in the equations (13), for a closed ϵ-neighborhood of the points of the strip (17) in the interior of R, then the solutions (14) are defined at least over an interval $|u| \leq \epsilon/M$, and the absolute values of A and B can be taken at least as great as this constant without disturbing the continuity properties desired for the functions (20) in the region R_{uv}. The condition (22) is satisfied for the values $u = 0$, $v_1 \leq v \leq v_2$ because of the first two of equations (13) and the third of the relations (18); and the region R_{uv} can therefore be chosen so that the determinant is different from zero everywhere in it.

By an argument similar to that used in proving the theorem of § 4 it can be shown that A and B can be restricted still further, if necessary, so that no two distinct points (u', v'), (u'', v'') in the region R_{uv} define the same point (x, y) by means of equations (20). The boundary of the region R_{uv} is transformed then by the first two of equations (20) into a simply closed regular curve in the xy-plane which bounds a portion R_{xy} of the xy-plane. The equations establish furthermore a one-to-one correspondence between the points of R_{uv} and those of R_{xy}, and the functions

(23) $$u = u(x, y), \quad v = v(x, y)$$

so defined are continuous and have continuous first derivatives in R_{xy}. The others of the equations (20) define then three functions

(24) $$z = z(x, y), \quad p = p(x, y), \quad q = q(x, y)$$

which are also continuous and have continuous first derivatives in R_{xy}, and which with the values (23) for u and v satisfy the equations (20) identically in x, y.

The functions (20) satisfy the relations

(25) $$\begin{aligned} PX_u + QY_u - Z_u &= 0, \\ PX_v + QY_v - Z_v &= 0, \\ F(X, Y, Z, P, Q) &= 0, \end{aligned}$$

identically in u, v. The first and third of these follow at once from the equations (15), (16), the second of the equations (18), and (21). The expression

$$\Omega(u, v) = PX_v + QY_v - Z_v$$

has the initial values

(26) $$\Omega(0, v) = \pi \xi_v + \kappa \eta_v - \zeta_v = 0,$$

which vanish on account of the first of equations (18). Furthermore

$$\Omega_u = P_u X_v + Q_u Y_v + P X_{uv} + Q Y_{uv},$$

and from the first of equations (25),

$$0 = P_v X_u + Q_v Y_u + P X_{uv} + Q Y_{uv}.$$

By subtracting the last expression from that for Ω_u and using the equations (13) which the functions (20) satisfy, it follows that

$$Q_u = P_u X_v + Q_u Y_v - P_v X_u - Q_v Y_u = -\Omega F_z - \frac{\partial F}{\partial v},$$

in which the arguments of the derivatives of F are the functions (20). Hence with the help of the third of equations (25) and the initial values (26),

$$\Omega_u = -\Omega F_z, \quad \Omega = \Omega(0, v) e^{\int_0^u F_z du} = 0.$$

The single-valued function $z(x, y)$ defined above over the region R_{xy} has the derivatives

$$Z_x = \frac{\begin{vmatrix} Z_u & Y_u \\ Z_v & Y_v \end{vmatrix}}{\begin{vmatrix} X_u & Y_u \\ X_v & Y_v \end{vmatrix}} = p(x, y), \quad Z_y = \frac{\begin{vmatrix} X_u & Z_u \\ X_v & Z_v \end{vmatrix}}{\begin{vmatrix} X_u & Y_u \\ X_v & Y_v \end{vmatrix}} = q(x, y),$$

found by substituting the functions (23), (24) in the equations (20), differentiating the resulting identities, and applying the first two of the relations (25). It satisfies the differential equation $F = 0$ on account of the third of the equations (25). Furthermore

$$x, \quad y, \quad z(x, y), \quad p(x, y), \quad q(x, y)$$

reduce to $\xi, \eta, \zeta, \pi, \kappa$ at any point of the strip (17), since at such a point $u(\xi, \eta) = 0$ and the relations (21) are satisfied.

It has been proved therefore that there is a single-valued

function
(27) $$z = z(x, y),$$

defined over a region R_{xy} of the xy-plane, which is continuous and has continuous first and second derivatives, contains the initial strip (17), and satisfies the differential equation $F = 0$.

There is no other surface

(28) $$z = z_1(x, y)$$

defined over the region R_{xy} and having these properties. If there were such a one, it would have to contain all of the points of the strips defined by equations (20). To prove this, suppose that (x', y', z', p', q') is an element belonging to one of the strips (20) for values (u', v'), and also to the surface (28). The equations

(29) $$\frac{dx}{du} = F_p(x, y, z_1, p_1, q_1), \quad \frac{dy}{du} = F_q(x, y, z_1, p_1, q_1),$$

where p_1 and q_1 are the derivatives of z_1, have a unique solution

(30) $$x = x_1(u), \quad y = y_1(u)$$

reducing to x', y' for the initial value $u = u'$ and defined over an interval $u' - \epsilon \leqq u \leqq u' + \epsilon$. The corresponding equations

(31) $$x = x_1(u), \quad y = y_1(u), \quad z = z_1(u),$$
$$p = p_1(u), \quad q = q_1(u),$$

found by substituting the functions (30) in z_1, p_1, q_1, define a characteristic strip. For on the surface (28) the equations

$$F_x + F_z p_1 + F_p r_1 + F_q s_1 = 0,$$
$$F_y + F_z q_1 + F_p s_1 + F_q t_1 = 0$$

are identities in x, y, where r_1, s_1, t_1 are the three second derivatives of $z_1(x, y)$. As a result of these identities and the equations (29),

FUNDAMENTAL EXISTENCE THEOREMS. 105

(32)
$$\frac{dz_1}{du} = p_1 \frac{dx}{du} + q_1 \frac{dy}{du} = p_1 F_{p_1} + q_1 F_{q_1},$$
$$\frac{dp_1}{du} = r_1 \frac{dx}{du} + s_1 \frac{dy}{du} = - F_x - p_1 F_{z_1},$$
$$\frac{dq_1}{du} = s_1 \frac{dx}{du} + t_1 \frac{dy}{du} = - F_y - q_1 F_{z_1},$$

where the arguments of the derivatives of F are the functions (31). The equations (29) and (32) show that the strip (31) is a characteristic strip. Its initial element for $u = u'$ is (x', y', z', p', q'), the same as that for the strip (20) corresponding to $v = v'$. Hence the two must coincide on the interval $u' - \epsilon \leq u \leq u' + \epsilon$ on which both are defined.

The initial element (21) of any one of the strips (20) is by hypothesis on the surface (28). According to the last paragraph all of the elements of the strip in an interval $|u| \leq \epsilon$ must also lie on the surface, and it follows that there can be no upper bound except B for the values of u for which this is true. If $u' < B$ were such a limiting value, the element (x', y', z', p', q') corresponding to u' on the characteristic strip would also belong to the surface, on account of the continuity of $z_1(x, y)$ and its derivatives; and the interval of coincidence would therefore be necessarily longer than $0 \leq u < u'$.

For any point (x, y) in the region R_{xy} there is but one set of values (u, v) solving the first two of equations (20), and the corresponding value of z from the third equation belongs to both of the surfaces (27) and (28). The two surfaces must therefore coincide throughout.

Suppose now that an initial curve of the form (19) is given instead of the initial strip (17). If to any value v_0 defining a point (ξ_0, η_0, ζ_0) of the curve there corresponds a direction $\pi_0 : \kappa_0 : -1$ satisfying the relations (18), and such that $(\xi_0, \eta_0, \zeta_0, \pi_0, \kappa_0)$ is interior to R, then there will be a strip of elements of the form (17) along the curve containing these initial values for $v = v_0$. For the first two equations (18) have the solution (v_0, π_0, κ_0)

when their first members are regarded as functions of v, π, κ, and on account of the third relation (18) their functional determinant for π, κ does not vanish at these values. According to the fundamental theorem of § 1 there is therefore a pair of functions $\pi(v)$, $\kappa(v)$ defined over an interval $v_1 \leq v \leq v_2$ containing v_0 and satisfying, with $\xi(v)$, $\eta(v)$, $\zeta(v)$, the relations (18).

The results of the preceding paragraphs may be summarized as follows:

Suppose that

(C) $\qquad x = \xi(v), \quad y = \eta(v), \quad z = \zeta(v)$

is a continuous and differentiable curve, at some point (ξ_0, η_0, ζ_0)
$= (\xi(v_0), \eta(v_0), \zeta(v_0))$ *of which there is a normal* $\pi_0 : \kappa_0 : -1$
satisfying the equation

$$F(\xi_0, \eta_0, \zeta_0, \pi_0, \kappa_0) = 0.$$

Suppose furthermore that

$$\begin{vmatrix} \xi_v(v_0) & F_p(\xi_0, \eta_0, \zeta_0, \pi_0, \kappa_0) \\ \eta_v(v_0) & F_q(\xi_0, \eta_0, \zeta_0, \pi_0, \kappa_0) \end{vmatrix} \neq 0,$$

and that the initial element $(\xi_0, \eta_0, \zeta_0, \pi_0, \kappa_0)$ *lies in a region R of points* (x, y, z, p, q) *in which F is continuous and has continuous first and second derivatives. Then there is a strip of the form*

(S) $\quad x = \xi(v), \quad y = \eta(v), \quad z = \zeta(v), \quad p = \pi(v), \quad q = \kappa(v)$

$$(v_1 \leq v \leq v_2)$$

containing $(\xi_0, \eta_0, \zeta_0, \pi_0, \kappa_0)$ *for* $v = v_0$, *and such that all of its elements have the properties ascribed above to this initial one. If the projection C_{xy} of C in the xy-plane does not intersect itself, the characteristic strips of the differential equation*

$$F(x, y, z, p, q) = 0$$

which pass through the elements of S simply cover a region R_{xy} of *the xy-plane and envelop a single-valued surface*

$$z = z(x, y).$$

This surface is continuous and has continuous first and second derivatives in R_{xy}, contains the strip S, and satisfies the differential equation $F = 0$. There is no other surface over the region R_{xy} which has these properties.

DIFFERENTIAL-GEOMETRIC ASPECTS OF DYNAMICS

BY

EDWARD KASNER

CONTENTS

	Pages
INTRODUCTION	1

CHAPTER I

TRAJECTORIES IN AN ARBITRARY FIELD OF FORCE

1–8.	Trajectories in the plane	7
9.	Actual and virtualt rajectories	16
10–15.	Trajectories in space	17
16–25.	The inverse problem of dynamics: a method of geometric exploration	22
26–27.	Tests for a conservative field	31

CHAPTER II

NATURAL FAMILIES: THE GEOMETRY OF CONSERVATIVE FIELDS OF FORCE

28.	Origin and application of the natural type	34
29–31.	Characteristic properties A and B	37
32.	General velocity systems	42
33.	Reciprocal systems	44
34.	Character of the transformation T	46
35–44.	The converse of Thomson and Tait's theorem	47
45–53.	Wave propagation in an isotropic medium: properties of wave sets	57
54–61.	A second converse problem connected with the Thomson-Tait theorem	61
62–67.	Geometric formulation of some curious optical properties	65
68–72.	The so-called general problem of dynamics	69

CHAPTER III

Transformation Theories in Dynamics

73–81.	Projective transformations	73
82–91.	Conformal transformations	81
92–94.	Contact transformations	87
95–97.	A group of space-time transformations	89

CHAPTER IV

Constrained Motions in a Field. Generalization of the Trajectory Problem Including Brachistochrones and Catenaries

98–114.	Systems S_k defined by $P = kN$	91
115–116.	Curves of constant pressure	97
117–118.	Tautochrones	98
119.	Non-uniform catenaries	100

CHAPTER V

More Complicated Types of Force

120–122.	Motion in a resisting medium	102
123–126.	Particle on a surface	104
127–130.	The general field in space of n dimensions	107
131–132.	Interacting particles in the plane and in space	109
133–141.	Forces depending on the time. Trajectories and space-time curves	111

DIFFERENTIAL-GEOMETRIC ASPECTS OF DYNAMICS

BY

EDWARD KASNER

INTRODUCTION

The relations between mathematics and physics have been presented so frequently and so adequately in recent years, that further discussion would seem unnecessary. Mathematics, however, is too often taken to be analysis, and the role of geometry is neglected. Geometry may be viewed either as a branch of pure mathematics, or as the simplest of the physical sciences. For our discussion we choose the latter point of view: geometry is the science of actual physical or intuitive space. All physical phenomena take place in space, and hence necessarily present geometric aspects. We confine our discussion to mechanics, and consider the rôle of geometry in mechanics.

The fundamental concepts of mechanics are: space, time, mass, and force. Certain preliminary theories deal with some instead of all these concepts. Space by itself gives rise to pure geometry with all its subdivisions. According to Sir William Rowan Hamilton, algebra is the science of pure time; in fact time is the simplest one-dimensional manifold suggesting the notion of real number, the continuum, the foundation of analysis. Neither mass by itself, nor force by itself, gives rise to an independent theory, for these notions cannot be considered without considering space also.

Space and time together give rise to kinematics. If we do not consider velocities and accelerations, but only displacements (that is, initial and terminal positions without introducing continuous motion from one to the other), we obtain Ampère's "geometry of motion," which belongs to pure geometry rather than to kinematics.

Space and mass give rise to a separate discipline which may be called the geometry of masses. This deals with centers of gravity, moments of inertia, and moments of higher type, which have been studied extensively in recent years, especially by the Italian mathematicians.

Space and force are the essential concepts employed in rigid statics. Mass and time are not necessary in this theory, which deals essentially with the equivalence and reduction of systems of vectors. The remaining combinations, mass and time, force and time, mass and force, do not produce separate theories, since they can not be discussed without introducing also the concept of space.

Consider then space, time, and mass. The principal development along this line is Hertz's remarkable "geometry and kinematics of material systems," a theory entirely independent of the concept of force.

The other combinations of three of the four concepts have not produced separate developments.

Finally, we have the theory which involves all four concepts simultaneously, namely, kinetics.

Although the geometric aspects of the preliminary theories are very interesting and important, it is not our intention to review the progress which has been made in this line. We mention only Ball's theory of screws, Study's Geometrie der Dynamen, and the law of duality connecting kinematics and statics—a law which is not dynamical, but purely geometric.

The notion of vector is of course fundamental in many of these theories. We recall the fact that there are three distinct types of vector used in mechanics: the *free* vector, the *sliding* vector, the *bound* vector. These three types differ with respect to the definition of equivalence. In the first theory, two vectors are regarded as equivalent when they have the same length and direction (including sense). Such free vectors are employed in combining translations, or forces acting at a point. A free vector in space has three coordinates. The sum of any number of vectors is a vector.

In the second theory, dealing with sliding vectors, two vectors are equivalent only when they have the same line as well as the same length and sense. Such vectors are used in the statics of rigid bodies. The sliding vector in space has five coordinates. A system of these vectors can not usually be reduced to a single vector. The most general system depends in fact on six essential parameters: it is a new geometric element which may be represented either as a screw or a dyname.

Finally, in the third type of vector theory, two vectors are not equivalent unless they have the same initial point and same terminal point, that is the vector is completely bound. Such a vector in space depends on six coordinates. The most general system depends on twelve essential parameters. This is the theory required in the developments of *astatics*.

Statics and kinematics have given rise to very extensive geometric developments; but kinetics still is thought of almost exclusively as a matter of differential equations. Lagrange, in the famous preface to his Mécanique Analytique, stated that no diagrams would be found in his work: "Lovers of analysis will thank me for adding a new branch to that science." The special object of these lectures will be to point out some of the geometric aspects of kinetics, especially properties of the trajectories described in arbitrary fields of force. While the investigations connected with statics and kinematics are mainly of algebraic-geometric character, our kinetic discussions relate to infinitesimal properties, tangents, distribution of curvature, osculating conics, and so on: we shall deal chiefly with the *differential geometry of systems of trajectories*. It is essential to observe that the properties considered relate not to the individual curves, but to the infinite systems of curves.

To emphasize this point, consider the motion of a particle

in a plane field of force, the force depending only on the position of the point. For given initial conditions, the particle will move on a definite curve; taking all possible initial conditions, we shall obtain a triply infinite system of curves. A single curve obviously has no peculiarities, for a particle may be made to describe any given curve by selecting a proper force, varying from point to point of that curve. The system of curves, however, will have intrinsic peculiarities, for if a triply infinite system of curves is given at random, it will not usually be possible to find any field of force such that every particle moving in that field will describe one of the given curves; there is, for instance, no field of force which produces as its trajectories all the circles of the plane.

The simplest general property of the system of trajectories is as follows: If a particle is started at a given position in a given direction with all possible initial speeds into a field of force, a single infinity of trajectories will be obtained, one for each value of the speed; construct for each of these curves the parabola having four-point contact (osculating parabola); the foci of these parabolas will always lie on a circle passing through the given initial point. An equivalent statement is that the directrices of these parabolas will always be concurrent. In space we employ osculating spheres and find that the locus of the centers is a straight line.

A completely characteristic set of properties, for both the plane and space, is given in Chapter I. It is thus possible to tell when a given system of curves can serve as a system of dynamical trajectories. A method is obtained for constructing the field from its trajectories. If say a handful of particles is thrown into an unknown field (the force acting at any point depending only on the position of the point) and if a photograph of the totality of paths is taken, then, without any record of velocity or any observation of time, the field can be constructed. In particular it is possible, by simple geometric tests, to distinguish conservative from non-conservative fields.

Chapter II deals with the geometry of conservative forces. Here the energy equation allows us to group the trajectories into "natural families." Such a family is obtained most concretely as the totality of ∞^4 rays or paths of light in any medium where the index of refraction varies continuously from point to point. The geometric characterization is first given by two simple properties relating to circles of curvature; and then by a new converse of the theorem of Thomson and Tait. It is seen, for example, that if a candle is placed in the atmosphere or in any gas of variable density, the ∞^2 rays emitted by it, which may be curves of very complicated shape, will necessarily have these properties: (A) the circles of curvature constructed at the given source all meet at a second point; (B) three of these circles have four-point (instead of merely three-point) contact with their curves, and these three are mutually orthogonal; (C) the ∞^2 rays form a normal congruence, that is, admit ∞^1 orthogonal surfaces. Natural families are characterized either by (A) and (B), or by (A) and (C).

These results are applied to the propagation of waves in any isotropic medium. A second and more complicated converse question suggested by the Thomson-Tait theorem is discussed. Some interesting optical theorems are given a geometric formulation, but the converse problems are left unsettled. The final section deals with the "general problem of dynamics" in the sense of the French writers.

The third chapter deals with transformation theories. It is interesting to notice how the most important groups of geometry, the projective and the conformal, play essential rôles in dynamics, the former in connection with arbitrary fields, the latter in connection with conservative fields and natural families. The infinitesimal contact transformations of mechanics, and a new group of space-time transformations are also discussed.

The chief subject of Chapter IV is a simple problem in constrained motion, which includes, and hence serves to unify, the theories of trajectories, brachistochrones, catenaries, and velocity curves in an arbitrary field of force. Complete characterizations are given. Curves of constant pressure and tautochrones are treated only briefly.

Chapter V includes brief discussions of more complicated problems in motion, for example, the effect of a resisting medium on the geometric character of the system of trajectories; the motions of any number of interacting particles (the results being of course applicable to the problem of three bodies); finally, forces depending not only on position but also on the time, both trajectories and space-time curves being studied. The latter are constructed, in the sense of Minkowski, in the four-dimensional space (x, y, z, t), but the application made is to ordinary dynamics, not to electrodynamics or relativity theory.

The main results of the first two chapters (in particular the complete characterizations of general systems of trajectories and of natural families) were first given by the writer in a series of four papers published in the *Transactions* of this Society (1906–1910). Some of the other results are given in notes published in the *Bulletin*. The last two chapters, as well as many sections of the other chapters, deal with hitherto unpublished results.

CHAPTER I

TRAJECTORIES IN AN ARBITRARY FIELD OF FORCE

§§ 1–8. TRAJECTORIES IN THE PLANE

1. We consider first the motion of a particle in the plane under the action of any positional field of force. The general equations of motion are

$$m\frac{d^2x}{dt^2} = \varphi(x, y), \quad m\frac{d^2y}{dt^2} = \psi(x, y),$$

where m is the mass and φ, ψ are the rectangular components of the force acting at any point x, y. There is no loss of generality in assuming the mass of the particle to be unity, so we write*

(1) $$\ddot{x} = \varphi(x, y), \qquad \ddot{y} = \psi(x, y).$$

The particle may be started from any position (x_0, y_0) with any velocity $(\dot{x}_0, \dot{y}_0)$. A definite trajectory is then described. Since the same curve may be obtained by starting from any one of its ∞^1 points, the total number of trajectories, for all initial conditions, is ∞^3. The differential equation of the third order representing this system of trajectories, found by eliminating the time from (1), is

(2) $$(\psi - y'\varphi)y''' = \{\psi_x + (\psi_y - \varphi_x)y' - \varphi_y y'^2\}y'' - 3\varphi y''^2.$$

This is not an arbitrary differential equation of the third order. Hence the system of trajectories generated by an arbitrary field of force must have peculiar geometric properties, which translate the peculiar analytic form of (2).

---

* *The following notation is employed throughout these lectures:* Dots indicate total differentiation with respect to the time t; primes indicate total differentiation with respect to x; subscripts x and y indicate partial differentiation; finally, the subscript s indicates total differentiation with respect to the arc length s.

2. Before stating these, we remark that a more intrinsic basis for the discussion is obtained by decomposing the acting force into components normal and tangential to the path, instead of parallel to x and y axis as in (1). Denoting these components by N and T respectively, the equations of motion are

$$(3) \qquad v^2/r = N, \quad vv_s = T,$$

where v denotes the speed, s the arc length, and r the radius of curvature. By differentiating the first of these equations with respect to s, and comparing with the second equation, we can eliminate v, obtaining

$$(4) \qquad (rN)_s = 2T,$$

a relation which defines the trajectories and is equivalent to (2).

To reduce this to a more explicit form, we introduce an auxiliary vector, completely determined by the given field of force, namely the space derivative of the force (considered of course as a vector). The normal and tangential components of the force vector are

$$(5) \qquad N = \frac{\psi - y'\varphi}{\sqrt{1 + y'^2}}, \quad T = \frac{\varphi + y'\psi}{\sqrt{1 + y'^2}};$$

the corresponding components of the new vector are

$$(6) \qquad \mathfrak{N} = \frac{\psi_s - y'\varphi_s}{\sqrt{1 + y'^2}} = \frac{\psi_x + (\psi_y - \varphi_x)y' - \varphi_y y'^2}{1 + y'^2},$$

$$\mathfrak{T} = \frac{\varphi_s + y'\psi_s}{\sqrt{1 + y'^2}} = \frac{\varphi_x + (\varphi_y + \psi_x)y' + \psi_y y'^2}{1 + y'^2}.$$

While the new vector is the s derivative of the force vector, its components are obviously not the same as the s derivatives of the old components: the correct relations are found to be

$$(7) \qquad N_s = \mathfrak{N} - \frac{T}{r}, \quad T_s = \mathfrak{T} + \frac{N}{r}.$$

These formulas are sufficient for the discussion of trajectories.* By means of (7) we can reduce (4) to the form

(8) $$Nr_s = -\mathfrak{N}r + 3T.$$

This is the fundamental intrinsic representation of the system of ∞^3 trajectories connected with a given field of force.

From it we may derive very simply a number of geometric properties. But in dealing with the converse questions which arise, and in proving the completeness of the set obtained, it is more convenient to use the equivalent cartesian representation, that is, equation (2).†

3. *The Five Characteristic Properties in the Plane.*—The system of trajectories generated by any positional field of force in the plane has the following set of properties, and conversely, any system of ∞^3 curves which has these properties will be a system of dynamical trajectories.

---

* In some of the later discussions we shall need also the space derivatives of $\mathfrak{N}$ and $\mathfrak{T}$, which may be written in the form

$$\mathfrak{N}_s = \mathfrak{N}_1 + \frac{\mathfrak{N}_2}{r}, \qquad \mathfrak{T}_s = \mathfrak{T}_1 + \frac{\mathfrak{T}_2}{r},$$

where

$$\mathfrak{N}_1 = \frac{\psi_{xx} + (2\psi_{xy} - \phi_{xx})y' + (\psi_{yy} - 2\phi_{xy})y'^2 - \phi_{yy}y'^3}{(1+y'^2)^{3/2}},$$

$$\mathfrak{N}_2 = \frac{\psi_y - \phi_x - 2(\phi_y + \psi_x)y' + (\phi_x - \psi_y)y'^2}{1+y'^2},$$

$$\mathfrak{T}_1 = \frac{\phi_{xx} + (2\phi_{xy} + \psi_{xx})y' + (\phi_{yy} + 2\psi_{xy})y'^2 + \psi_{yy}y'^3}{(1+y'^2)^{3/2}},$$

$$\mathfrak{T}_2 = \frac{\phi_y + \psi_x + 2(\psi_y - \phi_x)y' - (\phi_y + \psi_x)y'^2}{1+y'^2}.$$

The functions ϕ, ψ depend only on the position of the particle; the auxiliary intrinsic functions $N, T, \mathfrak{N}, \mathfrak{T}, \mathfrak{N}_1, \mathfrak{N}_2, \mathfrak{T}_1, \mathfrak{T}_2$, defined above, depend also upon the direction of motion; finally, $N_s, T_s, \mathfrak{N}_s, \mathfrak{T}_s$ depend upon the curvature of the path. Cf. *Bull. Amer. Math. Soc.*, vol. 15 (1909), p. 475.

† Cf. *Trans. Amer. Math. Soc.*, vol. 7 (1906), pp. 401–424. The result contained in property IV of § 3 gives this simple, but apparently overlooked, dynamical theorem: If a particle starts from rest, the initial curvature of the path described is one third of the curvature of the line of force through the initial position.

I. If for each of the ∞^1 trajectories passing through a given point in a given direction we construct the osculating parabola, at the given point, the locus of the foci of these parabolas is a circle passing through that point.

II. The circle that corresponds, according to property I, to a lineal element, is so situated that the element bisects the angle between the tangent to the circle and a certain direction fixed for the given point (the direction of the force acting at the given point).

III. In each direction at a given point there is one trajectory which has four-point contact with its circle of curvature: the locus of the centers of the ∞^1 hyperosculating circles constructed at the given point is a conic passing through that point in the fixed direction described in property II.

IV. With any point O there is associated a certain conic passing through it as described in property III. The normal to the conic at O cuts the conic again at a distance equal to three times the radius of curvature of the line of force passing through O. (The lines of force are defined geometrically by the fact that the tangent at any point has the direction associated with that point in accordance with property II.)

V. When the point O is moved, the associated conic referred to above changes in the following manner. Take any two fixed perpendicular directions for the x direction and the y direction; through O draw lines in these directions meeting the conic again at A and B respectively. Also construct the normal at O meeting the conic again at N. At A draw a line in the y direction meeting this normal in some point A', and at B draw a line in the x direction meeting the normal in some point B'. The variation property referred to takes the form

$$\frac{\partial}{\partial x}\frac{1}{AA'} + \frac{\partial}{\partial y}\frac{1}{BB'} + \frac{\omega\omega_{xy} - \omega_x\omega_y}{3\omega^2} = 0,$$

where AA' and BB' denote distances between points, and where ω denotes the slope of the lines of force relative to the chosen

x direction. This is true for any pair of orthogonal directions,

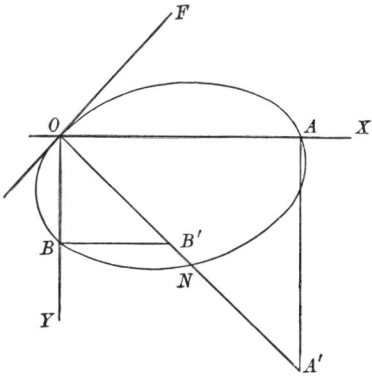

Fig. 1.

and therefore really expresses an intrinsic property of the system of curves.

4. The most general system of ∞^3 curves in the plane is represented by an arbitrary differential equation of the third order

(F_0) $\qquad y''' = f(x, y, y', y'').$

It thus involves *one* arbitrary function of *four* arguments.

A system of dynamical trajectories, on the other hand, is represented by an equation of the particular form

(F_v) $\quad (\psi - y'\varphi)y''' = \{\psi_x + (\psi_y - \varphi_y)y' - \varphi_y y'^2\}y'' - 3\varphi y''^2,$

and thus involves *two* arbitrary functions of *two* arguments. These are the only systems having all five properties I–V.

It is interesting to notice just how the successive imposition of the properties gradually narrows down the general form (F_0) to the particular form (F_v).

5. *The most general system having property I is found to be*

(F) $\qquad y''' = G(x, y, y')y'' + H(x, y, y')y''^2.$

It thus involves *two* arbitrary functions of *three* arguments. This type of course includes the dynamical type as a very special case. It arises in a number of different geometric and physical investigations. It has therefore its own interest. The characteristic property may be stated in various ways, all of course equivalent to the original form: (I) The osculating parabolas of the trajectories passing through a given point in a given direction have the foci situated on a circle passing through the given point. Five equivalent forms are as follows:

I (2). The directrices of the osculating parabolas form a pencil It follows that there exists a point (the vertex of this pencil) from which all the parabolas subtend an angle of 90°.

I (3). If for each of the trajectories considered, we construct the center of curvature of its evolute, the locus of the centers thus obtained is a parabola passing through O, and having its axis parallel to the given initial direction.

I (4). For each of the trajectories, construct the osculating equiangular spiral. The locus of the centers of the poles of these spirals is a circle passing through O.

I (5). Construct for each of the trajectories the axis of deviation, that is the line bisecting the chords of the curve which are parallel and infinitesimally close to the tangent. The tangent of the angle between the varying axis of deviation and the fixed normal is a linear function of the radius of curvature.

I (6). The derivative of the radius of curvature with respect to the arc length is a linear integral function of the radius of curvature. This is practically a restatement of (5), since for any curve the derivative of the radius of curvature is known to be equal to three times the tangent of the angle of deviation. But in this form it has the advantage of being valid, not only in the plane, but in space of three and in fact any number of dimensions.

If in addition to property I, we impose property II, the function $H(x, y, y')$ is specialized to

$$H = \frac{3}{y' - \omega(x, y)},$$

Thus the most general system with properties I and II is

$$(F_{\text{II}}) \qquad (y' - \omega)y''' = (y' - \omega)Gy'' + 3y''^2$$

where G is any function of x, y, y', and ω is any function of x, y. The type thus involves *one* arbitrary function of *three* arguments and one arbitrary function of two arguments.

6. *Systems with Properties I, II, III.*—Imposing also property III, we find that $G(x, y, y')$ must be of the special form

$$G = \frac{\lambda y'^2 + \mu y' + \nu}{y' - \omega}.$$

Thus the most general system of ∞^3 curves having properties I, II, III is represented by

$$(F_{\text{III}}) \qquad (y' - \omega)y''' = (\lambda y'^2 + \mu y' + \nu)y'' + 3y''^2,$$

involving *four* arbitrary functions ω, λ, μ, ν each of the *two* arguments x, y.

This type may be characterized by the following properties which are then equivalent to I, II, III.

I (2). For a given lineal element, the directrices of the ∞^1 osculating parabolas pass through a common point D.

II (2). When the lineal element turns about the given point O, the point D describes a straight line passing through O.*

III (2). The correspondence between the range of points D and the pencil of elements through O is one-to-two of the special form

$$\frac{3}{2d} = \lambda \sin^2 \theta + \mu \sin \theta \cos \theta + \nu \cos^2 \theta,$$

where d denotes the distance OD, and θ is the angle between the element and fixed direction of OD.

---

* In the dynamical case this line OD is perpendicular to the force vector acting at O. For certain special fields the point D may remain fixed: this happens only when the components of the force are conjugate harmonic functions, that is when the field is of the type termed "analytic" by Lecornu.

7. If now we add the properties IV and V, the four functions ω, λ, μ, ν appearing in (F_{III}) must obey the relations

(F_{IV}) $\qquad \lambda\omega^2 + \mu\omega + \nu + \omega_x + \omega\omega_y = 0,$
(F_{V}) $\qquad (\omega_y + \lambda\omega + \mu)_y - \lambda_x = 0.$

Thus the general system having properties I–IV involves *three* arbitrary functions of x, y; while that having all five properties involves *two* such functions.

By integrating these relations, we may express the four functions in terms of two arbitrary functions φ, ψ as follows:

$$\omega = \frac{\psi}{\varphi}, \qquad \lambda = \frac{\varphi_y}{\varphi}, \qquad \mu = \frac{\varphi_x - \psi_y}{\varphi}, \qquad \nu = -\frac{\psi_x}{\varphi}.$$

These values, substituted in the type (F_{III}), actually give rise to the type

$$(\psi - y'\varphi)y''' = \{\psi_x + (\psi_y - \varphi_y)y' - \varphi_y y'^2\}y'' - 3\varphi y''^2,$$

and thus the proof is completed that the set of five properties characterizes the dynamical type.

In connection with the statements I (2), II (2), III (2), property IV may be formulated as follows:

IV (2). In the correspondence described in III (2), if the element approaches the direction of the force the corresponding distance OD has for its limiting value 3/2 the radius of curvature of the line of force passing through O. It is to be remembered that the direction of the force, and hence also the lines of force, are defined purely geometrically in terms of the given triply infinite system of curves by the fact that at any point O in the plane the "direction of the force" is perpendicular to the line described as the locus of D in the above equivalent II (2) of property II.

In the same line of ideas it would be possible to find an equivalent for property V (thus completing the characterization), but the result V (2) cannot be put into simple form. The original form V may be criticized as inelegant because in it reference is made to a system of cartesian axes. Of course the result expresses an intrinsic property since it is true for all systems of axes. It would certainly be desirable to restate the result in intrinsic language. This can be done, for instance, by introducing the distances cut off by the conic described in IV, not only on the normal ON, but also on the two lines inclined at an angle of 45° to that normal. However it does not seem possible to obtain a statement which is both simple and intrinsic in form.

8. Of course many other properties of trajectories may be obtained, either by reasoning synthetically from the five fundamental properties, or by reasoning analytically from the fundamental differential equation. We state only a few samples.

If we shoot particles from a given position in a given direction with variable speed, the center of curvature of the resulting trajectories describes a straight line (the normal) and the focus of the osculating parabola simultaneously describes a circle (by property II), in such a way that the two ranges (one linear, the other circular) are homographically related; furthermore the given point, which is on both ranges, corresponds to itself.

If we shoot from the same position in a direction perpendicular to that previously employed, the new focal circle will be tangent (at the given point) to the former focal circle. Conversely if two focal circles, for the same point, are tangent, the initial directions to which they correspond will be perpendicular to each other.

We shall make use of the following properties which describe the disposition of the ∞^1 focal circles constructed at a given point. The two results which follow are geometrically equivalent, and either may be substituted for property III in the fundamental set.

III (3). If for each of the elements at a given point we construct the corresponding focal circle, the locus of the centers of the ∞^1 circles thus obtained is a conic with one focus at that point.

III (4). The envelope of the ∞^1 focal circles is always a circle.*

---

* This enveloping circle is in general position: it does not usually have its center at the given point. This simple position arises only when the force is of the Lecornu type.

§ 9. Actual and Virtual Trajectories

9. If we consider the motion of a cannon ball in a given vertical plane under the action of gravity assumed constant, the triply infinite system of trajectories consists of parabolas with vertical axes. We do not, however, obtain all vertical parabolas, represented by the differential equation of the system of trajectories, which is here $y''' = 0$, but only those whose concavity is directed downwards. The other vertical parabolas, with concavity directed upwards, satisfy the same differential equation, and it is therefore convenient to include them in the system studied. We thus have a distinction of *actual* and *virtual* trajectories. The latter are the actual trajectories corresponding to gravity reversed in direction.

In an arbitrary field of force the same distinction arises. The complete system of trajectories is composed of the actual trajectories corresponding to the given force, and the virtual trajectories which are the actual trajectories corresponding to the reversed field. It is obvious that the system of trajectories is not changed if the force acting at every point is multiplied by a constant. If we were considering only actual trajectories, it would be necessary to restrict this constant to positive values, but as we include both actual and virtual, the constant factor may also be negative. (Of course the constant must not be zero, since then the force would vanish and we should obtain the trivial system of straight lines.)

It is easy to show that the virtual trajectories corresponding to the given field may be found by giving the initial speed of the particle a pure imaginary value. The cannon ball could be made to describe a parabola with its concavity directed upwards if only some kind of powder could be invented which would cause its initial speed to be imaginary!

In discussing the general geometric properties of trajectories, we had in mind of course the complete system as defined by the differential equation. Consider for example property I: for any given lineal element the locus of the foci of the parabolas osculating the corresponding trajectories is a circle through the given point. The question arises, what part of this circle corresponds to the actual trajectories. It is easily found to be the arc of the circle cut off by the initial direction line (the common tangent of the trajectories considered) on that side which is indicated by the force vector. Thus, if we confined our discussion to actual trajectories, the focal locus would be, not a *circle*, but an *arc of a circle*, the arc running from the given point O to a certain terminal point A. If we consider all elements through O the locus of the corresponding terminal point A is found to be a conic passing through O in the direction of the force vector.*

For a given element, the point A, which separates the actual from the virtual, may be defined as the limiting position of the focus of the osculating parabola as the initial speed becomes infinitely large. The osculating parabola in this limiting case becomes a straight line, but the focus has a definite limiting position.

An analogous distinction, into actual and virtual, presents itself also in the theories of brachistochrones, catenaries, and tautochrones. The differential equations of the systems of curves are satisfied by both the actual and virtual curves, and it is the complete systems that we refer to in all our discussions unless the contrary is explicitly mentioned.

§§ 10–15. Trajectories in Space

10. Consider the motion of a particle, which we may take to be of unit mass, in an arbitrary positional field of force. The equations of motion are

(1) $\quad \ddot{x} = \varphi(x, y, z), \qquad \ddot{y} = \psi(x, y, z), \qquad \ddot{z} = \chi(x, y, z).$

The particle may be started from any position, in any direction, with any speed: its motion is then determined by the field of

---
* This conic is not the same as the conic arising in property III.

force, and it describes a definite trajectory. The totality of trajectories constitutes a definite system of ∞^5 curves. (We exclude the case where the force vanishes at every point, the trajectories then being merely the ∞^4 straight lines.)

What are the properties of such quintuply infinite systems of curves? Obviously an arbitrary system of space curves cannot be obtained as the totality of trajectories connected with any field of force. In fact the most general system of ∞^5 curves (assuming that ∞^1 curves pass through any point of space in any direction) would be represented by a pair of differential equations, one of the third order and one of the second order, of the general form

(2) $\quad y''' = f(x, y, z, y', z', y''), \quad z'' = g(x, y, z, y', z', y''),$

thus involving two arbitrary functions of *six* arguments; while the dynamical type involves merely three arbitrary functions of *three* arguments. The differential equations representing the dynamical type, obtained by eliminating the time from the equations of motion, may be written in the form

(3)
$$(\psi - y'\varphi)y''' = \begin{vmatrix} 1 & \varphi_x + y'\varphi_y + z'\varphi_z \\ y' & \psi_x + y'\psi_y + z'\psi_z \end{vmatrix} y'' - 3\varphi y''^2,$$
$$(\psi - y'\varphi)z'' = (\chi - z'\varphi)y''.$$

The question is to express the peculiar form of these equations in simple geometric language.

The interpretation of the second equation is obvious: the osculating plane of the path passes not only through the given initial direction $1 : y' : z'$, but also through the fixed direction $\varphi : \psi : \chi$; that is, the osculating plane always passes through the direction of the force acting at the given point. The other properties are not obvious;* they take into account the form of the differential equation of third order.

---

* The simplest of these, property II below and certain consequences, were first stated in the author's note published in the *Bull. Amer. Math. Soc.*,

We cannot now, as in the case of the plane discussion, employ osculating parabolas, since our curves are twisted. Three consecutive points of a curve determine an osculating circle. What do four consecutive points determine? No simple type of osculating curve is available, so we shall make use of the osculating sphere. The results are therefore quite different in form from those obtained in the two-dimensional theory.

11. *The Four Properties in Space.*—In order that a system of ∞^5 space curves, of which ∞^1 pass through each point in each direction, shall be identifiable with the system of trajectories generated by a positional field of force, it is necessary and sufficient that it shall have the following four purely geometric properties:

I. The osculating planes of the ∞^3 curves passing through a given point form a pencil; that is, all the planes pass through a fixed direction.

II. The osculating spheres of the ∞^1 curves passing through a given point in a given direction form a pencil; their centers thus lie on a straight line.

III. The straight lines which correspond, in accordance with II, to all the ∞^2 directions at a given point, form a congruence (of order one and of class three) consisting of the secants of a twisted cubic curve; which cubic furthermore passes through the given point in the direction fixed by property I.

IV. The associated plane systems S', determined by the given space system in the manner described below, have the five geometric properties characteristic of a system of plane dynamical trajectories. Consider the given system of ∞^5 space curves in connection with any plane p. Through any point of p there pass ∞^2 curves of the given system which are tangent to the plane. Project the differential elements of the third order belonging to these space curves orthogonally upon p, thus obtaining ∞^2

vol. 12 (1905), pp. 71–74. Somewhat simplified proofs were then given by Cesàro, in a paper published shortly before his death, in the *Memorie di Torino* (1905). The complete set of properties appeared in the *Trans. Amer. Math. Soc.*, vol. 8 (1907), pp. 121–140.

plane differential elements of the third order at the selected point. Applying this process to all points of p, we have a defined set of ∞^4 differential elements of the third order. These elements define a certain differential equation of the third order, and thus determine a system of ∞^3 integral curves. This we term the associated system in the plane p. The space system has the property that every one of these plane systems associated with it is a system of dynamical trajectories, and therefore has the five properties stated in § 3, which we here denote by I_p–V_p in order to avoid confusion with the four spatial properties.

These four properties are ordinally independent: no one can be derived from those which precede it. The question of absolute independence is left open: it is quite probable that IV is sufficiently strong to furnish a complete characterization by itself.

12. The most general system having property I involves one arbitrary function of six arguments besides two functions of three arguments. These systems have the following properties, which are of course consequences of property I.

The ∞^1 curves passing through a given point in a given direction have not only the same osculating plane, but also the same *torsion*.

If the torsion is given the corresponding initial directions form a quadric cone. In particular such a cone defines the directions of those curves, through the given point, which admit hyper-osculating planes.

If for each of these curves we construct, at the common point, the *related helix** (that is the helix which agrees with the curve in osculating plane, curvature, and torsion), the axes of the helices so obtained generate a cylindroid.

13. The most general system with properties I and II involves two arbitrary functions of five arguments, besides two functions of three arguments. Two further statements, each equivalent to II, are as follows:

---

*An osculating helix, that is, one having four-point contact with the curve, does not in general exist.

If for each of the ∞^1 curves defined by a given lineal element we construct the osculating circle and the osculating sphere, the distance between the center of the circle and the center of the sphere varies as a linear integral function of the radius of curvature.

For the same set of ∞^1 curves, the derivative of the radius of curvature with respect to the arc length can be expressed as a linear integral function of the radius of curvature.

This last form has the advantage of being valid in space of two or any number of dimensions. On this basis, however, it would be difficult to formulate equivalents for the higher properties, so as to obtain a complete characterization.

14. Property III is perhaps the most interesting result obtained. The most general system having this property in addition to I and II is represented by a pair of differential equations involving ten arbitrary functions of three arguments.

One may ask what is the significance of the cubic curve (we denote it by Γ), which arises in connection with III. To each point O of space there is related a certain cubic Γ. If we shoot from O in every direction with every speed, we obtain ∞^3 trajectories. Each of these has an osculating sphere with a definite center C. To each of the trajectories there corresponds one center C. Usually the center C determines the trajectory. However if C lies on the curve Γ, there are ∞^1, instead of one, corresponding trajectory: in this case in fact the initial direction may be any direction perpendicular to the line joining O and C.* Thus the curve Γ may be defined as the locus of those points which may serve as centers for more than one trajectory through the given point O.

A simple consequence of III is that the locus of the centers of the osculating spheres of the ∞^2 trajectories touching a given plane at a given point is a quadric surface.

---

* Two trajectories through O have the same osculating sphere only if the initial speed is the same, and if the line through O perpendicular to the initial elements meets the cubic Γ.

If the plane varies, the given point being held fixed, the ∞^2 quadrics obtained form a linear system.*

The properties so far considered relate to the curves through a given point O. If we have ∞^3 curves passing through a point O, ∞^1 in each direction, and if, at that point, properties I, II, III are fulfilled, it will not usually be possible to generate the curves as trajectories in any field of force. All that follows is that the relations between y', z', y'', z'', y''', z''' are of precisely the same form as those holding for trajectories; and therefore it is possible to find (in infinitely many ways) a field of force such that each of the ∞^3 trajectories passing through the given point shall have contact of the third order with some one of the given curves.

15. In order to cause our system to be of the dynamical type, it is necessary to restrict the ten arbitrary functions involved in the type characterized by I, II, III so that only three arbitraries remain, namely, the components φ, ψ, χ defining the field of force. This is the rôle of property IV, which states that in any plane p the associated system S is of the plane dynamical type. An equivalent statement is as follows:

IV (2). If the ∞^2 space curves touching any plane p at any point O are projected orthogonally upon p, the plane curves thus obtained possess the properties I_p, II_p, III_p; when the point O varies in p, the direction associated with it by II_p, and the conic associated with it by III_p, vary in accordance with the restrictions expressed in IV_p and V_p.

It may be remarked that the first half of this statement holds for all space systems having properties I, II, III; in fact all such systems have also property IV_p. The real restriction is in V_p. It is also sufficient to consider, instead of all planes p, merely those of a triply orthogonal set.

§§ 16–25. The Inverse Problem of Dynamics: A Method of Geometric Exploration

16. The usual direct and inverse problems arising in dynamics are: first, given the force acting on a particle, to find its motion;

---

* On the other hand if we vary the given point, keeping the plane fixed, no simple result is obtained: the ∞^2 quadrics constitute an arbitrary family.

and second, given the motion of a particle, to find the force acting on it. The first problem is solved by integrating the differential equations of motion. The second is solved by differentiating the coordinates of the point with respect to the time.

Suppose, however, that we are given only the path described by the particle but have no information about the motion along the curve. If merely a single curve is given, the problem of finding the acting force would of course be indeterminate. But if all the trajectories, described by starting particles in a field of force from all initial positions in all directions with all speeds, are given, then the field of force is essentially determined (that is, up to a constant factor). *Hence if we were given a photograph of the entire system of curves generated by some (positional) field of force, without any record of motion or time, it ought to be possible to find the law of the field of force.* This is easily seen to be true analytically; but we wish also a purely geometric solution which will enable us to pass from the given curves to the vector representing the force at each point of the plane (taking first the two-dimensional case). The result gives what may be described as a method for the *geometric exploration of a field of force.*

17. First consider two trajectories passing through the same point O in the same direction. Construct the two osculating parabolas. The circle passing through the point O and the foci of these parabolas will, according to property I, be the focal circle corresponding to the given point and the given direction. Then, according to property II, the direction of the force acting at O will be symmetric to the tangent to this circle at O with respect to the common tangent of the two curves. An equivalent of this construction is to join O to the intersection of the directrices of the osculating parabolas: this line is perpendicular to the direction of the force acting at O.

If we are given two trajectories passing through O in different directions, then the direction of the force at O is not determined. The same is true if we are given three curves with distinct tangents.

18. *If, however, we are given four trajectories with distinct tangents, the force direction is (in general) uniquely determined.*

Consider an arbitrary direction at O, and let us see if it can be the direction of the force acting at that point. Take the image of this direction in the tangent to the first of the given curves; then pass a circle through O in the direction so obtained and through the focus of the corresponding osculating parabola. Doing this for each of the four curves, we obtain four focal circles. *If there exists a circle touching these four, the direction tested is correct.* This follows from property III (4) of § 8. We have then a purely geometric problem: to find a direction at O such that the four circles constructed by means of it shall admit a common tangent circle. We may simplify this problem by inverting the configuration considered with respect to O. We then have, instead of the four focal circles, four straight lines which are to be concyclic. As we change the direction tested, these rotate simultaneously through equal angles about four fixed points, namely, those obtained by inverting the four foci.

Take an arbitrary oriented direction for trial; construct for each of the four inverse foci, a direction parallel to the image of the tangent to the focal circle with respect to the tangent to the corresponding trajectory. We thus obtain four oriented lineal elements, one at each of the inverse foci. The problem is then to rotate these through the same angle α, so that the new elements shall have concyclic lines.* In this position the image of the direction of any one of the four elements in the corresponding tangent at O will give the required direction of the force. The only ambiguity, in general, will be in the sense (arrowhead) of the force: this, however, may be determined separately for actual† trajectories by considerations of concavity and convexity.

19. The direct analytical treatment is as follows. The differential equation of the ∞^3 trajectories of any positional field

---

* A simple ruler and compass solution of this problem was suggested to the author by Professor Wedderburn.

† See § 9.

of force is of the form

$$(y' - \omega)y''' = (\lambda y'^2 + \mu y' + \nu)y'' + 3y''^2,$$

where λ, μ, ν, ω are functions of x, y (and have therefore fixed numerical values so far as we deal with the ∞^2 curves passing through a given point O), the latter quantity ω representing the slope of the acting force. Each of the four given curves C_1, C_2, C_3, C_4 through the point O determines certain values of the derivatives y', y'', y'''; that is we are given the differential elements of third order

$$y_i', \quad y_i'', \quad y_i''' \qquad (i = 1, 2, 3, 4).$$

Substituting these values we have four linear equations

$$(y_i' - \omega)y_i''' = (\lambda y_i'^2 + \mu y_i' + \nu)y_i'' + 3y_i''^2 \quad (i = 1, 2, 3, 4),$$

from which we can find the values of λ, μ, ν, ω at the given point. *The required direction of the acting force is determined by the slope*

$$\omega = \frac{\begin{vmatrix} 1 & y_i' & y_i'^2 & \dfrac{y_i' y_i''' - 3y_i''^2}{y_i''} \end{vmatrix}}{\begin{vmatrix} 1 & y_i' & y_i'^2 & \dfrac{y_i'''}{y_i''} \end{vmatrix}},$$

where numerator and denominator are determinants of the fourth order.

20. By any of these methods we may determine the direction[*] of the vector representing the force acting at any point O of the plane. How shall we determine the magnitude of the vector? The determination cannot be absolute, since, as already remarked, two fields that differ by a constant factor have identical trajectories. The magnitude of the vector at any one point may be taken at random, and then the field is completely determined.

This depends upon the simple fact that if we know the *path*

---

[*] Of course if *all* the trajectories were given, the direction of the force would be determined immediately by the fact that the curves in that direction have zero curvature.

of a particle and also the *direction* of the force acting at each of its points, then assuming the *magnitude* arbitrarily at one point, it is completely determined at all points. This is an integration problem. We know the force vector at the initial point O, and may decompose it into components N and T, normal and tangent to the given curve. Assuming the mass to be unity, the initial speed is given by

$$v_0^2 = rN,$$

where r is the known radius of curvature. Then from

$$vv_s = T,$$

we may find v_s, the rate of variation of the speed for unit of arc. The speed at any point P of the curve is thus found in the form

$$v = v_0 e^{\int \frac{T}{rN} ds},$$

where all the quantities in the right-hand member are geometrically given. (The integrals throughout are calculated from point O to point P.) If we denote by θ the inclination of the force to the curve, so that $\tan \theta = N/T$, the speed is

$$v = v_0 e^{\int \frac{\cot \theta}{r} ds}.$$

Since the speed, that is the motion, is now known, the magnitude of the force is also known. The components at any point P are

$$N = N_0 e^{\int \frac{2\cot \theta - r_s}{r} ds}, \quad T = N \cot \theta.$$

21. We see that the construction of the field may be carried out without knowing all the ∞^3 trajectories. So far as the direction of the force is concerned, it is sufficient to know at each point of the plane either two trajectories with a common tangent, or four trajectories with distinct tangents. So far as

the magnitude is concerned it is then sufficient to know ∞^1 trajectories through one point O, one for each direction, since we can then integrate from this point to any point of the plane* along some one of the curves.

A field of force is in general determined, and may be constructed, if we know $4\infty^1$ *out of the totality of* ∞^3 *trajectories,* each of the four systems of ∞^1 curves covering the plane (or the region considered) simply, that is, one passing through each point of the plane.

The complete system of ∞^3 trajectories is thus determined in general by four systems of ∞^1 trajectories. Further reduction is possible. In general $3\infty^1$ curves determine the totality, but no simple constructions are then available. If two simply infinite systems of curves (that is, a net of curves) are assigned arbitrarily, a corresponding complete system can be found in a large infinitude of ways: the corresponding field of force is not determined up to a constant, but involves arbitrary functions.

The first and most interesting example of the geometric exploration of a field of force arose in Bertrand's discussion of Kepler's laws. The first of these laws (every planet describes an ellipse having the sun for a focus) is geometric, while the second and third are kinematic (involving the areal velocity and the period). The first law determines all the trajectories, and therefore determines the field of force.† Hence the newtonian law of gravitation can be deduced from the first law alone, instead of, as usual, from all three. Bertrand thus concludes that the other two laws are consequences of the first. If Kepler had been a mathematician of the twentieth century, he would have stopped his laborious observational inductions after noting his first law, and deduced the other two analytically.

The first law, in Bertrand's discussion, is of course to be taken ideally: not only the actual planets describe conics with a focus at the sun, but every particle starting from any position with

---

* That is, in some region of the plane—in some neighborhood of O.

† It is assumed, of course, that the force depends only upon the position of the planet.

any velocity describes such a conic. From what has been stated above it is sufficient to limit the observations to four simply infinite systems of conics in "general" position.

On account of the last phrase, it is easily possible to commit errors in the application of the result. It would be possible to give $4\infty^1$ or even ∞^2 conics in certain special ways, so that the field is *not* determined. (See § 23.)

22. This raises the general question: *How many trajectories may be common to two distinct fields of force?*

The first field, defined by its components φ, ψ, has a system of ∞^3 trajectories with a differential equation

$$(y' - \omega)y''' = (\lambda y'^2 + \mu y' + \nu)y'' + 3y''^2;$$

the second field, with components φ_1, ψ_1, has a system of trajectories given by an analogous equation

$$(y' - \omega_1)y''' = (\lambda_1 y'^2 + \mu_1 y' + \nu_1)y'' + 3y''^2.$$

If there are any solutions in common,* they must satisfy the equation of second order

$$3(\omega - \omega_1)y'' = (y' - \omega)(\lambda_1 y'^2 + \mu_1 y' + \nu_1)$$
$$- (y' - \omega_1)(\lambda y'^2 + \mu y' + \nu).$$

Two systems of trajectories cannot have more than ∞^2 curves (one through each point in each direction) in common without coinciding. If they have ∞^2 curves in common the differential equation of the second order defining these curves must be of the cubic form†

$$y'' = Ay'^3 + By'^2 + Cy' + D,$$

where the coefficients are functions of x, y.

Usually the solutions of the equation of the second order will

---

* In addition to straight lines, $y'' = 0$, which are common to *all* systems.

† This form is characterized by the fact that the locus of the centers of curvature of the curves passing through a given point is a special type of cubic curve. Cf. *Amer. Jour. Math.*, 1908, p. 207.

not satisfy either equation of the third order and the two systems will have no curves in common. An example showing that the two systems may actually have ∞^2 curves in common is given by the fields

$$\varphi = x, \quad \psi = 4y; \qquad \varphi_1 = x^{-3}, \quad \psi_1 = 1,$$

where the equation of second order,

$$xy'' = y',$$

defines ∞^2 curves $y = ax^2 + b$, which are trajectories in both fields.

23. A fortiori $4\infty^1$ curves, or any number of simple systems, *may* belong to two distinct fields. If the four simple systems are given in the form

$$y' = f_i(x, y) \qquad (i = 1, 2, 3, 4),$$

the field, if it exists, will be *uniquely* defined provided not all the determinants of fourth order in the matrix

$$\|f_i', \; f_i f_i', \; f_i^2 f_i', \; f_i'', \; 3f_i'^2 - f_i f_i''\|$$

vanish identically. Here the primes denote complete differentiation with respect to x, so that

$$f' = f_x + f f_y,$$
$$f'' = f_{xx} + 2f f_{xy} + f^2 f_{yy} + f_x f_y + f f_y^2.$$

This is the exact formulation of the result stated previously "in general."

24. Consider the simplest of all fields, gravity assumed constant. If a cannon ball is projected in any way into the field it describes a vertical parabola. Conversely if every path in an unknown field is a vertical parabola, it follows that the acting force is vertical and constant in intensity. *How many cannon ball experiments would have to be made in order to arrive at this conclusion?*

We confine the discussion for simplicity to a fixed vertical

plane, taken as the xy-plane, so that the equations of motion are
$$\ddot{x} = 0, \qquad \ddot{y} = 1$$
and the trajectories are the ∞^3 parabolas
$$y = ax^2 + bx + c.$$
Suppose first the cannon is kept in one place, say the origin, and the ball is fired in all directions with all initial speeds, giving in all ∞^2 parabolas
$$y = ax^2 + bx.$$
This would not be sufficient to prove that the field is uniform. Another possible field, for example, is
$$\ddot{x} = x^{-5}, \qquad \ddot{y} = yx^{-6}.$$
In fact there are ∞^2 distinct fields each consistent with the given set of ∞^2 parabolas.

The same is true if we confine our geometric experiments to the ∞^2 parabolas $y = ax^2 + c$ found by shooting horizontally from every point in the axis of ordinates with variable initial speed. The differential equation of this family is $xy'' = y'$, precisely the one given at the end of § 22, and so the two forces there given are consistent with the experiments, just as much as ordinary gravity.

If however the shots are fired from all points in the axis of abscissas, with all initial speeds, at the fixed inclination of 45°, producing as trajectories the ∞^2 vertical parabolas whose foci are on the axis of abscissas, *the field must be uniform gravity*. The only possible field is in fact $\ddot{x} = 0$, $\ddot{y} =$ constant.

The same is true if we fix the amount of powder, that is the initial speed, and shoot from every point on the ground (the axis of x), at every angle. This gives ∞^2 parabolas with a common directrix.

As an example of a set of $4\infty^1$ observations that would be sufficient, we mention only the case of shooting from four*

---

* It may be that three stations are sufficient, but this requires a separate discussion. Two stations would certainly not suffice.

ASPECTS OF DYNAMICS. 31

stations on the ground, pointing the cannon at the angle 45°, and using all initial speeds.

25. Consider very briefly the general inverse problem in space of three dimensions. The determination of the magnitude of the force involves the same considerations as in the plane case.

If we are given two trajectories through O in the same direction, the osculating planes must coincide. The force acts in this common plane; its direction is determined by projecting the given space curves orthogonally on this plane, and then using the plane construction described above.

If we are given two trajectories with distinct osculating planes, the initial directions will be necessarily distinct; the force-direction is then determined by the intersection of the osculating planes.

If we are given two trajectories through O in different directions, but with the same osculating plane, the direction of the force is not determined. We need in fact four such curves with the same osculating plane and different directions before the force-direction is determined: the requisite construction is again obtained by orthogonal projections of the curves of the common osculating plane, thus reducing the problem to that considered in the two-dimensional theory (cf. § 18).

§§ 26–27. Tests for a Conservative Field

26. Since the system of trajectories determines the field of force, it ought to be possible to find out from the trajectories, whether the field belongs to any special type, for example, whether the field is central or conservative.

The lines of force are determined geometrically by property I in the plane and property II in space. The field will be central if the lines of force are straight lines passing through a common point.

We now give a number of tests any one of which will distinguish a conservative from a non-conservative force. It is not possible to decide this from the lines of force alone.

1°. First consider the *plane theory*. Here there is for each point a certain conic determined by the trajectories in accordance with property III of § 3 as the locus of the centers of the hyperosculating circles. *For a conservative field (and for no other) this conic is always a rectangular hyperbola.*

2°. In connection with property III (3) of § 8 we have this test: The conic which there appears as the locus of the centers of the focal circles is in the conservative case merely a straight line. That is, the focal circles constructed at any point all have a second point in common.

3°. The focal circles corresponding to two perpendicular directions are, in any field, tangent to each other. In the conservative case the two circles coincide.

4°. In any field two trajectories through a given point O exist whose osculating parabolas have the same given focus. If for one given focus the trajectories are orthogonal at O, this will be true for any given focus. When this is the case for every point O, the force will be conservative.

27. In the *three-dimensional theory*, the lines of force in the conservative case necessarily form a normal congruence; but this is not a sufficient test. All the tests given below are both necessary and sufficient.

1°. First consider property III of § 11. In any field there corresponds to each point O a certain twisted cubic curve Γ. The conservative fields are distinguished by the fact that the cubic Γ is, for every point O, of the rectangular type.*

2°. An interesting kinematic test, connected with the theorem of Thomson-Tait, is the following. If from any point O we shoot with a given speed v_0 in every direction, ∞^2 trajectories will be obtained. If these form a normal congruence (that is admit a set of orthogonal surfaces), the same will necessarily be true for any other speed v_0. *The trajectories starting out from any point with*

---

* That is, the cubic intersects the plane at infinity in three mutually orthogonal directions. All the quadrics passing through the curve are then of the equilateral type.

a given speed form a normal congruence when, and only when, the field is conservative.

The necessity of this condition is included in the Thomson-Tait theorem discussed in the next chapter. Its sufficiency, of course, requires a separate discussion which is connected with the theory of velocity systems.

3°. In order to make the preceding test purely geometric, it is necessary to have a geometric method of assembling those trajectories which, starting from the same point, correspond to the same initial speed. Such a method is readily found from the fact that the square of the speed varies directly as the radius of curvature and directly as the normal component of the force. The ∞^2 trajectories corresponding to a given speed have circles of curvature intersecting each other at the same point on the line of the force vector; that is, the centers of curvature lie in a plane perpendicular to the direction of the force acting at the given point. In the conservative case, the ∞^2 trajectories so selected form a normal congruence.

4°. Among the ∞^2 trajectories considered there are, for any field, three which admit hyperosculating circles of curvature. The three initial directions thus determined will be mutually orthogonal when and only when the field is conservative.

Only test 1° is directly connected with the set of properties I–IV of page 19. The other three are suggested by the discussion of velocity systems (cf. § 32).

CHAPTER II

NATURAL FAMILIES: THE GEOMETRY OF CONSERVATIVE FIELDS OF FORCE

§ 28. ORIGIN AND APPLICATION OF THE NATURAL TYPE

28. We now consider the properties of the trajectories generated by conservative fields of force. The total system of trajectories will have the general properties previously considered for an arbitrary field of force, together with the additional properties stated in §§ 26, 27, peculiar to the conservative case.

An entirely new feature presents itself, due to the fact that the differential equations of motion admit an integral of the first order, namely, the energy equation. During any motion of the particle in the given field, the sum of the kinetic and potential energies is constant; thus each motion corresponds to a definite value of the constant h, representing the total energy. The motions may therefore be grouped according to the values of h. Those corresponding to a given value form what may be termed, following Painlevé, a *natural* family.

Thus, in space of two dimensions, the complete system of trajectories for a given conservative field of force consists of ∞^3 curves grouped into ∞^1 natural families, each composed of ∞^2 curves. For example, in the case of ordinary gravity the trajectories are the ∞^3 vertical parabolas (in a given vertical plane), and the natural families are formed by grouping together those parabolas which have the same (horizontal) line as directrix.

In space of three dimensions, the complete system contains ∞^5 trajectories grouped into ∞^1 natural families, each containing ∞^4 curves. Examples are the ∞^4 parabolas with vertical axes whose directrices are situated in a fixed horizontal plane; and the ∞^4 circles orthogonal to a fixed sphere. The simplest example, corresponding to the case of zero force, is the ∞^4 straight lines of space.

This grouping of the trajectories according to the values of the total energy constant, that is, into natural families, is fundamental in most dynamical investigations relating to conservative forces, in particular, those connected with the principle of least action and the developments of Hamilton and Jacobi. From this point of view, dynamical problems relating to the same field of force, but having distinct values of h, are considered as essentially distinct problems. Quoting Darboux: "This restriction is in accordance with the spirit of modern mechanics which attaches less importance to force than to energy, and which permits us to regard as distinct two problems in which the force function or work function is the same, but the total energy is different."

It therefore seems of interest to work out the purely geometric properties of natural families. According to the principle of least action, such a family is made up of the extremals defined by the variation problem

$$\int \sqrt{W + h} \, ds = \text{minimum},$$

that is, the curves which cause the first variation of the integral to vanish. This follows from the fact that the speed v, in the action integral $\int v \, ds$, is determined by the energy equation

$$v^2 = 2(W + h).$$

Abstractly, a natural family of curves may be defined as one which can be regarded as the totality of extremals connected with a variation problem of the form

$$\int F \, ds = \text{minimum},$$

where F is any point function, that is, any function of x, y, z in the three-dimensional case.

Such families arise not only in the discussion of trajectories, but also, for example, in the discussion of brachistochrones, catenaries, optical rays, geodesics, and contact transformations.

The brachistochrone problem for a conservative field with any work function W leads to the integral

$$\int dt = \int \frac{ds}{\sqrt{W+h}}.$$

Thus the complete system of brachistochrones is made up of ∞^1 natural families, one for each value of h.

When a homogeneous, flexible, inextensible string is suspended in the conservative field, the forms of equilibrium, which are termed catenaries in the general sense of the word, are obtained by rendering the integral

$$\int (W+h)ds$$

a minimum. Hence here also we have ∞^1 natural families, one for each value of h.*

Consider an isotropic medium in which the index of refraction ν varies arbitrarily from point to point. The paths of light in such a medium, according to Fermat's principle of least time, are determined by minimizing the integral $\int \nu ds$ and hence form a single natural family. This is the most concrete way of defining a natural family.

The connection with the theory of geodesics is obvious. Thus in the two-dimensional case the geodesics of the surface whose squared element of length (first fundamental form) is $\lambda(x, y)(dx^2 + dy^2)$ are found by minimizing the integral $\int \sqrt{\lambda}\, ds$, and hence the representing curves in the x, y plane constitute a natural family. Hence if any surface is represented conformally on a plane, the geodesics are pictured by a natural family of curves in that plane. The extension to more variables is evident:

---

* The complete systems of ∞^5 brachistochrones and ∞^5 catenaries have geometric properties distinct from each other and from those of the ∞^5 trajectories: no quintuply infinite system of curves can be at the same time the system of trajectories for some field and the system of brachistochrones or catenaries in either the same or a different field. The distinctive properties for an arbitrary field are given in § 107, p. 94. Cf. § 103.

any natural family in any space may be obtained by conformal representation from the geodesics of some other space.*

As a last application we consider the transformations which Sophus Lie has termed the infinitesimal contact transformations of mechanics. In the plane case, such a transformation is defined by a characteristic function of the special form $\Omega(x, y)(1 + y'^2)^{\frac{1}{2}}$ and is characterized by the fact that the lineal elements at each point are converted into the elements of a circle about that point as center. The path curves of every contact transformation of this category form a natural family.

§§ 29–31. Characteristic Properties A and B

29. *Osculating Circles—Property A.*—We now consider the general geometric properties of a natural family in ordinary space, that is, the totality of ∞^4 extremals connected with an integral of the form

$$(1) \qquad \int F(x, y, z) \sqrt{1 + y'^2 + z'^2}\, dx.$$

The differential equations of the family are then the corresponding Euler-Langrange equations

$$(2) \qquad \begin{aligned} y'' &= (L_y - y'L_x)(1 + y'^2 + z'^2), \\ z'' &= (L_z - z'L_x)(1 + y'^2 + z'^2), \end{aligned}$$

where

$$L = \log F.$$

Of the ∞^4 curves in this family ∞^2 pass through any given point p, one in each direction. Our first result is:

Theorem 1: *The ∞^4 curves in any natural family have this property: the circles which at any point p of space osculate the ∞^2 curves passing through that point, have a second point P in common and thus form a bundle.*

---

* A natural family on a given surface may be regarded as a family of pseudo-geodesics, that is, one which may be obtained as the conformal picture of the geodesics on some other surface.

This property we shall refer to as *property A*. In the discussion it is convenient to decompose it into these two statements, also relating to the ∞^2 curves through a given point:

(A_1) The osculating planes constructed at the common point form a pencil.

(A_2) The centers of curvature lie in a plane perpendicular to the axis of the pencil of osculating planes.

A proof of the theorem stated is easily obtained by regarding the family as made up of dynamical trajectories. Property A_1 results from the fact that the osculating plane of a trajectory always passes through the force vector. Property A_2 is proved by noting that those trajectories through a given point, which correspond to the same value of the total energy h, are all described with the same initial velocity v_0. The radius of curvature at the initial point is given by the formula

$$r = v_0^2/N,$$

where N denotes the component of the force along the principal normal. Since N is the orthogonal projection of a fixed vector, the locus of its terminal point will be a sphere through the initial point. The conclusion then follows from the fact that r varies inversely as N.

The following analytical discussion has the advantage of answering the converse question which naturally arises: Are there other systems with property A?

The differential equations of any system of ∞^4 space curves, one determined by each lineal element of space, may be assumed in the form

(3) $\quad y'' = f(x, y, z, y', z'), \quad z'' = g(x, y, z, y', z').$

Property A_1 requires that at each point there shall be a certain direction through which all the osculating planes at that point must pass. Let the direction in question be given by the ratios of three arbitrary point functions

(4) $\quad\quad \phi(x, y, z), \quad \psi(x, y, z), \quad \chi(x, y, z);$

then the requisite condition is

(5)
$$\frac{z''}{y''} = \frac{\chi - z'\phi}{\psi - y'\phi}.$$

Property A_2 requires that the centers of curvature shall lie in a plane perpendicular to the direction (4); hence

(6)
$$\phi X + \psi Y + \chi Z = 1,$$

where X, Y, Z denote the coordinates of the center relative to axes with the common point as origin. Using the general formulas for the center of curvature, and combining with (5), we find

THEOREM 2: *The differential equations of any system of curves possessing property A are of the form*

(7)
$$y'' = (\psi - y'\phi)(1 + y'^2 + z'^2),$$
$$z'' = (\chi - z'\phi)(1 + y'^2 + z'^2),$$

where ϕ, ψ, χ are arbitrary functions of x, y, z. The converse is valid also.

The equations (2) are seen to be included in this form, hence the result certainly holds for our natural systems, as stated in theorem 1.

30. *Hyperosculations—Property B.*—The circles of curvature at a given point, for any system of the form (7), constitute a bundle. We now inquire whether any of these circles correspond to four-point, instead of three-point, contact.

If a twisted curve is to have an hyperosculating circle of curvature at a given point, two conditions must be satisfied, namely,

(8)
$$\begin{vmatrix} 1 & y' & z' \\ 0 & y'' & z'' \\ 0 & y''' & z''' \end{vmatrix} = 0,$$

(9)
$$\frac{dr}{ds} = 0.$$

The first of these states that the osculating plane has four-point contact with the curve; the second, in which r denotes the radius of curvature, is the condition for the existence of an osculating helix, i. e., one with four-point contact. When both conditions hold the helix is simply the circle of curvature, which then has hypercontact.

Applying these conditions to the curves defined by (7), we find, from (8),

(10) $\quad (\psi - y'\phi)\chi' - (\chi - z'\phi)\psi' + (y'\chi - z'\psi)\phi' = 0;$

and, from (9),

(11) $\quad (1 + y'^2 + z'^2)\Sigma\phi\phi' - (\phi + y'\psi + z'\chi)$
$$\times \left\{ \begin{array}{l} (1 + y'^2 + z'^2)\Sigma\phi^2 + \phi' + y'\psi' + z'\chi' \\ - (\phi + y'\psi + z'\chi)^2 \end{array} \right\} = 0,$$

where the indicated summations extend over ϕ, ψ, χ and where ϕ', for example, denotes $\phi_x + y'\phi_y + z'\phi_z$.

Since we wish to discuss the ∞^2 curves through a given point, we may simplify our equations considerably by taking the axis of abscissas in the special direction (4). Then, at the selected point, ψ and χ vanish, and the above equations reduce to

(10′) $\qquad\qquad y'\chi' - z'\psi' = 0,$
(11′) $\qquad (y'^2 + z'^2)(\phi' - \phi^2) - (y'\psi' + z'\chi') = 0.$

Neglecting the trivial solutions for which $y'^2 + z'^2$ vanishes, we may reduce this pair of simultaneous equations to the form

(12) $\quad \dfrac{\psi_x + y'\psi_y + z'\psi_z}{y'} = \dfrac{\chi_x + y'\chi_y + z'\chi_z}{z'} = \phi_x + y'\phi_y + z'\phi_z - \phi^2.$

This set of equations for the determination of y', z' is of a familiar type, namely, that arising in the determination of the fixed points of a collineation, and is easily shown to admit three solutions.* Hence

---

* Of course in special cases some of these may coincide, or the number of solutions may become infinite. The theorem stated is true " in general " in so far as it omits these cases which are definitely assignable.

THEOREM 3: *The curves defined by equations of the form* (7) *are such that through each point there pass three with hyperosculating circles at that point.*

Since the form (7) is characterized by property A, it follows that the existence of three hyperosculating circles in each bundle is a consequence of property A.

We state two further properties, found by considering the conditions (10′) and (11′) separately.

The tangents to those curves of a system (7) *which pass through a given point and there have an hyperosculating plane form a quadric cone.* This cone passes through the special direction (4).

The tangents to those curves which have an osculating helix at the given point form a cubic cone. This cone passes through the special direction (4) and through the minimal directions in the plane normal to that direction.

These properties hold for natural families since they hold for all systems with property A. By comparing (7) with (2), we see that the functions ϕ, ψ, χ in the case of a natural family are

(13) $$\phi = L_x, \quad \psi = L_y, \quad \chi = L_z;$$

and hence are connected by the relations

(14) $$\psi_z - \chi_y = 0, \quad \chi_x - \phi_z = 0, \quad \phi_y - \psi_x = 0.$$

We now inquire what is the effect of these relations on the directions of the hyperosculating circles. Introducing, for symmetry,

(15) $$X : Y : Z = 1 : y' : z',$$

we may write our equations (12) in the homogeneous form

(16) $$\begin{aligned}&\chi_y Y^2 - \psi_z Z^2 + (\chi_z - \psi_y)YZ + \chi_x XY - \psi_x XZ = 0,\\&-\chi_x X^2 + \phi_z Z^2 + \phi_y YZ - \chi_y XY + (\phi_x - \phi^2 - \chi_z)XZ = 0,\\&\psi_x X^2 - \phi_y Y^2 - \phi_z YZ - (\phi_x - \phi^2 - \psi_y)XY + \psi_z XZ = 0.\end{aligned}$$

In virtue of (14), each of the quadric cones (16) is seen* to be

---

* The condition for such a cone is that the sum of the coefficients of X^2, Y^2, and Z^2 shall vanish.

of the rectangular type. Hence the three generators common to the cones must be mutually orthogonal. This gives

THEOREM 4: *In the case of any natural family the three hyperosculating circles which exist in any bundle are mutually orthogonal.*

We refer to this property as *property B*.

31. The relations (14) are seen to be necessary as well as sufficient for the orthogonality in question. Hence property B is the equivalent of (14), and serves to single out the natural families from the more general class defined by equations of form (7). The latter form was characterized by property A; hence we have our

FUNDAMENTAL THEOREM: *A system of ∞^4 curves, one for each direction at each point of space, will constitute a natural family when, and only when, it possesses properties A and B: that is, the osculating circles at any given point must form a bundle, and the three hyperosculating circles contained in such a bundle must be mutually orthogonal.*

§ 32. GENERAL VELOCITY SYSTEMS

32. The most general system with property A is represented by differential equations of the form

(7) $$\begin{aligned} y'' &= (\psi - y'\phi)(1 + y'^2 + z'^2), \\ z'' &= (\chi - z'\phi)(1 + y'^2 + z'^2), \end{aligned}$$

and thus involves three arbitrary functions. Only in the case where these functions are the partial derivatives of the same function is the system a natural one. We now point out a dynamical problem that leads to the general type (7): this justifies the term *velocity system* which we hereafter employ to denote any system of this type.

Consider a particle (of unit mass) moving in any field of force, the components of the force being ϕ, ψ, χ. The equations of motion are then

$$\ddot{x} = \phi(x, y, z), \qquad \ddot{y} = \psi(x, y, z), \qquad \ddot{z} = \chi(x, y, z).$$

If the initial position and the initial velocity are given the motion is determined. If only the initial position and direction of motion are given, the osculating plane will be determined but the radius of curvature r will depend for its value on the initial speed v. Hence, in addition to the usual formula

$$v^2 = \dot{x}^2 + \dot{y}^2 + \dot{z}^2,$$

there must be a formula expressing v^2 in terms of x, y, z, y', z', r. This is furnished by the familiar equation

$$v^2 = rN,$$

where N denotes the (principal) normal component of the force, so that

$$N^2 = \phi^2 + \psi^2 + \chi^2 - \frac{(\phi + y'\psi + z'\chi)^2}{1 + y'^2 + z'^2}.$$

The result may be written in the two (equivalent) forms

$$v^2 = \frac{(\psi - y'\phi)(1+y'^2+z'^2)}{y''} = \frac{(\chi - z'\phi)(1+y'^2+z'^2)}{z''}.$$

In the actual trajectory v varies from point to point. If now we replace v^2 in this result by some constant, say $1/c$, the resulting equations may be written

$$y'' = c(\psi - y'\phi)(1 + y'^2 + z'^2),$$
$$z'' = c(\chi - z'\phi)(1 + y'^2 + z'^2).$$

The curves satisfying these differential equations—they are not in general trajectories—we define as *velocity curves*. For any field a curve is a velocity curve corresponding to the speed v_0, provided a particle starting from any lineal element of the curve with that speed describes a trajectory osculating the curve. In a given field of force there are ∞^5 trajectories and ∞^5 velocity curves.* If c is given we have ∞^4 velocity curves. In particular

---

* The properties of a complete system of ∞^5 velocity curves are analogous to, but distinct from, those of a complete system of trajectories. Cf. p. 94.

if c (and hence v) is taken to be unity, our equations become precisely (7).

Any system of ∞^4 curves possessing property A, that is, any system (7), may be regarded as the totality of velocity curves corresponding to unit velocity in some (uniquely defined) field of force.

Only when the field is conservative do the velocity systems for each value of v (or c) become natural systems. The trajectories also are in this case made up of ∞^1 natural families, one for each value of the energy constant h; but the two sets of natural families are distinct. The determination of a velocity system in one conservative field is equivalent to the determination of a trajectory system in another conservative field, and vice versa. We find in fact the following explicit result:

If two conservative fields with work functions W_1 and W_2 satisfy the relation*

$$W_2 = a e^{\frac{2W_1}{v_0^2}} - h,$$

then the ∞^4 velocity curves for the speed v_0 in the first field coincide with the ∞^4 trajectories for the constant of energy h in the second field.†

§ 33. Reciprocal Systems

33. With any velocity system S

$$(S) \qquad y'' = (\psi - y'\varphi)(1 + y'^2 + z'^2), \quad z'' = (\chi - z'\varphi)(1 + y'^2 + z'^2)$$

there is connected a definite point transformation T: for in virtue of property A to any point p corresponds a definite point P, the osculating circles constructed at the first point all passing through the second point. The transformation T is explicitly

$$(T) \qquad X = x + \frac{2\varphi}{\varphi^2 + \psi^2 + \chi^2}, \quad \cdots.$$

---

* We note that if W_1 is left unaltered and v_0 varied, W_2 takes quite distinct forms. The ∞^1 velocity systems in a given field do not constitute the complete system of ∞^5 trajectories in any field whatever.

† It is seen that the two fields have the same equipotential surfaces and therefore the same lines of force. (Central fields therefore correspond to central fields.)

It is thus entirely general. To an arbitrary transformation*
corresponds a definite velocity system. In particular, to the
inverse transformation T^{-1} there corresponds a certain system
S', which we define as reciprocal to S.

*Hence to a general† velocity system S, that is, any system possessing
property A, there corresponds a definite reciprocal velocity system
S'. The osculating circles of those curves of system S which pass
through any point p are at the corresponding point P the osculating
circles of the curves of the system S' passing through P.*

Consider the bundle of circles determined by two corresponding
points p and P. We know that three of these circles have
hypercontact with S-curves at p, and three have hypercontact
with S'-curves at P. It is not obvious that the circles so obtained really coincide. Omitting the rather long proof, we
merely state the result.

*Reciprocal velocity systems have the same hyperosculating circles:
the three circles hyperosculating curves of the given system S at
any point p also hyperosculate curves of the reciprocal system S'
at the corresponding point P.*

It follows at once that if S possesses property B (that is
mutually orthogonal hyperosculating circles) the same will be
true of S'. This means that whenever system S is natural so is S'.

The reciprocal of a natural family is always a natural family.

We may restate this in optical terms as follows: With any
isotropic medium, defined by its index of refraction $\nu(x, y, z)$,
there is connected a certain *reciprocal medium* with an index of
refraction $\bar{\nu}(x, y, z)$: the rays of light in this second medium,
namely, the extremals of

$$\int \bar{\nu}(x, y, z) ds = \text{minimum},$$

form the system reciprocal to that formed by the rays of light

---

* It may even degenerate but must not be merely the identical transformation. We however exclude systems with degenerate T's from the rest
of the discussion: we assume that the jacobian does not vanish, so that the
inverse transformation exists.

† See preceding footnote.

in the given medium, namely, the extremals of

$$\int \nu(x, y, z)ds = \text{minimum}.$$

The actual calculation of $\bar{\nu}$ from ν requires only operations that are *performable* in the Lie sense, namely, eliminations and differentiations. See *Transactions of the American Mathematical Society*, volume 10 (1909), page 213.

§ 34. Character of the Transformation T

34. The transformation T (from point p to point P) associated with the most general system possessing property A is, as we have seen, entirely arbitrary. The question arises what is the peculiarity of T if the given system is of the natural type. The answer to this will furnish an equivalent of property B, and will thus make it possible to characterize natural families without introducing hyperosculating circles.

The problem is to describe geometrically the class of transformations of the form

$$X = x + \frac{2L_x}{L_x^2 + L_y^2 + L_z^2}, \quad Y = y + \frac{2L_y}{L_x^2 + L_y^2 + L_z^2},$$

$$Z = z + \frac{2L_z}{L_x^2 + L_y^2 + L_z^2},$$

depending on one arbitrary function L of x, y, z, instead of three independent functions required in a general point transformation,

$$X = \Phi(x, y, z), \quad Y = \Psi(x, y, z), \quad Z = \mathrm{X}(x, y, z).$$

For a general (analytic) point transformation the bundle of lineal elements at any point is converted linearly into the bundle at the corresponding point. Are there any elements which go over into parallel elements? It is well known that there are three. If in particular these three elements are mutually perpendicular (for every point of space), we obtain a certain category of transformations which may be termed Darboux* transforma-

---

* See *Proceedings of the London Mathematical Society*, 1900.

tions or deformations. They are analytically of the form

$$X = f_x, \qquad Y = f_y, \qquad Z = f_z,$$

involving one arbitrary function. Obviously this is not the class we desire.

We next ask whether in the general transformation there are any elements at a given point p each of which is turned into a *cocircular* element at the corresponding point P. This is, in a way, a case correlative to the Darboux case: for whether two elements in space are parallel or cocircular they have in common the properties that they are coplanar and equally inclined to the line pP joining their points. It is found that there are always three such elements at any point. If we require these to be mutually orthogonal, we obtain precisely the transformations connected with natural families.

A system of ∞^4 space curves possessing property A will form a natural family when and only when the associated transformation T (from point p to point P) has the following property: the three lineal elements at p each of which is converted into a cocircular element at P are mutually orthogonal.

We have thus obtained an equivalent for property B. It may be shown synthetically that the three directions just described (cocircular elements) always coincide with the directions of the hyperosculating circles. The orthogonality of the one triple amounts to the same thing as the orthogonality of the other.

It may be remarked that the class of transformations connected with all natural systems do not form a group. It is obvious however that the inverse of any member of the class is contained in that class. *This is the essence of the law of reciprocity for natural systems, previously obtained by a different method.*

§§ 35–44. THE CONVERSE OF THOMSON AND TAIT'S THEOREM

35. It is well known that if straight lines are drawn orthogonal to any given surface they will necessarily be orthogonal to an

infinitude of surfaces (namely the surfaces parallel to the given surface). Thomson and Tait in their Natural Philosophy showed that this property of the ∞^4 straight lines of space holds for the ∞^4 trajectories described in any conservative field with the same total energy, that is, for any natural family. The writer has proved that no other families of curves have the property: it is entirely characteristic of the natural type.* We first state the original theorem in connection with the general theory of the calculus of variations, and then take up the converse theorem. Later a second converse question is discussed.

35'. *Thomson and Tait's Theorem.*—We have seen that a natural family of curves in space may be regarded as the totality of extremals of a variation problem of the particular form

(1) $$J \equiv \int F(x, y, z) ds,$$

where F is a point function, ds is the element of length

$$ds = \sqrt{dx^2 + dy^2 + dz^2} = \sqrt{1 + y'^2 + z'^2}\, dx,$$

and the integral is taken between fixed end points.

It is easily shown that for integrals of this form,† and for no others, the relation of *transversality*, in the sense of the calculus of variations, amounts merely to *orthogonality*. This suffices to distinguish our type among variation problems of the general form

(2) $$\int f(x, y, z, y'\, z') dx.$$

But of course it does not serve as a complete geometric test for a natural family. What is the geometric character of the systems of ∞^4 extremals connected with any variation problem (2)? This is an unsolved question in the calculus of variations.‡

---

* At least in the case of space of three dimensions. Cf. *Trans. Amer. Math. Soc.*, vol. 11 (1910), pp. 121–140.

† Cf. Bolza, Variationsrechnung, p. 691; also p. 146 for the two-dimensional problem due to Hedrick.

‡ See the author's paper, "Systems of extremals in the calculus of variations," *Bull. Amer. Math. Soc.*, vol. 13 (1908), pp. 289–292.

We are concerned here only with the integrals of special form J, defining natural families. Applying Kneser's fundamental theorem on transversals,* we have this well-known result: If from the points of any surface Σ we construct the extremals orthogonal to the surface, and on each lay off an arc so that the integral J takes some constant value, then the locus of the end points is a surface which is also orthogonal to the extremals.

36. This is known as the *theorem of Thomson and Tait*. It was obtained by them in connection with the dynamics of a particle moving in a conservative field—the first interpretation of a natural family considered in § 28. Here $F(x, y, z)$ represents the speed v, as determined by the energy equation

$$v^2 = 2(W + h),$$

where W denotes the work function (negative potential), and the mass is assumed to be unity. Of course h has a fixed value. We quote the original statement of the theorem:

" If from all points of an arbitrary surface particles not mutually influencing one another be projected normally with the proper velocities [so as to make the sum of the kinetic and potential energies have a given value]; particles which they reach with equal actions lie on a surface cutting the paths at right angles."

The integral J, in this case, represents the action

$$\int v\,ds = \int \sqrt{2(W + h)}\,ds.$$

The ∞^1 surfaces cutting the curves orthogonally thus appear as surfaces of equal action.

The corresponding statement for brachistochrones is sometimes called the *theorem of Bertrand*:† From the points of any surface draw the brachistochrones normal to the surface and on each lay off lengths so that the time of transit is equal to a given quantity; then the locus of the end points will be another surface orthogonal

---

* Bolza, pp. 131 and 691.

† Cf. Routh, Dynamics of a Particle (1898), p. 376. According to Appell, Mécanique rationelle, vol. 1 (1909), p. 466, this result was indicated by Euler.

to the brachistochrones. Here the integral J represents the time

$$\int dt = \int \frac{ds}{v} = \int \frac{ds}{\sqrt{2(W+h)}},$$

so that the orthogonal surfaces appear as surfaces of equal time.

Corresponding statements may be made, of course, for the other interpretations leading to natural families. The most concrete aspect is obtained by using the language of optics. Here the integrand function is simply the index of refraction $\nu(x, y, z)$, varying from point to point in any (isotropic) medium, and the integral $\int \nu ds$ is proportional to the time. The paths of light in such a medium form a (single) natural family, and every natural family may be obtained in this way. The ∞^2 rays (in general curved) starting out normally from any surface admit ∞^1 orthogonal surfaces. These present themselves as surfaces of equal time. We shall describe them as a *set of wave fronts or wave surfaces*.

37. *The geometric part of the theorem of Thomson and Tait may be stated as follows*: In any natural family of ∞^4 space curves, the ∞^2 curves which meet any surface orthogonally always form a normal congruence.

Is this geometric property, which we shall refer to as the *Thomson-Tait property*, characteristic? This is in fact the case. We shall prove, namely, the following

CONVERSE THEOREM. *If a quadruply infinite system of curves in space is such that ∞^2 of the curves meet an arbitrarily given surface orthogonally* and always form a normal congruence (that is, admit an infinitude of orthogonal surfaces), then the system is of the natural type, that is, it may be identified with the extremal system belonging to an integral of the form $\int F(x, y, z) ds$.*

38. The result is simple but the proof is rather long. We give the essential steps.

Consider an arbitrary quadruply infinite system of curves in

---

* This means the same as requiring that one curve of the system passes through each point of space in each direction.

space, assuming that one passes through each point in each direction. Such a system may be defined by a pair of differential equations of the second order

(1) $\quad y'' = F(x, y, z, y', z'), \quad z'' = G(x, y, z, y', z'),$

where F and G are uniform functions which we assume to be analytic in the five arguments. Denoting the initial values of x, y, z, y', z', which may be taken at random, by x, y, z, p, q respectively, and employing X, Y, Z as current coordinates, we may write the solutions of (1) in the form

(2) $\quad Y = y + p(X - x) + \tfrac{1}{2}F(X - x)^2 + \tfrac{1}{6}M(X - x)^3 + \cdots,$
$\quad Z = z + q(X - x) + \tfrac{1}{2}G(X - x)^2 + \tfrac{1}{6}N(X - x)^3 + \cdots.$

Here F and G are expressed as functions of x, y, z, p, q; and M and N, found by differentiating (1), are given by

(3) $\quad M \equiv F_x + pF_y + qF_z + FF_p + GF_q,$
$\quad N \equiv G_x + pG_y + qG_z + FG_p + GG_q.$

The terms of higher order will not be needed in our discussion. Equations (2) involve five arbitrary parameters but of course represent only ∞^4 curves.

Consider now an arbitrary surface Σ

(4) $\quad\quad\quad\quad\quad z = f(x, y).$

At each point of this surface and normal to it a definite curve of the given family (1) may be constructed. A certain congruence will thus be determined. We wish to express the condition that this shall be of the normal type, that is, that the ∞^2 curves shall admit a family of orthogonal surfaces.

The direction normal to the surface Σ at any point is given by

$$1 : p : q = f_x : f_y : -1,$$

so that

(5) $\quad\quad p = P(x, y), \quad q = Q(x, y),$

where

(5') $\quad\quad P \equiv f_y/f_x, \quad Q \equiv -1/f_x.$

These functions are connected by the relation

(5″) $$PQ_x - QP_x - Q_y \equiv 0.$$

The equations of the ∞^2 curves corresponding to the given initial conditions may now be written

(6) $$\begin{aligned} X &= x + t, \\ Y &= y + Pt + \tfrac{1}{2}\overline{F}t^2 + \tfrac{1}{6}\overline{M}t^3 + \cdots, \\ Z &= f + Qt + \tfrac{1}{2}\overline{G}t^2 + \tfrac{1}{6}\overline{N}t^3 + \cdots, \end{aligned}$$

where t takes the place of $X - x$ in (2), and where the bars indicate that the substitution (4), (5) has been carried out, so that, for example,

(7) $$\overline{F}(x, y) \equiv F(x, y, f, P, Q).$$

The coefficients of the powers of t in (6) are thus functions of the two parameters x, y.

The general condition for a normal congruence given in parametric form is*

(8) $$(Y'XY) - (Z'ZX) + Y'(Z'YZ) - Z'(Y'YZ) = 0,$$

where the parentheses denote jacobians taken with respect to t, x, y, and Y', Z' denote the derivatives of Y, Z respectively with respect to t.

Expanding (8) in powers of t in the form

(9) $$\Omega_0 + \Omega_1 t + \Omega_2 t^2 + \cdots,$$

we find that Ω_0 vanishes in consequence of (5″). This is as it should be, since our ∞^2 curves are orthogonal to Σ by construction.

The terms containing the first power of t give

(10) $$\begin{aligned}(1 + P^2 + Q^2)(P\overline{G}_x - Q\overline{F}_x - \overline{G}_y) \\ + 2\overline{F}\{(P^2+Q^2)(Q_x - PQ_y) + 2\overline{G}fPP_y - (P^2+Q^2)P_x\} = 0.\end{aligned}$$

---

* We may also use the convenient form due to Beltrami. Cf. Bianchi-Lukat, Differentialgeometrie, p. 340.

From (6') we find

$$\overline{F}_x = F_x + F_z f_x + F_p P_x + F_q Q_x,$$

with corresponding results for $\overline{G}_x$ and $\overline{G}_y$. Substituting these values, and observing from (5) and (5') that

$$f_x = -1/Q, \quad f_y = -P/Q, \quad Q_y = PQ_x - QP_x,$$

we may reduce (10) to

(10')
$$2F\{PQP_x + Q^2 Q_x\} + 2G\{PP_y - (P^2 + Q^2)P_x\} - (1 + P^2 + Q^2) \\ \times \{QF_x - F_z - PG_x + G_y + (QF_p - QG_q - PG_p)P_x \\ + G_p P_y + QF_q Q_x\} = 0.$$

This is then a necessary condition in order that the ∞^2 curves belonging to the quadruply infinite system (1) and orthogonal to the surface (4) shall form a normal congruence. The result is to hold in virtue of (4) and (5).

It is of course not a sufficient condition. It merely expresses the fact that the curves orthogonal to Σ are also orthogonal to some consecutive surface, that is, that the congruence is approximately normal to the first degree.

Our main problem is to find all systems (1) which have the orthogonality property with respect to *every* base surface Σ. It is then necessary that (10') should be true for an arbitrary function $f(x, y)$. The function can be so selected that for any chosen values of x and y the quantities f, P, Q, P_x, P_y, Q_x, shall take arbitrary numerical values; for the only relation to be fulfilled is (5'') and this merely determines Q_y. The condition (10') must therefore hold identically. Arranging it in the form

(10'') $\quad (1 + P^2 + Q^2)C_0 + QC_1 Q_x + C_2 P_y - C_3 P_x = 0,$

and equating coefficients to zero, we find

(11)
$$\begin{aligned}
C_0 &\equiv qF_x - F_z - pG_x + G_y = 0, \\
C_1 &\equiv (1 + p^2 + q^2)F_q - 2qF = 0, \\
C_2 &\equiv (1 + p^2 + q^2)G_p - 2pG = 0, \\
C_3 &\equiv (1 + p^2 + q^2)\{qG_q + pG_p - qF_p\} \\
&\quad + 2pqF - 2(p^2 + q^2)G = 0.
\end{aligned}$$

Integration of the second and third of these partial differential equations gives

$$F = f_1(p, x, y, z)(1 + p^2 + q^2), \qquad G = g_1(q, x, y, z)(1 + p^2 + q^2),$$

where f_1 and g_1 denote unknown functions of the four arguments indicated. Substituting these values in the fourth equation, we find $f_{1p} = g_{1q}$, and therefore

$$f_1 = \psi - p\phi, \qquad g_1 = \chi - q\phi,$$

where ϕ, ψ, χ are functions of x, y, z only. The general solution of the last three equations of the set (11) is therefore

(12) $\qquad F = (\psi - p\phi)(1 + p^2 + q^2), \qquad G = (\chi - q\phi)(1 + p^2 + q^2).$

We have still to satisfy the first equation of (11), which now reduces to
(13) $\qquad \psi_z - \chi_y + p(\chi_x - \phi_z) + q(\phi_y - \psi_x) = 0.$

The functions ϕ, ψ, χ must therefore satisfy the equations

(13′) $\qquad \psi_z - \chi_y = 0, \qquad \chi_x - \phi_z = 0, \qquad \phi_y - \psi_x = 0,$

and hence are expressible as the derivatives of a single function in the form
(13″) $\qquad \phi = L_x, \qquad \psi = L_y, \qquad \chi = L_z.$

The solutions of the set (11) are therefore

(14) $\qquad \begin{aligned} F &= (L_y - pL_x)(1 + p^2 + q^2), \\ G &= (L_z - pL_x)(1 + p^2 + q^2), \end{aligned}$

involving an arbitrary function L of x, y, z. The resulting system (1) is thus recognized to be a natural family. This gives our fundamental converse theorem.

39. In the above discussion use has been made, not of the complete condition for a normal congruence, but only of condition (10′) derived from the terms of the first order in t. We may therefore state a stronger converse result as follows:

The only systems of ∞^4 curves which have the property that the curves orthogonal to any surface are always orthogonal to some infinitesimally adjacent surface are those of the natural type.

If a congruence of curves meets two neighboring surfaces orthogonally it need not meet ∞^1 surfaces orthogonally, and therefore it approximates to, but need not coincide with, a normal congruence. The above theorem shows however that if the weak requirement of approximate normal character be imposed on *all* the congruences obtained from the given quadruply infinite system, they will *all* be exactly normal.

40. We may further strengthen our theorem by demanding the orthogonality property for some instead of all surfaces. Our fundamental equations (11) resulted from the fact that x, y, z, f, P, Q, P_x, P_y, Q_x might receive arbitrary numerical values. It will therefore be sufficient to take a manifold of surfaces sufficiently large to leave these quantities, or the equivalent quantities

(15) $\qquad x, \quad y, \quad z, \quad z_x, \quad z_y, \quad z_{xx}, \quad z_{xy}, \quad z_{yy},$

unrestricted. Since these quantities define a differential surface element of the second order, we may state the result as follows:

The converse theorem remains valid if, instead of considering all base surfaces, we employ a manifold of surfaces sufficiently large to include all the ∞^8 possible differential elements of the second order.

41. The Thomson-Tait theorem holds of course even when the base Σ shrinks to a curve or a point: there will still be a normal congruence orthogonal to the curve or point (in the latter case orthogonality means simply passage through the point). We state a number of results obtained in this connection.

If for an arbitrary curve as base the corresponding ∞^2 orthogonal curves of a given quadruply infinite system always form a normal congruence, the given system is necessarily natural.

If we require each of the congruences here considered to be of approximately normal character, a more general type of system is obtained, namely the *velocity type* of § 32. The velocity type is thus characterized by the fact that those curves of the system which meet an arbitrary curve orthogonally are orthogonal to some infinitesimally adjacent (of course tubular) surface. We may even restrict ourselves to the case where the base is a curve of the given system, or the case where it is any straight line.

42. Suppose next that the base is an arbitrary point. Are natural families the only families of ∞^4 curves such that the ∞^2 curves passing through any point form a normal congruence? A discussion shows that this is not the case. There exist families not of the natural type, for example, that defined by the differential equations
$$y'' = y'^2, \qquad z'' = 0,$$
with the restricted property stated. To find all such systems would be a rather difficult, but certainly an interesting, undertaking. The result would of course include the natural type as a special case.

43. It will not however be the velocity type. It may be shown in fact that the only velocity systems for which the curves passing through an arbitrary point constitute always a normal congruence are those of the natural type. Recalling the fact that the velocity type is characterized by property A, we may give a new characterization of the natural type as follows:

Natural families are the only quadruply infinite systems of curves in space such that the ∞^2 curves through an arbitrary point admit an infinitude of orthogonal surfaces, and such that the osculating circles constructed at the common point form a bundle.

44. It may also be shown that if for every point and every straight line as base the corresponding congruence is normal, the system will be natural. To have a velocity system it is sufficient to demand that the congruence corresponding to an arbitrary straight line shall be approximately normal. To have a natural system it is sufficient to demand approximate normality for the congruences corresponding to arbitrary straight lines and planes.

§§ 45–53. Wave Propagation in an Isotropic Medium: Properties of Wave Sets

45. The optical interpretation of a natural family and the Thomson-Tait property suggest certain sets of surfaces which we shall now study.

Consider a given medium defined by its index of refraction $\nu(x, y, z)$ given as a function of position. The rays (in general curved lines) are the ∞^4 extremals of

(1) $$\int \nu(x, y, z) ds = \text{minimum};$$

they form the natural family, whose differential equations are

(2) $$y'' = (L_y - y'L_x)(1 + y'^2 + z'^2),$$
$$z'' = (L_z - z'L_x)(1 + y'^2 + z'^2),$$

where

(2′) $$L \equiv \log \nu.$$

The ∞^2 rays orthogonal to any selected surface Σ form a normal congruence, that is, are orthogonal to a set of ∞^1 surfaces. A disturbance originating in the medium on the surface Σ will be propagated in the medium through this set of surfaces, which we term a *set of wave fronts*. In the given medium an arbitrary surface belongs to one and only one of these wave sets. A single surface is thus of arbitrary character, but the sets of surfaces

(3) $$f(x, y, z) = \text{constant}$$

that may be wave sets are restricted by the Hamilton-Jacobi equation

(4) $$f_x^2 + f_y^2 + f_z^2 = \nu^2.$$

The given medium defines also a certain set of *level surfaces*

$$\nu(x, y, z) = \text{constant}.$$

This, it should be noticed, is not usually a wave set—the only exception arising when the level surfaces are parallel. For a given medium the number of wave sets is ∞^∞, since there is one for each surface. Each of these sets is cut by the level surfaces in the equidistant curves of the wave set; that is, along any one of these curves the distance between consecutive wave surfaces remains the same.*

46. A single set of wave fronts has no geometric peculiarity. That is, given any set of surfaces $f(x, y, z) =$ constant, it will always be possible to find a medium in which that set will serve as a wave set. In fact there are ∞^∞ such media. For in equation (4), the given function f, without altering the given surfaces, may be replaced by an arbitrary function $\Omega(f)$ of itself, and this gives ∞^∞ distinct values for v.

When will two sets of wave fronts be consistent? Two arbitrary sets of surfaces $f =$ constant, $f_1 =$ constant cannot usually be regarded as wave sets in any single medium. The requisite condition is

$$\frac{f_x^2 + f_y^2 + f_z^2}{f_{1x}^2 + f_{1y}^2 + f_{1z}^2} = \frac{\Omega(f)}{\Omega_1(f_1)},$$

where Ω, Ω_1 may be any functions. An equivalent condition is that it must be possible to chose parameters for the two sets in such a way that

$$\frac{df}{dn} = \frac{df_1}{dn_1},$$

where dn and dn_1 denote the normal distance between consecutive surfaces.

47. But a clearer answer may be given in terms of the geometric properties A and B. If a set of surfaces is to be a wave set, the ∞^2 orthogonal curves must be members of the natural family of ∞^4 rays. If two sets of wave fronts are given, we have then two congruences of curves. The question then is, when can two normal congruences of curves be regarded as belonging to a natural family?

---

* This follows immediately from (4). It is to be remarked, however, that this property is not characteristic of wave sets.

Take any point p in space, and consider the two curves, one from each of the congruences, passing through it. The circles of curvature at p must intersect again at some point P (by property A). This condition makes sure that the two congruences belong to some velocity system. If now this is to be a natural system, we must also add property B or rather, since no hyperosculating circles are directly defined, the equivalent restriction (see page 47) relating to the transformation from p to P. The final answer may then be given as follows:

Two sets of wave surfaces belong to the same optical medium when and only when they satisfy the following geometric conditions:

(A') *At any point p of space the circles of curvature of the orthogonal trajectories of the two sets of surfaces, passing through that point, intersect again at some point P.*

(B') *The point transformation from p to P has the property that the three lineal elements of p each of which corresponds to a cocircular element at P are mutually orthogonal.*

48. Two sets of surfaces taken at random will not belong, as wave sets, to any medium. On the other hand, as we have said, one set belongs to ∞^∞ distinct media. The question then arises, just what will uniquely determine a medium.

A natural family is uniquely determined if we are given one set of wave fronts and a single extra trajectory. This means a trajectory not belonging to the congruence defined as the orthogonal trajectories of the wave set.

49. The *extra curve* however cannot be taken at random; it must be related in a certain way to the wave set. If the wave set is $f(x, y, z) = $ constant, then the condition on the curve is that it satisfy the Monge equation of second order

(3) $$\begin{vmatrix} 2\Delta y'' - (1 + y'^2 + z'^2)(\Delta_y - y'\Delta_x) & f_y - y'f_x \\ 2\Delta z'' - (1 + y'^2 + z'^2)(\Delta_z - z'\Delta_x) & f_z - z'f_x \end{vmatrix} = 0,$$

where

(3') $$\Delta \equiv f_x^2 + f_y^2 + f_z^2.$$

Here f, and hence Δ, are given, and y and z are unknown functions of x. The interpretation is obvious from property A.

In order that an extra curve shall be consistent with a given wave set (that is, in order that both shall belong to a single medium) it is necessary and sufficient that the curve shall cross the surfaces (of course obliquely) in such a way that at any point of intersection the circle of curvature of the extra curve shall intersect the circle of curvature of the curve orthogonal to the surfaces. When the curve satisfies this restriction, it defines with the given wave set a unique natural family.

50. If we are merely given one wave set, the number of possible media is ∞^∞ (since ν involves arbitrary functions). Each of these has ∞^4 rays (forming a natural family). The totality of media give rise to a totality of ∞^∞ rays, namely the solutions of the Monge equation of second order (3). This equation is of the type
$$Ay'' + Bz'' + C = 0$$
(where the coefficients are functions of x, y, z, y', z'), which the author has shown to be characterized by the *Meusnier property:** Those curves which pass through a given point in a given direction have circles of curvature (constructed at the common point) generating a sphere.

51. The inverse problem connected with natural families, namely, given the ∞^4 trajectories to construct the generating field of force, is solved immediately in connection with property A. The force acting at any point p acts in the line joining that point to the corresponding point P, and its intensity is proportional to the reciprocal of the distance between the two points.† This construction may be carried out if we know a sufficient number of trajectories, without knowing the whole system.

52. The greatest number of rays which two distinct media

---

* Kasner, *Bull. Amer. Math. Soc.*, vol. 14 (1908), pp. 461–465. The result includes the extension of Meusnier's theorem made by Lie, and is in fact the largest generalization possible.

† The determination of the potential function $W(x, y, z)$ or, what is equivalent, the index of refraction $\nu(x, y, z)$, requires a quadrature.

can have in common is ∞^2 (one through each point of space). If two media have that many in common, it is easily shown that the resulting congruence is necessarily normal. Any normal congruence can be obtained in this way, for, as stated above, it belongs, not only to two, but to ∞^∞ distinct media.

53. We mention only one special problem: the determination of those media in which disturbances are propagated by Lamé families of surfaces; that is, every wave set is to be of the Lamé type (thus forming part of a triply orthogonal family of surfaces). The index of refraction is found to vary inversely as the *power* of the point with respect to a fixed sphere; the rays then are the ∞^4 circles orthogonal to that sphere. Since the radius of the sphere may be zero, real, or imaginary, these media yield well known interpretations of parabolic, hyperbolic, and elliptic geometries. (See *Transactions of the American Mathematical Society*, volume 12 (1911), pages 70–74.)

§§ 54–61. A Second Converse Problem Connected with the Thomson-Tait Theorem

54. Consider the general conservative field, defined by its work function $W(x, y, z)$. With any motion of the particle there is associated a definite value of the constant of total energy

$$\tfrac{1}{2}v^2 - W = h.$$

If h is not assigned the complete system of trajectories is made up of ∞^5 curves.

Consider now an arbitrary surface, which we term the base surface,

(Σ) $\qquad\qquad z = f(x, y).$

From each of its points we may draw normal to the surface ∞^1 trajectories since the initial value of the speed v is arbitrary. We thus have in all ∞^3 trajectories normal to Σ. In order to have a congruence we must assign the value of v at each point of Σ, that is, we must give a law of distribution of the initial speed. The question arises: What form of law will make the corresponding congruence a normal congruence? Of course for any law the congruence will be orthogonal to the base surface, but usually it admits no other orthogonal surfaces.

The Thomson-Tait theorem (in its complete dynamical form) gives *one* such law: it states that if the initial speed is selected so as to make h have the same value at all the points of Σ, the congruence will be normal. It thus gives a plan for constructing ∞^1 normal congruences for a given base, one for each value of h. We shall refer to any one of these as "constructed according to the Thomson-Tait law."

Is this the only answer to our question? If ∞^2 trajectories are drawn orthogonal to Σ and if they form a normal congruence, does it follow that the distribution of values of the initial speed is precisely such that the sum of the kinetic and potential energies has the same value at all points of Σ?

The requisite discussion is not simple. We shall merely state the results we have obtained.

55. The answer to our question is "in general" in the affirmative. The first converse theorem, discussed in § 37, is true without exception. The present is true with exceptions—which may be definitely limited.

For a "general" base surface Σ in a given conservative field of force, the only congruences, formed by ∞^2 trajectories orthogonal to Σ (one drawn at each point), which admit an infinitude of orthogonal surfaces, are those constructed according to the Thomson-Tait law (so that the total energy has a constant value).

56. To make this precise we must of course limit the class of *exceptional surfaces* connected with a given field. These appear in the analytical discussion as the solutions of a certain partial differential equation of the second order*

$$\begin{vmatrix} W_x + pW_y & \overline{W}_x + p\overline{W}_y \\ W_x + qW_z & \overline{W}_x + q\overline{W}_z \end{vmatrix} = 0,$$

---

* The expanded result is of the form

$$P_1 r + P_2 s + P_3 t + P_4 = 0,$$

where r, s, t denote the derivatives of second order of $z = f(x, y)$.

where W is the given work function, and

$$\overline{W} \equiv \frac{pW_x + qW_y - W_z}{\sqrt{1 + p^2 + q^2}}.$$

This differential equation defines a class of surfaces which is seen to depend only on the equipotential surfaces

$$W(x, y, z) = \text{constant}.$$

The result may be put into geometric form and stated as follows:

The only surfaces Σ which may be exceptional in the theorem of § 55 (that is, which may give rise to normal congruences not included in the Thomson-Tait law) are those with this property: along each of the equipotential lines* of the surface the component of the acting force normal to the surface is constant.

57. Observe that it is not stated that the surfaces described, which exist in any field, actually give rise to additional normal congruences. To understand the situation more precisely, it is necessary to observe that in the analytic discussion the condition for a normal congruence is developed in the form

$$t\,\Omega_1 + t^2\Omega_2 + \cdots = 0,$$

where t is the parameter which varies along the curve, starting with the value zero on the surface Σ, and the coefficients Ω are functions of the two parameters defining the initial points on Σ. By assumption the congruence is orthogonal to Σ, so the term Ω_0, independent of t, will not appear. For a normal congruence all the coefficients Ω must vanish. If only a certain number vanish the congruence may be described as *approximately normal* (the approximation being of degree n if $\Omega_1 = \Omega_2 \cdots = \Omega_n = 0$): the curves are then orthogonal not only to Σ but also to one or more (infinitesimally) adjacent surfaces.

58. If now we impose on the congruence of trajectories normal to Σ the condition $\Omega_1 = 0$, we find that this may be fulfilled for

---

* The equipotential lines of any surface are the lines cut out by the equipotential surfaces $W = $ const.

any surface: the restriction is merely on the law of initial speed and means that the total energy must be the same, not necessarily over the entire surface, but along each equipotential line of the surface.*

59. If we further impose the condition $\Omega_2 = 0$, then for a "general surface" the law of speed must be the Thomson-Tait law, but for an "exceptional surface" the law is the more general one just stated.

60. The discussion of the higher conditions $\Omega_3 = 0$, etc., we have not completed. It is therefore not known precisely in which cases normal congruences (in the exact sense) may arise. For central and parallel fields it may be shown that the exceptional surfaces† actually give rise to normal congruences (in addition to those included in the Thomson-Tait theory): for such fields the vanishing of the higher coefficients follows from the vanishing of the first two.

61. The principal results of the converse problem may be formulated as follows:

If ∞^2 trajectories (of a conservative field), meeting a surface Σ orthogonally, are also orthogonal to an infinitesimally adjacent surface, then the total energy along each equipotential line of Σ is constant.

If ∞^2 trajectories, selected from the complete system of ∞^5, form a normal congruence, then in general they will all belong to the same natural family (that is, the total energy will be the same for all the curves); except possibly when the ∞^1 orthogonal surfaces‡ are exceptional in the sense defined in § 56 (the additional congruences then and only then are normal to at least the second degree of approximation).

Normal congruences not of the Thomson-Tait type (that is, not

---

* If, in particular, the surface is one of the equipotential surfaces, the distribution of speed is thus entirely arbitrary.

† In the case of ordinary constant gravity the exceptional surfaces are those termed *moulure* surfaces by Monge: they are generated by rolling the plane of any plane curve about a vertical cylinder of arbitrary cross section.

‡ If one of these surfaces is exceptional, all will be.

selected from within a natural family) actually arise for central and parallel fields.

§§ 62–67. Geometric Formulation of Some Curious Optical Properties

62. In Thomson and Tait's Natural Philosophy* the characteristic function of Hamilton is applied to the motion of a particle in a conservative field of force, and certain results are obtained which we shall try to restate as purely geometric properties of a natural family of trajectories. To what extent these properties are characteristic is not settled. We quote the principal passages referred to.

"Let two stations, O and O', be chosen. Let a shot be fired with a stated velocity, V, from O, in such a direction as to pass through O'. There may clearly be more than one natural path by which this may be done; but, generally speaking, when one such path is chosen, no other, not considerably diverging from it, can be found; and any infinitely small deviation in the line of fire from O, will cause the bullet to pass infinitely near to, but not through, O'. Now let a circle, with infinitely small radius r, be described round O as center, in a plane perpendicular to the line of fire from this point, and let—all with infinitely nearly the same velocity, but fulfilling the condition that the sum of the potential and kinetic energies is the same as that of the shot from O—bullets be fired from all points of this circle, all directed infinitely nearly parallel to the line of fire from O, but each precisely so as to pass through O'. Let a target be held at an infinitely small distance, a', beyond O', in a plane perpendicular to the line of the shot reaching it from O. The bullets fired from the circumference of the circle round O, will, after passing through O', strike this target in the circumference of an exceedingly small ellipse, each with a velocity (corresponding of course to its position, under the law of energy) differing infinitely little from V', the common velocity with which they pass through O'. Let now a circle, equal to the former, be described round O',

---
* Part I (Cambridge, 1903), pp. 355–359.

in the plane perpendicular to the central path through O', and let bullets be fired from points in its circumference, each with the proper velocity, and in such a direction infinitely nearly parallel to the central path as to make it pass through O. These bullets, if a target is held to receive them perpendicularly at a distance $a = a'V/V'$, beyond O, will strike it along the circumference of an ellipse equal to the former and placed in a "corresponding" position; and the points struck by the individual bullets will correspond; according to the following law of "correspondence":—Let P and P' be points of the first and second circles, and Q and Q' the points of the first and second targets which bullets from them strike; then if P' be in a plane containing the central path through O' and the position which Q would take if its ellipse were made circular by a pure strain; Q and Q' are similarly situated on the two ellipses."

63. The second passage is as follows: "The most obvious optical application of this remarkable result is, that in the use of any optical apparatus whatever, if the eye and the object be interchanged without altering the position of the instrument, the magnifying power is unaltered." . . . "Let the points O and O' be the optic centers of the eyes of two persons looking at one another through any set of lenses, prisms, or transparent media arranged in any way between them. If their pupils are of equal size in reality, they will be seen as similar ellipses of equal apparent dimensions by the two observers. Here the imagined particles of light, projected from the circumference of the pupil of either eye, are substituted for the projectiles from the circumference of either circle, and the retina of the other eye takes the place of the target receiving them, in the general kinetic statement."*

---

* This fact and many other applications are included in the following general proposition. "The rate of increase of any one component momentum, corresponding to any one of the coordinates, per unit of increase of any other coordinate, is equal to the rate of increase of the component momentum corresponding to the latter per unit increase or dimension of the former coordinate, according as the two coordinates chosen belong to one configuration of the system, or one of them belongs to the initial configuration and the other to the final."

64. The statement in the first passage is not purely geometric; for it involves not only the curves described, but also the speeds V and V' at the points O and O'. We therefore try to formulate the part of the theorem which is really geometric.

We have a natural family made up of ∞^4 curves in space, one for each initial lineal element (point and direction) of space. Select any one of these curves c and any two points O and O' upon it. Construct the planes p and p' normal to this curve at O and O'.

For each direction through O, a curve of our family is determined; this strikes the plane p' at a definite point. We thus have a certain correspondence between the bundle of directions through O and the points of p'. For directions infinitesimally close to the direction of c at O, and for points close to O', this correspondence is linear; and by a proper selection of cartesian axes at O and O', we may write the correspondence in the canonical form

$$\xi = \alpha_1 x', \quad \eta = \beta_1 y',$$

where (x', y') denote the coordinates of the point in the plane p', and the corresponding direction at O has direction cosines proportional to $(\xi : \eta : 1)$.

In an entirely analogous way, by considering the curves of the natural family which go through O', and the points of intersection with the plane p, we obtain a second linear correspondence which may be reduced to the form

$$\xi' = \alpha_2 x, \quad \eta' = \beta_2 y,$$

where (x, y) is the point in the plane p and $(\xi' : \eta' : 1)$ gives the corresponding direction at O'.

If we were dealing with an arbitrary family of ∞^4 curves, instead of a natural family, these linear correspondences would still exist; but the choice of axes in the second canonical form would be different from that required in the first, and the two constants appearing in the second form would be independent of those appearing in the first. *The peculiarity of the natural type may be stated in the following form: First, the canonical axes for the two correspondences coincide; second, the ratio of the characteristic constants has the same value for both correspondences.*

This is the essential geometric content of the long statement quoted above from Thomson and Tait. Is this characteristic of the natural type? We do not know.

64′. A statement in more concrete terms is of interest. If we start out from O in directions equally inclined (the fixed angle is of course assumed infinitesimal) to the direction of c, that is, along a cone of revolution having for axis the tangent of c, the resulting trajectories forming a sort of curvilinear cone, we strike points on p' located on an ellipse with O' as center. By changing the angle of the cone we obtain a family of similar and similarly situated ellipses. The principal axes of these ellipses are the canonical directions referred to above for the first correspondence, and the ratio of the diameters is equal to the ratio of the canonical constants ($\alpha_1 : \beta_1$). By starting from the other point O' along cones of revolution having for axis the tangent to c, we strike the plane p in a second set of homothetic ellipses. *The two sets of ellipses thus obtained, one in the plane p, and the other in the plane p', are similar.* This is part of the property stated, but not the whole. It should be observed that it has no meaning to say that the two sets are similarly situated, since they are in different planes.

65. We may, however, obtain two sets in the same plane as follows: If we start along the cone of revolution from O, we hit p' in an ellipse. If we wish to hit p in a circle, we must start at O' along a certain elliptical cone: the sections of this cone by planes parallel to p', projected orthogonally on p', give a set of homothetic ellipses. We thus have in the plane p', two sets of ellipses, the first set being obtained from cones of revolution at O, and the second set being obtained from elliptical cones at O' by orthogonal projection of parallel sections. If we were dealing with an arbitrary family of curves, the two sets thus obtained would be unrelated: *for a natural family, however, the two sets coincide.*

ASPECTS OF DYNAMICS. 69

66. Of course we could also construct two sets in the plane p and these would coincide; but this would not give an additional property. In the statement quoted, certain pairs of congruent instead of merely similar ellipses appear, but that is due to the introduction of kinematics: namely, use is made of the velocities V and V' at the points O and O'. "If O and O' are regarded as optic centers of the eyes of two persons looking at one another through any optical apparatus, and if their pupils are of equal size in reality, they will be seen as similar ellipses of equal apparent dimensions by the two observers." It should be observed, however, that the dimensions will be *equal* only under the assumption that the two eyes are at positions for which the velocities V and V', or, what is equivalent, the indices of refraction ν and ν', are equal. In the most general case of an isotropic medium, the ellipses will not have equal apparent dimensions, but the ratio of the dimensions will be equal to the ratio of the two velocities.

67. Two converse questions remain unanswered. First: Find all systems of ∞^4 curves in space such that circles about O and O' appear as similar ellipses.

Second: Find all systems such that the set of ellipses in the plane p' formed by starting from O along cones of revolution, and the set of ellipses found by orthogonal projection upon p' of the sections cut out by planes parallel to p' of those (elliptical curvilinear) cones at O' which strike plane p in circles,—such that these two sets of ellipses shall coincide.

§§ 68–72. THE SO-CALLED GENERAL PROBLEM OF DYNAMICS

68. Consider any material system (particles or rigid bodies) with n degrees of freedom, so that its position at each instant is determined by n independent coordinates denoted by $x_1, x_2, \cdots, x_n$. The kinetic energy T will be represented by a quadratic form

$$2T = \sum a_{ik}\dot{x}_i\dot{x}_k,$$

where the coefficients a are functions of the coordinates, and the dots denote time derivatives. If the acting forces are conservative, there will exist a force function $W(x_1, x_2, \cdots, x_n)$, which is assumed to be independent of the time, and the equation of energy

$$T - W = h$$

asserts that in any given motion the sum of the kinetic and potential energies is constant.

The so-called general problem of dynamics requires the determination of the motions when we are given the form T, the function W, and the constant h. The possible trajectories are then given by the Jacobi principle of least action as the extremals of the integral

$$\int \sqrt{W + h} \sqrt{\Sigma a_{ik} dx_i dx_k}.$$

This defines the *most general natural family*. The integral is of the form $\int F ds$, where F is any point function and ds is the length-element in a general n-dimensional variety V_n defined by

$$ds^2 = \Sigma a_{ik} dx_i dx_k.$$

69. Such a family consists of $\infty^{2(n-1)}$ curves, in the space V_n, one passing through each point in each direction. A *complete characterization* is given by J. Lipke, in his doctor's dissertation,[*] as follows:

(A_1) The locus of the centers of geodesic curvature of the ∞^{n-1} curves passing through any point of V_n is a flat space of $n-1$ dimensions S_{n-1}.

(A_2) The osculating geodesic surfaces (two-dimensional varieties) at the given point form a bundle of surfaces, all containing a fixed direction (and hence the geodesic line in that direction) which is normal to the S_{n-1} of property A_1.

(B) The n directions at any point, in which, as a consequence of the preceding properties, the osculating geodesic circles (circles

---

[*] *Trans. Amer. Math. Soc.*, vol. 13 (1912), pp. 77–95.

of constant geodesic curvature) hyperosculate the curves of the given family, are mutually orthogonal.

70. This gives the generalization of properties A and B stated in §§ 29–31. The simpler results there given for ordinary space apply to a euclidean space of any dimensionality and also to spaces of constant curvature. In the general space of variable curvature, the geodesic circles constructed at a given point do not all meet at a second point, and so no analogue of the law of reciprocity of natural families presents itself.

71. The theorem of Thomson and Tait remains valid for any space.* The converse questions connected with it have not been settled. In all probability the Thomson-Tait geometric property is characteristic in any space (flat or curved) of dimensionality greater than two. Obviously in the case of two dimensions the geometric converse is not valid, since any system of ∞^1 curves admits ∞^1 orthogonal curves.

72. The systems characterized by property A (meaning A_1 together with A_2) are the most general velocity systems in V_n. The case $n = 2$ presents a peculiar feature: for then, included in the velocity type, we have, in addition to the natural type, another special type of interest (geometric, rather than dynamic), namely the isogonal type† (systems formed by the ∞^2 isogonal trajectories of an arbitrary simply infinite system of curves). In the case of the plane (or any surface of constant curvature) the reciprocity construction for velocity systems is available, and each of the species, natural and isogonal, is self-reciprocal. The only families common to the two species are those formed by the isogonals of an isothermal system, or, what is the same, by velocity systems generated by Laplacian fields of force.‡

---

* Cf. Darboux, Leçons, vol. 2, last chapter, where references to the memoirs of Lipschitz and Beltrami are given.

† Scheffers introduced the systems of plane curves $y'' = (\psi - y'\varphi)(1 + y'^2)$ in connection with the theory of isogonals, and obtained a law of reciprocity for isogonal systems. Cf. *Leipziger Berichte*, 1898, 1900; *Mathematische Annalen*, vol. 60.

‡ Cf. the author's note, "Isothermal systems in dynamics," *Bull. Amer. Math. Soc.*, vol. 14 (1908), pp. 169–172.

We note finally this characteristic distinction between the two noteworthy species:

For both natural and isogonal families in the plane, the circles of curvature constructed at any point p have another point P in common. The point transformation T (from p to P) in the natural case is such that the two lineal elements at any point, each of which is converted into a cocircular element, are orthogonal; while in the isogonal case the two elements, each of which is converted into an element normal to a cocircular element, are orthogonal.

If the transformation T connected with a velocity system is required to be (direct) conformal, the corresponding field must be Laplacian. Such fields are distinguished from all others by the fact that each of the infinitude of systems of velocity curves is then expressible linearly in the two parameters involved.

CHAPTER III

TRANSFORMATION THEORIES IN DYNAMICS

§§ 73-81. Projective Transformations

73. The general object of a transformation theory is to relate new problems to old problems, and so to proceed from the solution of the latter to the solution of the former. The most important geometric transformations are the projective and the conformal. Both groups play important rôles in dynamics, the former in connection with general fields, and the latter in connection with conservative fields.

74. The importance of projective transformations in dynamics was brought out by Appell in 1889. Given any positional field of force in the plane, the corresponding equations of motion are of the form

$$\text{(1)} \qquad \frac{d^2x}{dt^2} = \varphi(x, y), \qquad \frac{d^2y}{dt^2} = \psi(x, y).$$

If an arbitrary point transformation, unaccompanied by any change in the time, is applied, the new differential equations will usually involve not only x and y, but also the velocity components dx/dt, dy/dt. In fact the only exception is where the point transformation is merely affine:

$$x_1 = ax + by + c, \qquad y_1 = a'x + b'y + c'.$$

Appell showed that if a general collineation

$$\text{(2)} \qquad x_1 = \frac{ax + by + c}{a''x + b''y + c''}, \qquad y_1 = \frac{a'x + b'y + c'}{a''x + b''y + c''}$$

is accompanied by a change of the time of the form

$$\text{(2')} \qquad dt_1 = \frac{dt}{k(a''x + b''y + c'')^2},$$

the new differential equations will be of the original form

$$\text{(3)} \qquad \frac{d^2 x_1}{dt_1^2} = \varphi_1(x_1, y_1), \quad \frac{d^2 y_1}{dt_1^2} = \psi_1(x_1, y_1),$$

and therefore define motion in some new positional field of force. The relation between the new field and the original field is explicitly as follows

$$\text{(4)} \quad \begin{aligned} \varphi_1 &= k^2(a''x + b''y + c'')^2 \{C'(x\psi - y\varphi) + B'\varphi - A'\psi\}, \\ \psi_1 &= k^2(a''x + b''y + c'')^2 \{-C(x\psi - y\varphi) - B\varphi + A\psi\}, \end{aligned}$$

where the capital letters denote minors in the determinant $|ab'c''|$ of (2).

74'. The trajectories of the original field are converted by the collineation into the trajectories of the new field. Also the directions of forces of the two fields are projectively related. It must not be thought, however, that the force vector acting at a given point (x, y) in the first plane is projected into the new force vector acting at the point (x_1, y_1) of the second plane: the initial points of the two vectors will correspond, of course, by the given collineation, but the terminal points will not. The question therefore arises, *what is the geometric relation between the new vector field and the old vector field?*

To answer this question we take our rectangular axes so that the collineation takes its metrical normal form. (Affinities of course require a separate discussion.) The canonical formulas for our transformation are

$$\text{(5)} \qquad \begin{aligned} x_1 &= \frac{\gamma \gamma_1}{x}, \quad y_1 = \frac{\gamma_1 y}{x}, \\ \varphi_1 &= -k^2 \gamma \gamma_1 x^2 \varphi, \quad \psi_1 = k^2 \gamma_1 x^2 (x\psi - y\varphi), \end{aligned}$$

together with

$$\text{(5')} \qquad dt_1 = \frac{dt}{kx^2}.$$

To each collineation between the two planes corresponds a definite vector transformation. The vectors are here of the third type (*bound* vectors) described in the Introduction, requiring four coordinates for their determination. The original vector is defined by the four numbers (x, y, φ, ψ), the first two defining the initial point, and the last two giving the components of the vector. The coordinates of the new vector are $(x_1, y_1, \varphi_1, \psi_1)$.

The vector transformation induced by the given collineation is not projective. The new vector has the same initial point and the same direction as the projection of the old vector, but has a different length. The ratio λ between the actual length of the new vector and the length of the projected vector is

(5'') $$\lambda = k^2 x^3 (x + \varphi).$$

Noting that in the canonical form x and $x + \varphi$ denote the distances from the initial and terminal points of the original vector to the vanishing line in the first plane, we may state this result.

Any given (non-affine*) collineation (2) induces a certain vector transformation (determined up to the factor k) defined analytically by (2) and (4), and geometrically as follows: If PQ is any bound vector in the first plane, and if the collineation converts the initial point P into P_1 and the terminal point Q into Q_1, then the transformed bound vector is not $P_1 Q_1$, but $P_1 Q_1'$, where Q_1' is the point on the line joining $P_1 Q_1$ such that the ratio $\lambda = P_1 Q_1' / P_1 Q_1$ equals k^2 times the cube of the distance from P to the vanishing line times the distance from Q to that vanishing line.

The transformation converts the ∞^4 bound vectors of the first plane, represented by the independent coordinates (x, y, φ, ψ), into the ∞^4 bound vectors of the new plane.† In the dynamical application, φ and ψ are given as functions of x, y, that is, we have a field of ∞^2 vectors, one for each initial point: the result of

---

* In the case of an affine collineation, the induced vector transformation is, except for the constant factor k, merely the result of applying the affinity to both ends of the vector. It is thus linear.

† The vector transformations induced by inverse collineations are inverse to each other. The four-dimensional transformations are therefore Cremona transformations.

the transformation is a new field, φ_1 and ψ_1 being expressible in terms of x_1, y_1. The ∞^3 trajectories of the first field are converted by the collineations into the ∞^3 trajectories of the new field; it is to be noticed however that, during any corresponding motions, positions which correspond according to the collineation will usually not correspond to the same instant of time; in fact from (2')

$$t_1 = \int \frac{dt}{k^2(a''x + b''y + c'')^2}.$$

75. If X, Y denote the velocity components at the position x, y and if the corresponding velocity in the second plane is X_1, Y_1, acting at the position x_1, y_1, then we find, from the canonical form (5),

(6) $\quad x_1 = \gamma\gamma_1/x, \qquad y_1 = \gamma_1 y/x,$
$\quad\quad X_1 = -k\gamma\gamma_1 X, \quad Y_1 = k\gamma_1(xY - yX).$

Thus we have a different vector transformation which may be termed the *phase* transformation* (in distinction from the *force transformation* of § 74): it gives the relation between the corresponding phases in the two planes.

If we speak of points and vectors which correspond in the two planes according to the given collineation as projectively related, then the result may be stated in this form:

The new phase vector does not coincide with the projection of the given phase vector: it has the same initial point, but the ratio of the actual length to the length of the projected vector is k^2 times the product of the distances from the ends of the original vector to the vanishing line of the collineation.

76. Having studied the Appell transformation and its geometric interpretation in terms of force vectors and phase vectors, we now ask whether other more general transformations can play a like rôle. Appell proved the following converse theorem:

---

* The phase of a particle at any instant, in the sense of Gibbs, is its position together with its velocity: it is defined by the four numbers $(x, y, \dot{x}, \dot{y})$.

The only transformations of the form

$$x_1 = \Phi(x, y), \qquad y_1 = \Psi(x, y), \qquad dt_1 = \mu(x, y)dt$$

which convert every set of differential equations

(1) $$\frac{d^2x}{dt^2} = \varphi(x, y), \qquad \frac{d^2y}{dt^2} = \psi(x,y),$$

into one of the same form are those defined by (2), (2').

77. By eliminating the time from (1), giving the differential equation of the trajectories in the form (page 7)

(7) $$(\psi - y'\varphi)y''' = \{\psi_x + (\psi_y - \varphi_x)y' - \varphi_y y'^2\}y'' - 3\varphi y''^2,$$

the author proved that the only point transformations which convert every trajectory system (of a positional field) into a trajectory system are the collineations. This remains valid even in the domain of all contact transformations, as we now proceed to show.

We first consider the class of differential equations (cf. page 11)

(8) $$y''' - G(x, y, y')y'' + H(x, y, y')y''^2$$

including (7) as a special case, and characterized geometrically by the possession of property I (that is, the focal locus for each element is a circle through the given point). We prove this theorem:

The only contact transformations which convert every equation of type (8) (that is, every system of curves with property I) into one of the same type are collineations and correlations.

That no other transformations are possible is seen as follows. If a contact transformation is to convert type (8) into itself, it must convert the part common to all systems of that type into itself. The curves defined by $y'' = 0$, that is, straight lines, obviously satisfy (8) for every form of G and H. It is obvious that no other (proper) curves satisfy all such equations. But since we are dealing with contact transformations and not merely point transformations, we must replace the concept *curve* by the concept *union*. In the plane the only unions which are not (proper) curves are points. A point is regarded as made up of ∞^1 lineal elements; so x is constant, y is constant, y' is arbitrary, and therefore y'' and y''' are infinite. Point unions are to be regarded then as solutions of all equations (8). The common part thus consists of the ∞^2 straight lines and the ∞^2 points of the plane. If this is to go into itself, either points go into points and lines into lines, or else points go into lines and lines into points. We thus obtain only collineations and correlations.

That the collineations actually leave type (8) unchanged is easily verified analytically.* The work for correlations is simplified by observing that every correlation may be reduced, by means of collineations, to the form of Legendre's transformation

(9) $\qquad x_1 = -y', \qquad y_1 = xy' - y, \qquad y_1' = -x,$

(which is simply polarity with respect to the conic $x^2 + 2y - 1 = 0$). Extending (9), we find

(9') $\qquad\qquad y_1'' = \dfrac{1}{y'''}, \qquad y_1''' = \dfrac{y'''}{y''^2}.$

This converts equation (8) into one of the same form

(10) $\qquad y_1''' = G_1(x_1, y_1, y_1')y_1'' + H_1(x_1, y_1, y_1')y_1''^2,$

the new coefficient functions being related to the old as follows:

(10') $\qquad\begin{aligned} G_1 &= H(-y_1', x_1y_1' - y_1, -x_1), \\ H_1 &= G(-y_1', x_1y_1' - y_1, -x_1). \end{aligned}$

This completes the proof of the theorem stated on the previous page.

78. If we impose property II on the system (8), that is, if we consider the subclass in which

(11) $\qquad\qquad H = \dfrac{3}{y' - \omega(x, y)},$

---

* *Trans. Amer. Math. Soc.*, vol. 7 (1906), p. 420.

the correlations are no longer available. That collineations actually convert this subclass into itself is readily verified. The same is true for the still narrower class, characterized by properties I, II, and III, in which the differential equation is of the form (cf. page 13)

$$(12) \qquad (y' - \omega)y''' = \{\lambda y'^2 + \mu y' + \nu\}y'' + 3y''^2.$$

79. We pass now to the case of dynamical trajectories, defined by type (7), and state the fundamental result:

Collineations are the only contact transformations of the plane which convert every system of ∞^3 dynamical trajectories (belonging to an arbitrary positional field of force) into such a system.

The only possibilities here also are collineations and correlations. The former actually have the required property. The latter have not, as is seen by observing that the application of the Legendre transformation (9) to a dynamical equation (8) will result in a new equation, which, while still of the general form (8), will not usually be of the dynamical form.*

80. Systems of trajectories are characterized by the set of five geometric properties of page 10. Therefore projective transformation will convert any system of curves having these properties into a system having the same properties. So, in spite of the fact that the properties as stated involve metric ideas (osculating parabolas, angles, circles of curvature, etc.), the set is actually projectively invariant. It ought to be possible therefore to *restate the geometric characterization in projective language.*

We shall not attempt to carry out this idea completely, and merely restate properties I and II as follows:

Consider the ∞^1 trajectories passing through a given point O in a given direction whose slope is y'. For each of these trajectories construct the conic which has four-point contact at O and touches the line determined by two arbitrarily selected

---

* We see from (10') that the coefficients G and H, which are rational with respect to y', are converted into coefficients which are not usually rational.

points* A and B (which remain fixed in the following statements); through A and B draw tangents to the conic (in addition to the fixed line) and join the points of contact. *The lines thus constructed, one for each of the ∞^1 trajectories, will form a pencil* (property I).

As the initial direction (that is y') varies about O, the vertex of the pencil just described will move along a straight line† *passing through O* (property II).

The other properties, especially the fifth, are much more complicated.

81. In conclusion we point out another way in which the projective group enters in dynamics. If an arbitrary point transformation

$$x_1 = \Phi(x, y), \qquad y_1 = \Psi(x, y)$$

is applied to the differential equations

$$\ddot{x} = \varphi(x, y), \qquad \ddot{y} = \psi(x, y),$$

defining motion under a purely positional force, the new differential equations, of the more general form

$$\Phi_x \ddot{x} + \Phi_y \ddot{y} + \Phi_{xx} \dot{x}^2 + 2\Phi_{xy} \dot{x}\dot{y} + \Phi_{yy} \dot{y}^2 = \varphi(\Phi, \Psi),$$

$$\Psi_x \ddot{x} + \Psi_y \ddot{y} + \Psi_{yy} \dot{y}^2 + 2\Psi_{xy} \dot{x}\dot{y} + \Psi_{yy} \dot{y}^2 = \psi(\Phi, \Psi),$$

will usually define a motion due to a positional force together with a force depending on the velocity $\dot{x}$, $\dot{y}$. If this latter force is to be absent the transformation will be affine, as already remarked (§ 74). If, instead, we demand that the latter force shall act in the direction of the velocity (and thus be in the nature of a resistance), we find that the transformation may be any collineation.

More generally, *projective transformations are the only point*

---

* In the original metric statements these are of course the circular points at infinity.

† The force direction will be determined projectively as the harmonic of this line with respect to the lines joining O to A and B.

ASPECTS OF DYNAMICS. 81

transformations which leave invariant the type

$$\ddot{x} = \varphi(x, y) + \dot{x}R(x, y, \dot{x}, \dot{y}),$$
$$\ddot{y} = \psi(x, y) + \dot{y}R(x, y, \dot{x}, \dot{y}),$$

defining motion of a particle under any positional force together with any resistance term acting in the direction of motion.

§§ 82–91. Conformal Transformations

82. The importance of conformal transformation is well known in connection with the theory of the potential. Geometric inversion or transformation by reciprocal radii, for example, yields the method of electric images due to Sir William Thomson. In connection with dynamics, the importance of general conformal transformations has been emphasized by Larmor, Goursat, and Darboux.*

83. Consider any conformal representation of the points of two surfaces S and S_1. The first fundamental forms of the surfaces may be taken to be

$$ds^2 = Edu^2 + 2Fdudv + Gdv^2,$$
$$ds_1^2 = \lambda(Edu^2 + 2Fdudv + Gdv^2),$$

where corresponding points have the same parameters u, v. *The principal theorem is that every natural system on one surface becomes by the conformal representation a natural system on the other.* This is obvious if we remember that natural systems are obtained by minimizing an integral in which the integrand is the element of length multiplied by any point function. Hence

The only point transformations (in any space) which convert every natural family into a natural family are the conformal.

84. Consider now the ∞^3 dynamical trajectories on S produced by a conservative field of force, the work function being W. These consist of ∞^1 natural families, one for each value of the

---

* Cf. the discussion in Routh, Dynamics of a Particle, Nos. 628–635 (method of inversion and conjugate functions).

constant of total energy h. It will be convenient to refer to the particular natural system produced in the given field W for a particular value h, as the family due to $W + h$.

The corresponding family on S_1 is due to

$$\frac{W + h}{\lambda}.$$

Hence the ∞^1 related natural families on S, found by varying h, go over by the conformal representation into ∞^1 natural families which are *not* usually related, that is, do not form the complete system of trajectories belonging to a conservative field. The only case in which the new families are related arises when

$$W = \lambda,$$

for then the new systems are due to the work function

$$W_1 = 1/\lambda.$$

We then reach the conclusion that *in any conformal representation (excluding the trivial homothetic case*) there is a unique conservative force whose complete system of ∞^3 dynamical trajectories is converted into the complete system of some (usually distinct) conservative force. The work function of the force in question is defined by the squared ratio of magnification,*

$$W = \lambda = \frac{ds_1^2}{ds^2}.$$

85. Similar statements may be made for brachistochrones. Every system of ∞^2 brachistochrones due to any work function and a given value of h of course becomes such a system, for any natural family may be regarded as a family of brachistochrones. But *there is only one complete system of ∞^3 brachistochrones which is converted into a complete system, namely, that defined by the work*

---

* It is obvious that in this case *every* complete system of trajectories becomes a complete system. The same holds for brachistochrones and catenaries.

function
$$W = 1/\lambda.$$

For any other work function the ∞^1 families of brachistochrones, due to $W + h$, become ∞^1 non-related natural families on S_1 due to
$$\lambda(W + h).$$

86. In the case of catenaries due to $W + h$, the ∞^1 usually non-related natural families corresponding on S_1 are due to
$$\frac{W + h}{\sqrt{\lambda}}.$$

Hence *the only complete system of catenaries which is turned into a complete system is defined by the work function**
$$W = \sqrt{\lambda}.$$

87. Consider, for example, the conformal representation of the plane
$$z = x + iy = re^{i\theta}$$
on the plane
$$z_1 = x_1 + iy_1 = r_1 e^{i\theta_1}$$
defined by
$$z_1 = z^n,$$
where n is neither 0 nor 1.

Here the squared ratio of magnification is
$$\lambda = r^{2(n-1)} = r_1^{\frac{2(n-1)}{n}}$$

---

* The three physical cases mentioned may be included in one general discussion by considering the extremals of
$$\int v^m ds \equiv \int (W + h)^{m/2} ds = \text{minimum};$$
when $m = 1$, we have least action and trajectories; when $m = -1$, least time and brachistochrones. For every value of m we obtain, by varying h, a system of ∞^3 curves. Cf. the general discussion of the systems S_k defined (for arbitrary fields) in Chapter IV.

Applying the theorems stated above, we find that the trajectories generated by
$$W = r^{2(n-1)}$$
go over into the trajectories of a new field
$$W_1 = r_1^{2\left(\frac{1}{n}-1\right)}.$$
For brachistochrones the corresponding fields are
$$W = r^{-2(n-1)}, \qquad W_1 = r_1^{-2\left(\frac{1}{n}-1\right)};$$
and for catenaries
$$W = r^{n-1}, \qquad W_1 = r_1^{\frac{1}{n}-1}.$$

The particular transformation $z_1 = z^2$, that is, $n = 2$, gives rise to simple fields. Stating the results in terms of the law of the central forces obtained, instead of the corresponding work functions, we have:

The trajectories of a central force varying as r (that is, the conics described about the center of force as center) become the trajectories of a central force varying as r_1^{-2} (that is, the conics described about the center of force as focus).

The brachistochrones of a central force varying as r^{-3} become the brachistochrones of a central force of constant intensity.

The catenaries of a central force of constant intensity become the catenaries of a central force varying as $r_1^{-3/2}$.

88. Returning to the general conformal representation, we observe that ∞^1 natural families forming a complete system of trajectories can never become a complete system of brachistochrones. For the trajectories on S due to $W + h$ become ∞^1 natural families on S_1, which, when regarded as brachistochrones, are due to $\lambda/(W + h)$; and there is no work function which reduces this expression to the form of a function of u, v plus a constant depending only on h. Thus for a given (non-homothetic) conformal transformation there is *one* system of trajectories

which is converted into a system of trajectories, and *one* system of brachistochrones which is converted into a system of brachistochrones, but there is *no* system of trajectories which is converted into a system of brachistochrones. The same is true for any two of the three types trajectories, brachistochrones, catenaries or of the infinite number of types described in the preceding footnote (page 83).

89. As another application, consider the *velocity curves* connected with a plane field of force whose work function is $W(x, y)$. For a given speed v_0, we obtain ∞^2 such curves, defined by the property that the curvature at each point and direction equals the curvature of a free particle starting out from that point and direction with the speed v_0. The differential equation of this velocity system is

$$y'' = \frac{(W_y - y'W_x)(1 + y'^2)}{v_0^2}.$$

This is recognized as a natural family; it corresponds to the geodesics of the surface whose first fundamental form is

$$e^{\frac{2W}{v_0^2}}(dx^2 + dy^2).$$

By varying v_0 we obtain the ∞^1 velocity systems belonging to the given field; they are pictured by the geodesics of ∞^1 surfaces.

Consider now a conformal representation of the xy-plane upon itself. This converts $dx^2 + dy^2$ into

$$e^H(dx^2 + dy^2),$$

where $H(x, y)$, by known theory, is a harmonic function. We thus obtain ∞^1 new natural families corresponding to the geodesics of the ∞^1 surfaces

$$e^{\frac{2}{v_0^2}W + H}(dx^2 + dy^2).$$

These ∞^1 natural families cannot usually be regarded as related velocity systems for some new field: the requisite condition is that W shall be the same as H except for a constant factor.

Hence for a given conformal transformation of the plane (which is not merely a similitude), there is a unique complete velocity system belonging to a conservative field of force which is converted into a complete system. The unique work function is

$$W = H = \log \lambda,$$

where λ denotes the squared ratio of magnifaction in the given conformal representation. The fields obtained are *Laplacian*, that is, satisfy the condition

$$W_{xx} + W_{yy} = 0.$$

As an example, the transformation $z_1 = \log z$ converts the ∞^3 velocity curves of the field $W = \log r$ (in which the force varies inversely as the distance from the origin) into the ∞^3 velocity curves of the field $W_1 = x_1$ (force vertical and constant).

90. It was shown above that conformal transformations are the only point transformations which convert every natural family into a natural family. Natural families are characterized by properties A and B of § 31. It is of interest to notice that property A by itself is conformally invariant. The most general system having this property (that osculating circles constructed at any point have another point in common) is what we have termed a velocity system. We now prove that

The only point transformations which convert every velocity system into a velocity system are the conformal transformations.

Consider say the three-dimensional case, where the general velocity system is

$$y'' = (\psi - y'\varphi)(1 + y'^2 + z'^2), \quad z'' = (\chi - z'\varphi)(1 + y'^2 + z'^2).$$

The only curves which are common to all such systems must

satis y

$$1 + y'^2 + z'^2 = 0, \quad y'' = 0, \quad z'' = 0,$$

and are therefore the minimal straight lines of space. Since the only transformations converting minimal lines into minimal lines are conformal, we have the result stated. That conformal transformations actually leave the velocity type invariant is easily verified analytically*. The result is obvious synthetically (in the case of more than two dimensions) since the conformal group converts circles into circles and bundles of circles into bundles. Hence if the original system possesses property A, the same will be true of the transformed system.

91. It may be shown that, for any given non-conformal transformation, there exists one and only one velocity system which is converted into a velocity system.

§§ 92–94. Contact Transformations

92. With each natural family, or, what is the same, with each isotropic medium, there is associated a definite infinitesimal contact transformation. This connection, which appears implicitly in Hamilton's fundamental memoir of 1835, was worked out in detail by S. Lie.†

If the index of refraction is $\nu(x, y, z)$, the associated contact transformation has the characteristic function

(1) $$\nu(x, y, z)\sqrt{1 + p^2 + q^2},$$

where x, y, z, p, q are considered as the coordinates of a surface element. If the one-parameter group generated is applied to an arbitrary surface the resulting ∞^1 surfaces form a wave set. The trajectories or rays appear as the path curves of this group. Lie showed that the category of transformations which thus

---

* Cf. *American Journal of Mathematics*, vol. 27 (1906), p. 213, for the two-dimensional case.

† "Die infinitesimalen Berührungstransformationen der Mechanik," *Leipziger Berichte* (1889), pp. 145–153. A very elegant discussion, with new results, is given by Vessiot, *Bull. Soc. math. de France*, vol. 34 (1906), pp. 230–269.

appears, with a characteristic function of type (1), and which he termed " the infinitesimal contact transformations of mechanics," is distinguished geometrically by the fact that the so-called* *transversality* relation reduces to orthogonality.

93. The following simple and easily proved theorem appears to be new.

The alternant (or Klammerausdruck of Lie) of the contact transformations associated with any two media is always a point transformation.

94. Here we are dealing with two natural families in the same three-dimensional space. In connection with the most general problem of dynamics (page 70), spaces of any dimensionality must be considered, with arbitrary variable curvature. The space depends on the quadratic form defining the kinetic energy: this determines the quadratic expression appearing under the radical in the generalization of (1). The potential† determines the factor ν which may be any point function. The general theorem is then as follows:

The alternant of the contact transformations associated with two dynamical problems (or natural families) will be a point transformation when, and only when, the two expressions for the kinetic energy are either the same or differ by a factor (which may be any point function); the two potential energies remain entirely arbitrary.

In particular, if any two natural families are constructed in the same space (which space is entirely arbitrary), the alternant will be a point transformation.

For a detailed discussion of the two-dimensional case, including a number of converse results, the reader is referred to the author's paper, cited in the first footnote below.

---

* Lie does not use this term. The author borrows it from the closely connected problem in the calculus of variation. See "The infinitesimal contact transformations of mechanics," *Bull. Amer. Math. Soc.*, vol. 16 (1910), pp. 408–412.

† Here considered as including the energy constant h, which is fixed, since we are dealing with a natural family.

§§ 95–97. A Group of Space-time Transformations

95. In the fundamental transformation of the relativity theory, known as the Lorentz transformation, the position coordinates x, y, z and the time coordinate t are merged: the new position and the new time appear as functions of both the original position and the original time. The Lorentz group is composed of the linear transformations of the four variables x, y, z, t which leave invariant the quadric

$$x^2 + y^2 + z^2 - c^2 t^2 = 0.$$

Its importance is due to the fact that it leaves unaltered the form of the Maxwell equations.

We consider in this section an entirely different group of space-time transformations, depending on arbitrary functions instead of arbitrary constants. It arises in connection with ordinary (newtonian) dynamics in the theory of forces depending on the time as well as position.

We confine the discussion for the sake of simplicity to the case of two dimensions. What transformations of the three variables x, y, t will convert any set of equations of the form

(1) $$\frac{d^2x}{dt^2} = \varphi(x, y, t), \qquad \frac{d^2y}{dt^2} = \psi(x, y, t)$$

into another set of the same form? An arbitrary transformation would produce equations representing a force depending, not only on x, y, t, but also on the velocity $dx/dt, dy/dt$. The problem is to find those peculiar transformations which do not introduce the velocity in the final equations. The result is as follows:

The only space-time transformations which convert every space-time field of force into a space-time field are those of the form

(2) $$t_1 = f(t), \quad x_1 = (ax + by)\sqrt{f'(t)} + g(t),$$
$$y_1 = (cx + dy)\sqrt{f'(t)} + h(t).$$

The group thus involves three arbitrary functions $f(t), g(t), h(t)$ as well as four arbitrary constants a, b, c, d.

96. Another representation of the same group, which has the advantage of avoiding radicals, is

(3)
$$\frac{dt_1}{dt} = \{\lambda(t)\}^2, \quad x_1 = (ax + by)\lambda(t) + \mu(t),$$
$$y_1 = (cx + dt)\lambda(t) + \nu(t).$$

When such a transformation is applied to equations (1), the new equations are found to be

$$\lambda^5 \ddot{x}_1 = (\lambda\ddot{\lambda} - 2\dot{\lambda}^2)(ax + by) + \lambda^2(a\varphi + b\psi) + \lambda\ddot{\mu} - 2\dot{\lambda}\dot{\mu},$$
$$\lambda^5 \ddot{y}_1 = (\lambda\ddot{\lambda} - 2\dot{\lambda}^2)(cx + dy) + \lambda^2(c\varphi + d\psi) + \lambda\ddot{\nu} - 2\dot{\lambda}\dot{\nu}.$$

Of course the original variables x, y, t are here to be replaced by their values in the new variables x_1, y_1, t_1.

97. The transformation converts the space-time curves of the original force into the space-time curves of a new force. Of course it is not a point transformation of the xy-plane, so it does not, as was the case for the Appell transformation (page 76), convert trajectories into trajectories. These remarks apply even in the special case where the force is positional. Consider, as a simple example, the transformation

$$t_1 = \tfrac{1}{2}e^{2t}, \quad x_1 = xe^t, \quad y_1 = ye^t,$$

applied to the equations

$$\ddot{x} = x, \quad \ddot{y} = y.$$

The transformed equations are found to be

$$\ddot{x}_1 = 0, \quad \ddot{y}_1 = 0.$$

The first field is central, the force varying directly as the distance, so that the trajectories are ∞^3 conics with the same center. The second force is everywhere zero, so the trajectories are merely ∞^2 straight lines.

CHAPTER IV

CONSTRAINED MOTIONS IN A FIELD. GENERALIZATION OF THE TRAJECTORY PROBLEM INCLUDING BRACHISTOCHRONES AND CATENARIES

§§ 98–114. Systems S_k Defined by $P = kN$

98. In connection with a field of force, the only curves usually studied are the lines of force and the trajectories. In the plane the lines of force form a simply infinite system, and the trajectories a triply infinite system. The former system has no peculiar properties, since any set of ∞^1 curves may be regarded as the lines of force in some field, in fact in an infinite number of different fields. The triply infinite system of trajectories has peculiar properties which have been discussed in Chapter I. Other noteworthy systems of curves are connected with the field, for example, brachistrochrones, catenaries, velocity curves, and tautochrones.

99. Omitting the tautochrones, *the other three systems named, together with the trajectories, may all be obtained as special cases of this simple general problem*: to find curves along which a constrained motion is possible such that the pressure is proportional to the normal component of the force.

100. If an arbitrary curve is drawn in the plane field of force, and the particle, of say unit mass, is started along it from one of its points with a given speed, the constrained motion along the given curve is determined. The acceleration along the curve is given by T, the tangential component of the force vector. So the speed at any point is determined by

(1) $$v^2 = \int T ds.$$

The pressure P (of course normal to the curve, since the curve is considered smooth) is given by the elementary formula

(2) $$P = \frac{v^2}{r} - N.$$

If we increase the initial speed, the effect is to increase v^2 by a constant c; and hence P changes by the addition of a term of the form c/r.

101. If the given curve is a trajectory, the initial speed may be so chosen that the pressure vanishes throughout the motion; that is, trajectories may be defined as curves of no constraint. Of course, if a different initial speed is used, P will be of the form c/r; but, as regards the curves, they are completely characterized by $P = 0$.

102. If the given curve is a brachistochrone and if the motion along it is brachistochronous, Euler proved (assuming the force to be conservative) that the pressure was double the normal component of the acting force and opposite to it in direction, that is, $P = -2N$. If the force is not conservative, the real brachistochrones, as defined by a problem of the calculus of variations, form a quadruply infinite system. The curves defined by the property $P = -2N$ then form a triply infinite system of what should be called pseudo-brachistochrones. These curves are really brachistochrones only in the conservative case. No ambiguity however will arise by terming the system here considered brachistochrones instead of pseudo-brachistochrones.

103. *The general problem suggested is to find curves such that P shall be proportional to N. So $P = kN$. To a given value of k there correspond ∞^3 such curves: the system so obtained will be denoted by S_k. The four special cases of physical interest are as follows*:

$k = 0$ gives S_0, the system of *trajectories*;

$k = -2$ gives S_{-2}, the system of *brachistochrones*;

$k = 1$ gives S_1, the system of *catenaries*;

$k = \infty$ gives S_∞, the system of *velocity curves*.

104. The last case requires a justification in terms of limits which is easily carried out analytically.

105. The third case follows from the known fact that when an inextensible flexible homogeneous string is suspended in any field of force, the resulting form of equilibrium, called a catenary in the general sense of the term, has the dynamical property that when a particle, started out with the proper initial velocity, rolls along the curve, the pressure at any point equals the normal component of the force: that is, catenaries are defined by $P = N$, corresponding to $k = 1$.

106. Of course a triply infinite system S_k exists for any value of the parameter k. The differential equation of the system, in intrinsic form, is easily obtained by eliminating v from the equations
(3) $$v^2/r = (k+1)N, \qquad vv_s = T.$$
The result is
(4) $$Nr_s = (n+1)T - r\mathfrak{N},$$
where
(4') $$n = 2/(k+1).$$

We may readily find various properties from this intrinsic equation, but in order to obtain a complete set it is necessary to have recourse to the equivalent equation in cartesian coordinates
(5) $$(\psi - y'\varphi)y''' = \{\psi_x + (\psi_y - \varphi_x)y' - \varphi_y y'^2\}y'' \\ - \left\{3 + \frac{(n-2)(\varphi + y'\psi)}{1 + y'^2}\right\}y''^2.$$

This obviously reduces to the familiar trajectory equation of §1 when $n = 2$, corresponding to $k = 0$. Brachistochrones correspond to $n = -2$, catenaries to $n = 1$, velocity curves to $n = 0$.

107. We now state the characteristic properties of a system of the above type for any value of n, that is, any value of k.

Characteristic Properties of the System S_k

Property 1.—For any given element (x, y, y') the foci of the osculating parabolas of the single infinity of curves determined by the given element lie on a circle passing through the given point.

Property 2.—At any point O the tangent of the angle which the focal circle makes with the given element is to the tangent of the angle which the given element makes with a certain direction fixed at O (the direction of the acting force) as 3 is to $n + 1$, that is, as $3k + 3$ is to $k + 3$.

Property 3. Through a given point there pass a single infinity of curves admitting hyperosculating circles of curvature; the centers of these circles lie on a conic passing through the given point in the direction of the force vector.

Property 4.—The normal at the given point O cuts the conic described in property 3, at a distance equal to $n + 1$, that is $(k + 3)/(k + 1)$, times the radius of curvature of the line of force passing through O.

Property 5.—This is of the same form as property V (§ 3) obtained in the discussion of trajectories, the number 3 being replaced by the number $n + 1$. In the notation of page 11

$$\frac{\partial}{\partial x}\frac{1}{AA'} + \frac{\partial}{\partial y}\frac{1}{BB'} + \frac{\omega\omega_{xy} - \omega_x\omega_y}{(n + 1)\omega^2} = 0.$$

108. The special case where n equals -1, that is, the system S_{-3}, is exceptional and requires a separate discussion; but as we do not need the results, this case is omitted.

109. While the properties corresponding to different values of k are analogous, they are of course not identical. The first property is common to all the systems. But the second property involves the parameter k. Thus, while for trajectories the constant ratio that appears is 1 (bisection), it is -3 for brachistochrones, $3/2$ for catenaries, and 3 for velocity curves. Not only are the triply infinite systems S_k, corresponding to different values of k, distinct in any given field of force, but also no two

systems arising in two distinct fields can ever coincide. For example, if a certain system of ∞^3 curves arises as trajectories in one field, it cannot also arise as catenaries in either the same or another field.

110. If we combine all the systems S_k, in a given field (φ, ψ), we obtain a quadruply infinite system which we now proceed to study. The differential equation of the fourth order defining this system is readily obtained by eliminating k from the equation of S_k. It is more convenient to carry this out in terms of intrinsic quantities, using either the radius of curvature and its first and second derivatives with respect to the arc, quantities denoted by r, r_s, r_{ss}, or else the radius of curvature together with the radii of the first and second evolute, quantities which we denote by r, r_1, r_2. The two sets of quantities are equivalent, being connected by the relations $r_1 = rr_s$, $r_2 = r^2 r_{ss} + rr_s^2$. The equation of the quadruply infinite system may then be put, using the notation of § 2, into the form

$$\begin{vmatrix} Nr_s + r\mathfrak{N} & T \\ Nr_{ss} + \left(2\mathfrak{N} - \dfrac{T}{r}\right) r_s & \mathfrak{T} + \dfrac{N}{r} \end{vmatrix} 0 = .$$

This may be written in either of the forms

$$r_{ss} = (\beta_1 + \beta_2 r^{-1})r_s + \beta_3 r + \beta_4,$$
$$r_2 = r^{-1} r_1^2 + (\beta_1 r + \beta_2) r_1 + \beta_3 r^3 + \beta_4 r^2,$$

where the β's are functions of x, y, y'.

111. We notice first that r_2 is quadratic with respect to r_1. Hence for given values of x, y, y', r, that is for a given curvature element, the ∞^1 curves of the system have the property that the locus of the third center of curvature is a parabola with axis parallel to the fixed radius of curvature, that is, perpendicular to the initial direction y'.

112. An equivalent statement is this: If for each of the curves we construct the osculating conic (five-point contact), the locus

of the centers of these conics is a conic passing through a given point in the given direction. It is perhaps worth while to restate this, so far as it concerns the four special cases of physical interest, as follows: In any plane field of force select any fixed element of curvature; corresponding to the initial values of x, y, y' and r so given, construct the unique trajectory, unique brachistochrone, unique catenary, the unique velocity curve, and the respective centers of the osculating conics; the four centers so found and the given point (x, y) will lie on a conic passing through the latter point in the given direction y'. (Cf. the first footnote on page 98.)

113. Keeping the curvature element fixed and varying the parameter k, the value of r_s or, what is equivalent, of r_1, varies linearly. As above, let n denote the fraction $2/(k+1)$; then if values of n forming an arithmetic progression are selected, the corresponding values of r_1 also form an arithmetic progression. The successive differences in the values of r_1 corresponding to the case of trajectory, brachistochrone, catenary, and velocity curve are proportional to $4, -3, 1$.

114. If in the system S_k we keep x, y, y' fixed and vary r, two limiting cases of interest arise. First, if r becomes infinite, then r_s is also infinite, and the limiting curve obtained is a straight line. In fact the ∞^2 straight lines of the plane form part of every system S_k.

On the other hand, if r approaches zero, then r_s approaches a definite limit
$$(n+1)T/N.$$
Remembering that the tangent of the angle of deviation is one third of r_s, we may state the result obtained as follows: In any system S_k if we take any lineal element and let r approach zero, the tangent of the corresponding angle of deviation is to the tangent of the angle which the force vector makes with the normal to the given element in the fixed ratio of $n+1$ to 3. The special values of this ratio for the four special systems of physical interest are respectively $1, -1/3, 2/3, 1/3$. In the case of trajectories, it is noteworthy that the limiting position of the axis

of deviation coincides with the direction of the force acting at the given point.

§§ 115–116. Curves of Constant Pressure

115. We now consider a second simple generalization of the problem $P = 0$, defining trajectories. We consider, namely, curves corresponding to $P = c$, where c denotes any constant. The curves obtained may be termed curves of constant pressure: only along such a curve is a constrained motion of a particle possible such that the pressure against the curve remains constant.

For a given value of c a system of ∞^3 such curves is obtained, whose intrinsic equation, found by differentiating the relation

$$P \equiv v^2/r - N = c,$$

is

$$(c + N)r_s = 3T - \mathfrak{N}.$$

We see that this system for any value of c retains property I of the system of trajectories. Omitting the discussion of the higher properties of these triply infinite systems we consider the quadruply infinite system whose differential equation, found by eliminating c, may be written in either of the intrinsic forms

$$(\mathfrak{N}r^2 - 3Tr)r_{ss} = (2r\mathfrak{N} - T)r_s^2 + [\mathfrak{N}_1 r^2 + (\mathfrak{N}_2 - 3\mathfrak{T})r - 3N]r_s,$$

$$r(\mathfrak{N}r - 3T)r_2 = (3r\mathfrak{N} - 4T)r_1^2 + [\mathfrak{N}_1 r^2 + (\mathfrak{N}_2 - 3\mathfrak{T})r - 3N]rr_1.$$

This gives the totality of ∞^4 curves of constant pressure defined by a given field.

As regards special cases of interest, we note, in addition to $c = 0$, giving trajectories, the case $c = \infty$ which gives $r_s = 0$, defining circles; hence for any field of force the ∞^4 curves of constant pressure include the ∞^3 circles of the plane, which arise in fact as curves of infinite pressure.

116. The quadruply infinite system which here arises, as well as that obtained in the previous problem $P = kN$, comes under

the category represented by a differential equation of the type*

$$y^{\text{IV}} = Ay'''^2 + By''' + C.$$

It therefore enjoys the property, previously stated in the other problem (§ 112), that the locus of the centers of the osculating conics corresponding to any element (x, y, y', y'') is a conic touching the element (x, y, y'). Of course, since the forms of A, B, C in the two problems are quite distinct, the systems are distinguished in their higher properties.

§§ 117–118. Tautochrones

117. Tautochrones are not included in either of the previous problems. They are not distinguished by any simple law of pressure.† The condition for a tautochrone is that the resulting constrained motion of a particle along the curve be harmonic, that is,

$$(1) \qquad T = k(s - s_0),$$

where k is a constant (which is negative for actual and positive for virtual tautochrones) and $s - s_0$ denotes the arc reckoned from a fixed point of the curve, the center of the tautochronous motion. From this

$$(2) \qquad T_{ss} = 0$$

and hence, by expansion, *the general equation of the system of ∞^3 tautochrones in any field is*‡

$$(3) \qquad Nr_s = \mathfrak{T}_1 r^2 + (\mathfrak{T}_2 + \mathfrak{N})r - T,$$

where the notation is that of § 2.

---

* This type (noteworthy in that it unifies many distinct mathematical and physical problems) first presented itself in the author's study of "Systems of extremals in the calculus of variations," *Bull. Amer. Math. Soc.*, vol. 13 (1907), p. 290: the extremals of any integral of the second order $\int f(x, y, y', y'')dx$ form a system of that type. In these lectures other physical problems leading to species included in this type are treated in §§ 110, 135, 137.

† It may be shown that during any tautochronous motion

$$P = k(s - s_0)^2/r - N.$$

‡ "Tautochrones and brachistochrones," *Bull. Amer. Math. Soc.*, vol. 15 (1909), pp. 475–483.

We see that r_s is a quadratic function of r, and not a linear function as in the case of trajectories and the other systems S_k. For a discussion of the geometric properties of tautochrones, we refer to the dissertation of H. W. Reddick.*

118. There is no field in which the tautochrones coincide with the trajectories, or with any of the systems S_k, in either the same or some other field, except for the case $k = -2$ corresponding to brachistochrones. The classical work of Huygens and J. Bernoulli showed that for a uniform field the system of tautochrones is identical with the system of brachistochrones. The author has shown that the only other field where such duplication occurs is that in which the force is central and varies directly as the distance. The only case of duplication in two distinct fields is as follows: The tautochrones of the field $\varphi = 0$, $\psi = y$ coincide with the brachistochrones of the field $\varphi = 0$, $\psi = y^{-3}$. The particular fields arising in this duplication problem are included in the interesting class of fields, involving eight parameters, characterized by the vanishing of the element function T_1. For such a field r_s, according to (3), becomes linear in r, and hence the ∞^2 straight lines of the plane are included in the system of tautochrones.†

118′. Each of the ∞^3 tautochrones in a given field has associated with it a certain time of oscillation, determined by the value of the constant k in (1). To each value of the period, that is, to each value of k, corresponds a certain family of ∞^2 tautochrones, whose differential equation, in implicit form, is

$$r(k - \mathfrak{T}) = N,$$

or, expanded,

$$(\psi - y'\varphi)y'' = k(1 + y'^2) - \{\varphi_x + (\varphi_y + \psi_x)y' + \psi_y y'^2\}.$$

We pass over the easy geometric interpretation; and note merely the special family, corresponding to the value $k = 0$, for which

---

* *Amer. Jour. of Math.*, vol. 33 (1911).

† The corresponding problem in space is treated in Reddick's paper and gives a class of fields involving twenty parameters.

the period is infinite. This separates the actual from the virtual tautochrones.

§ 119. NON-UNIFORM CATENARIES

119. It is a familiar fact that vertical parabolas appear in elementary dynamics in two distinct discussions; first, as trajectories of a cannon ball, and secondly as forms of equilibrium of a chain in which the mass (or load) of any element is proportional to the horizontal projection of that element. Here the force is ordinary gravity. The question arises whether any other fields of force give rise to a like duplication.

We first consider the following general problem of non-uniform catenaries. If a flexible string or chain, in which the mass of any element of length is proportional to some given function μ of x, y, y', is suspended in a positional field, the possible forms of equilibrium are defined by the equation

$$Nr_s = 2T - (1 + y'^2)N\bar{\mu}_{y'} - r\{\Re + (1 + y'^2)^{-\frac{1}{2}}N(\bar{\mu}_x + y'\bar{\mu}_y)\}.$$

This represents the ∞^3 non-uniform catenaries for a given field $\varphi(xy)$, $\psi(xy)$ and a given density law $\mu(x, y, y')$, where $\bar{\mu}$ denotes $\log \mu$.

On the other hand, the trajectories in the given field are defined by the equation

$$Nr_s = 3T - r\Re.$$

Our problem then is to find those fields for which the two systems described coincide. *The result obtained is that the field must be central or parallel.* The detailed result is as follows:

In any central field of force the ∞^3 trajectories may be also obtained as catenaries by loading the chain so that its density is proportional to the perpendicular dropped from the center to the tangent line. In the more special case where the field is parallel, the density is proportional to the sine of the angle between the element of the curve and the force.

It is easy to obtain analogous comparisons between brachistochrones and catenaries. In this case the density must vary inversely as the cube of the perpendicular dropped from the center (or of the sine of the angle referred to above). For example, in the case of gravity the vertical cycloids which appear as brachistochrones may be obtained as catenaries by causing the load applied to any element to vary inversely as the cube of its horizontal projection.

All the results may be included in a generalization found by comparing the non-uniform catenaries with the systems denoted by S_k in § 103. The density must vary as the $(n-1)$th power of the perpendicular, where n is the number defined on page 93. The field is necessarily central or parallel.

CHAPTER V

MORE COMPLICATED TYPES OF FORCE

§§ 120–122. Motion in a Resisting Medium

120. We consider the motion of a particle moving in the plane under a positional field of force and influenced by a resisting medium, the resistance acting in the direction of the motion and varying as some function of the speed v. The equations of motion will then be of the form

(1) $\quad \ddot{x} = \varphi(x, y) + \dot{x}f(v), \qquad \ddot{y} = \psi(x, y) + \dot{y}f(v),$

where the resistance R is equal to

$$R = vf(v).$$

The differential equation of the trajectories is found to be

(2) $\quad (\psi - y'\varphi)y''' = \{\psi_x + y'(\psi_y - \varphi_x) - y'^2\varphi_y\}$
$$- 3\varphi y''^2 - 2f\sqrt{\psi - y'\varphi}\, y''^{\frac{3}{2}},$$

where the argument v of f is to be expressed in terms of x, y, y', y'' by means of

$$v^2 = \frac{(\psi - y'\varphi)(1 + y'^2)}{y''}$$

Consider now the ∞^1 trajectories starting from a given element (x, y, y'). The focal locus, that is, the locus of the foci of the osculating parabolas, varies in shape with the function f, that is, with the law of resistance.

We know that, if there is no resistance, property I of § 3 holds, that is, the focal locus is a circle passing through the given point. Are there any resisting media for which this property is preserved? A simple discussion shows that there are, the appropriate media being those for which R is of the form $Av^2 + B$.

For such media, property II will not usually be fulfilled; in fact *the only medium preserving the properties I and II is that in which the resistance varies as the square of the speed.*

If we impose also property III, both A and B must vanish, that is, the resistance vanishes and the force is purely positional.

It is of interest to examine the case where the resistance varies as any power v^n of the speed. The differential equation of the trajectories is then of the form

$$y''' = ay'' + by''^2 + cy''^m,$$

where
$$m = \tfrac{1}{2}(4 - n).$$

The focal locus is a curve whose inverse with respect to the given point is
$$X = a_1 + b_1 Y + c_1 Y^{m-1}.$$

This becomes a straight line (as in the case of no resistance), when m is 1 or 2, that is, when n is 2 or 0.

The curve is a conic when m is 3 or 0 or 3/2, that is, when n has one of the values -2 or 4 or 1. When $n = -2$ the conic is a parabola with its axis parallel to the given element. When $n = 4$ it is a hyperbola, asymptotic to the line of the given initial element. When $n = 1$ it is a parabola touching the initial line (not at the given point).

121. We now state briefly the corresponding results in ordinary space. No matter what the law of resistance is, property I (of the set of four properties for space given in § 11) is fulfilled; for the osculating planes necessarily pass through the force vector. The only laws for which property II is preserved are those included in
$$R = Av^2 + B.$$

If property III is also to be preserved, the resistance must vanish.

122. The results may be derived easily from the intrinsic equations
(3) $$v^2 = rN, \qquad vv_s = T + R,$$

obtained by taking components of the acting forces along the normal and tangent to the trajectory. The geometric equation, resulting from the elimination of v, is of the form*

$$(4) \qquad Nr_s = -r\mathfrak{R} + 3T + 2R.$$

This gives the relation between r_s (the rate of variation of r with respect to s) and r (the radius of curvature). The resistance R, which is given as a function of v, is here to be expressed in terms of r by means of the first of the relations (3). If property I, of plane trajectories, is to hold, r_s must be a linear integral function of r; this will be the case not only when R vanishes, but also, as stated above, when it is of the form $Av^2 + B$.

§§ 123–126. Particle on a Surface

123. The motion of a particle on any constraining surface

$$x = \varphi(u, v), \qquad y = \psi(u, v), \qquad z = \chi(u, v)$$

under any positional forces may be investigated most simply by means of the Lagrangian equations

$$\frac{d}{dt}\left(\frac{\partial T}{\partial \dot{u}}\right) - \frac{\partial T}{\partial \dot{v}} = U, \quad \frac{d}{dt}\left(\frac{\partial T}{\partial \dot{v}}\right) - \frac{\partial T}{\partial v} = V,$$

where T is the kinetic energy

$$2T = E\dot{u}^2 + 2F\dot{u}\dot{v} + G\dot{v}^2$$

and U, V are the components of the force given as functions of u, v.† The explicit equations of motion are of the form

$$\ddot{u} = \Phi + A_0\dot{u}^2 + 2A_1\dot{u}\dot{v} + A_2\dot{v}^2,$$
$$\ddot{v} = \Psi + B_0\dot{u}^2 + 2B_1\dot{u}\dot{v} + B_2\dot{v}^2;$$

---

* From this we may obtain the following dynamical result: If a particle starts from rest, the initial radius of curvature of the trajectory is to the radius of curvature of the line of force passing through the initial point as $3T + 2R$ is to T. When R vanishes we have the simple result previously stated.

† See for example Whittaker, Analytical Dynamics, p. 390, and Hadamard, *Jour. de Math.* (5), vol. 3, p. 331.

where Φ, Ψ define the force and the A's and B's are functions of u, v depending only on the given surface.

124. We observe that here $\ddot{u}$, $\ddot{v}$ depend not only on the position u, v but also upon the velocity $\dot{u}$, $\dot{v}$. Hence the motion in the uv-plane corresponding to the actual motion on the surface is not usually generated by any positional force in that plane. The only exception arises when the A's and the B's vanish identically: this is the case only if the given surface is developable, and if its representation on the uv-plane differs from its development on the plane by at most an affine transformation.

Another problem including this as a special case is to determine when the motion in the uv-plane can be regarded as due to a positional force together with a resistance acting in the direction of the motion. The condition for this is

$$\frac{A_0\dot{u}^2 + 2A_1\dot{u}\dot{v} + A_2\dot{v}^2}{B_0\dot{u}^2 + 2B_1\dot{u}\dot{v} + B_2\dot{v}^2} = \frac{\dot{u}}{\dot{v}}.$$

Expanding, we find four conditions on the six functions A, B, which turn out to be precisely the conditions that the geodesics of the surface shall be pictured by straight lines, a result which may be proved directly. Hence the only case in which the motion on the surface is pictured in the uv-plane by a motion due to a positional force together with a resistance depending on the velocity components and acting in the direction of the motion, is that in which the surface has constant curvature and the representation is geodesic.

125. We proceed with the general equations of motion. If we eliminate the time, we obtain the differential equation of the third order defining the ∞^3 trajectories in the form

$$(\Psi - v'\Phi)v''' = \{\delta_0 + \delta_1 v' + \delta_2 v'^2 + \delta_3 v'^3 + \delta_4 v'^4 + \delta_5 v'^5\} \\ + \{\epsilon_0 + \epsilon_1 v' + \epsilon_2 v'^2 + \epsilon_3 v'^3\}v'' - 3\Phi v''^2,$$

where the coefficients are functions of u, v. We confine ourselves to the observation that the picture curves in the uv-

plane come under the type

$$v''' = F_0 + F_1 v'' + F_2 v''^2,$$

where the coefficients are lineal-element functions: the focal locus is thus not a circle, but a special quartic. Hence if we consider the ∞^1 trajectories on the surface obtained by starting a particle at a given point in a given direction with different speeds, the picture curves in the uv-plane have osculating parabolas at the common point whose foci lie on a special quartic curve.

126. What is the simplest property of the actual trajectories described on the surface? What is, in particular, the locus of the osculating spheres of the ∞^1 trajectories considered?

To answer this we take our surface not in parametric form, but in the explicit form

$$z = f(x, y).$$

We may take the given point as origin, the tangent plane as the xy-plane, and the fixed initial direction as that of the axis of x. We find, by differentiating the equation of the surface and making use of $y' = 0$, $z' = 0$, that

$$z'' = a, \qquad z''' = b + cy'',$$

where a, b, c are constants, equal respectively to the values of the partial derivatives f_{xx}, f_{xxx}, $4f_{xy}$ at the origin. Again, from the general equation of the trajectories, we have a relation of the form

$$y''' = \alpha + \beta y'' + \gamma y''^2.$$

The center of the osculating sphere of the trajectory is then

$$X = 0,$$

$$Y = \frac{z'''}{y''z''' - z''y'''} = \frac{b + cy''}{y''(b + cy'') - a(\alpha + \beta y'' + \gamma y''^2)},$$

$$Z = \frac{-y'''}{y''z''' - z''y'''} = \frac{-(\alpha + \beta y'' + \gamma y''^2)}{y''(b + cy'') - a(\alpha + \beta y'' + \gamma y''^2)}.$$

Here y'' enters as parameter, varying from curve to curve: eliminating it, we find the locus, lying in the plane $X = 0$, to be

$$\alpha Y^2 + \beta Y(1 - aZ) + \gamma(1 - aZ)^2 + Z\{bY + c(1 - aZ)\} = 0.$$

Hence for any positional force on any surface, the ∞^1 trajectories starting from a given lineal element of the surface have osculating spheres, at the common point, whose centers lie on a (general) conic in the plane normal to the element.

This conic passes through the center of curvature of the normal section of the surface determined by the given element. If the element is in one of the principal directions of the surface, the conic touches the normal to the surface.

§§ 127–130. THE GENERAL FIELD IN SPACE OF n-DIMENSIONS

127. Any dynamical system with n degrees of freedom may be represented by a particle in space of n dimensions. For example, an arbitrary rigid body in ordinary space is represented by a particle in six-dimensional space, and the astronomical problem of three bodies in the most general case leads to a representative particle in space of nine dimensions.

For conservative forces, or natural families, the general discussion for any dimensionality has already been given (§ 69). We shall not attempt a complete discussion for arbitrary positional forces (corresponding to that given in Chapter I for two and three dimensions). The equations of motion for an arbitrary field are

$$\ddot{x}_1 = \varphi_1(x_1, \cdots, x_n), \quad \cdots, \quad \ddot{x}_n = \varphi_n(x_1, \cdots, x_n).$$

We confine ourselves to the simplest questions. If the initial position and initial direction are kept fixed, and only the initial speed v is varied, what are the properties of the ∞^1 trajectories obtained? The simplest geometric result is that r_s (the rate of variation of the radius of curvature with respect to the arc length) varies as a linear function of r. The locus of the centers of the osculating spheres is a straight line, just as in the case where n is three.

128. A general curve in n-space has at each point an osculating plane, an osculating 3-flat, and so on up to an osculating $(n-1)$-flat. It is obvious that our ∞^1 trajectories have the same osculating plane since this is determined by the given initial direction and the direction of the force. It can be shown that the osculating 3-flat is also fixed; the 4-flat varies, generating a pencil; the 5-flat varies, generating a quadratic system; and so on, with more complicated variations.

129. Consider next the connection between the various curvatures and the speed.

In the plane ($n = 2$) there is only one curvature γ_1, and this varies inversely as the square of v.

In space ($n = 3$) the first curvature γ_1 varies as above, and the second curvature or torsion γ_2 remains fixed.

If $n = 4$, we have three curvatures. The laws for γ_1 and γ_2 are as above, while
$$\gamma_3 = c_1 + c_2 v^{-2},$$
where c_1, c_2 are constants (depending of course on the given initial lineal element).

If $n = 5$, we have $\gamma_1 = av^{-2}$, $\gamma_2 = b$ (these forms are valid for any dimensions) and
$$\gamma_3 = \sqrt{c_1 + c_2 v^{-2} + c_3 v^{-4}}, \qquad \gamma_4 = \frac{d_1 + d_2 v^2 + d_3 v^4}{d_4 + d_5 v^2 + d_6 v^4}.$$

If $n = 6$, γ_3 remains the same, the numerator in γ_4 is replaced by the square root of a polynomial involving v^8, and γ_5 is given by a rational formula.

It is easy to write down the general formulas for the $n-1$ curvatures in n space. All except the first, second, and the last are irrational. These results are to be regarded as generalizations of the elementary fact (included in the formula for centrifugal force v^2/r), that the ordinary curvature varies as v^{-2}.

130. By eliminating v from any two of the formulas, we can obtain purely geometric results. For example, in space of four dimensions, $\gamma_3 = A + B\gamma_1$, where A and B depend only on the

common initial element. But in higher spaces

$$\gamma_3 = \sqrt{A + B\gamma_1 + C\gamma_1^2}.$$

This is the form required in particular in the application to the problem of three bodies, since the representative space has nine dimensions.

§§ 131–132. Interacting Particles in the Plane and in Space

131. We consider the motion of $n + 1$ particles, denoted by $M, M_1, \cdots, M_n$, moving in the plane under the action of any forces depending on the position of the particles. The differential equations of motion are then of the form

$$\ddot{x} = \varphi(x, y, x_1, y_1, \cdots, x_n, y_n),$$
$$\ddot{y} = \psi(x, y, x_1, y_1, \cdots, x_n, y_n),$$
$$\ddot{x}_1 = \varphi_1(x, y, x_1, y_1, \cdots, x_n, y_n),$$
$$\ddot{y}_1 = \psi_1(x, y, x_1, y_1, \cdots, x_n, y_n),$$

and so on, where the masses—which cannot be assumed to be unity as in the case of a single particle—are absorbed with the forces in the right hand terms. From these equations the following properties may be deduced.

(1) Given the phases of $M_1, \cdots$, and the position and the direction of M, a set of ∞^1 trajectories of M is determined (one for each value of the speed). The foci of the osculating parabolas lie on a special quartic curve whose inverse with respect to the given point is a parabola tangent to the given initial line (the point of contact, however, is usually not the given point).

(2) If the speed of one of the remaining particles, say M_1, is varied, all the other initial conditions being unaltered, the parabolic locus just obtained varies. Its point of contact with the initial line remains fixed and all the ∞^1 parabolas, one for each value of the speed, are homothetic with respect to the point of tangency.

(3) The normal constructed at the common point of tangency cuts the parabola again at a distance d which varies in such a way that the square root of d can be expressed as a linear combination of the square roots of the radii of curvature of the corresponding trajectories described by the particles $M_1, \cdots, M_n$.

(4) If we preserve the phases of the particles $M_1, \cdots, M_n$, then, for each initial direction y' of M, we obtain, by (1), a certain parabolic locus. Consider the relation between the axis of this parabola and the initial direction. It is found that the initial direction y' always bisects the angle between the direction of the force acting at the given point and the direction of the axis of the parabola.

(5) Furthermore, the point where the parabola touches the initial line describes, when y' varies, a quartic curve whose inverse with respect to the given point is a conic passing through that point in the direction of the force.

It is to be observed that the statement (3) about the variation of d simplifies considerably in the case of *two* particles (that is, $n = 1$). In that case d varies directly as the radius of curvature of the trajectory described by M_1.

132. A few corresponding results for the case of any number of particles moving in space are as follows: If the speed of M is the sole arbitrary parameter, the ∞^1 trajectories of M have the same osculating plane; the torsion varies according to a linear integral function of the square root of the curvature; the locus of the centers of the osculating spheres is a cubic curve of special type.

If we assign the phases of all the particles except M_1 and assign the position and direction of M_1, then the speed of M_1, or, in consequence, the curvature of the trajectory described by M_1, is the only arbitrary parameter. There will then be ∞^1 corresponding trajectories described by M. These will of course start from the same point in the same direction with a common osculating plane and a common curvature, that is, they all have contact of the second order. The torsion varies and so does the center of the osculating sphere. The simultaneous variation is controlled by the law that the distance from the center of the osculating sphere to the fixed center of curvature varies as a linear integral function of the radius of torsion. An equivalent statement is that the rate of variation of the radius of curvature per unit of the arc is expressed by a linear integral function of the torsion.

All these results apply in particular to the three-body problem. The present application is more concrete than that indicated in § 130, since no higher space is here introduced.*

§§ 133–141. FORCES DEPENDING ON THE TIME. TRAJECTORIES AND SPACE-TIME CURVES

133. Hitherto the force has been assumed to be independent of the time; now we consider the generalization where the force depends in any way upon the time as well as the position. Take the case of a particle moving in the plane; the equations of motion are then of the form

(1) $$\ddot{x} = \varphi(x, y, t), \qquad \ddot{y} = \psi(x, y, t).$$

From these, by differentiation and elimination, we may derive

(2) $$y''' = Py'' + Qy''^2 + Ry''^{\frac{3}{2}},$$

where the coefficients are functions of x, y, y', t, namely,

$$P = \frac{\psi_x + y'\psi_y - y'(\varphi_x + y'\varphi_y)}{\psi - y'\varphi},$$

$$Q = \frac{-3\varphi}{\psi - y'\varphi}, \qquad R = \frac{\psi_t - y'\varphi_t}{(\psi - y'\varphi)^{\frac{3}{2}}}.$$

If we are given the initial time, position and direction, that is, the initial values of t, x, y, y', there will be a certain set of ∞^1

---

* Since the forces in the three-body problem are conservative, we may decompose the motions into natural families, and interpret each family in a flat space of eight dimensions. The circles of curvature at a given point will meet again; eight of them will be hyperosculating, and these will be mutually orthogonal. Cf. § 70.

trajectories, one for each value of the initial speed. The following properties are obtained:

(1) We find that the focal locus (that is, the locus of the foci of the ∞^1 osculating parabolas) is a quartic curve whose inverse with respect to the given point is a parabola which is tangent to the given direction line (the point of contact is not usually at the given point).

(2) As y' varies (x, y, t being held fixed) this point of contact describes a cubic curve whose inverse is a conic passing through the given point in the direction of the force.

(3) The initial direction of y' bisects the angle between the direction of the force and the direction of the axis of the parabola described in (1).

134. The total system of trajectories, for all initial conditions, consists of ∞^4 curves. Only in the case where the force does not depend upon the time does the system consist of ∞^3 trajectories. In the properties stated above, the initial time is kept fixed. In a certain sense then the results are not purely geometric: they would not appear in a photograph of the complete system of trajectories. This system will be represented by a certain differential equation of the fourth order; but it is not possible to carry out the requisite eliminations in explicit form, and hence the derivation of purely geometric properties involves essentially new difficulties. A complete characterization is however obtained, by projection from space curves, in §§ 136, 140.

135. There is an interesting special case in which the elimination can be carried out: namely, the problem of the motion of a *particle of variable mass* in a positional field of force. The time then appears only through the mass, so the equations of motion are of the form

(3) $$f(t)\ddot{x} = \varphi(x, y), \qquad f(t)\ddot{y} = \psi(x, y).$$

As the result of the elimination is complicated, we shall here consider only the case where the function $f(t)$, representing the mass, is of one of the special types t^4, t^2, e^t, $(\log t)^2$. The equation of the fourth order representing the trajectories is then found to be of the form

(4) $$y^{IV} = Ay'''^2 + By''' + C,$$

where A, B, C involve only x, y, y', y''.

We see that the fourth derivative is a quadratic function of the third derivative. This category of equations of the fourth order arises in a number of different connections, in particular in the inverse problem of the calculus of variations, as stated in § 116. The characteristic geometric property may in the present case be stated as follows:

If the particle, whose mass varies according to one of the four laws stated, is projected into a field of force from a fixed initial position in a fixed direction at different times, with the initial speed for each time so adjusted as to cause the initial curvature of the trajectory to have a fixed value, and if for each of the ∞^1 trajectories thus obtained we construct the osculating conic (having five-point contact), the locus of the centers of these conics is a conic passing through the given conic in the given direction.

Of course not every system of ∞^4 curves having this property can be regarded as a trajectory system corresponding to equations of motion of the form considered. We do not, however, attempt a complete characterization.

136. *Space-time Curves.*—When we integrate the equations of motion, either in the special case where the forces depend only on the position

(1') $$\ddot{x} = \varphi(x, y), \qquad \ddot{y} = \psi(x, y),$$

or in the general case where the force depends also on the time

(1) $$\ddot{x} = \varphi(x, y, t), \qquad \ddot{y} = \psi(x, y, t),$$

we obtain x and y expressed as functions of t and four constants of integration. If we represent t by an ordinate perpendicular to the xy-plane, thus considering x, y, t as rectangular coordinates in space, we obtain a certain system of ∞^4 curves in that space which we designate as *space-time curves*.*

If we project these curves orthogonally on the xy-plane, we obtain the trajectories. In the general case (1) there will be ∞^4 of these trajectories; but in the special case where the force is positional, only ∞^3 trajectories arise, since the system of space-time curves, whose number is still ∞^4, now admits the group of translations along the t-axis.

If we project the space-time curves orthogonally on the xt-plane and on the yt-plane, we obtain in each case a system of ∞^4 plane curves.

What are the properties of the system of ∞^4 space-time curves? The following two properties are characteristic:

(1). The osculating planes of the ∞^2 space-time curves through a given point go through a fixed line parallel to the xy-plane. (This line is parallel to the direction of the force acting at the projected point in the xy-plane.)

(2). If the ∞^2 space-time curves through the given point are orthogonally projected on any plane perpendicular to the xy-plane, the ∞^2 plane curves obtained are such that those which have the same tangent also have the same curvature.

Another complete characterization may be given as follows:

(3). If the ∞^2 space-time curves through a given point are orthogonally projected on either the xt-plane or the yt-plane, the ∞^2 plane curves obtained have their centers of curvature located on a special cubic of the form $t^3 = a(x^2+t^2)$ or $t^3 = b(y^2+t^2)$. A corresponding cubic locus will then necessarily arise by projection on any plane perpendicular to the xy-plane.

---

* It may be remarked that if, in problem (1), the force is multiplied by a constant c (or, what is equivalent, the mass of the particle is multiplied by $1/c$), a distinct system of ∞^4 space-time curves will be obtained. The totality of ∞^5 space curves, thus related to the ∞^1 plane problems

$$\ddot{x} = c\varphi(x, y, t), \qquad \ddot{y} = c\psi(x, y, t),$$

may be generated as trajectories in a three-dimensional positional field of force. The ∞^5 curves have the four characteristic properties of a space system (§ 11) and the further peculiarity that the direction of the force is parallel to the xy-plane.

137. Consider the ∞^4 curves in say the xt-plane. These are the curves representing graphically the relation between the abscissa x and the time t. By eliminating y from the set (1), we obtain a relation of the form

$$x^{\text{IV}} = A\dddot{x}^2 + B\dddot{x} + C,$$

where A, B, C involve only x, $\dot{x}$, $\ddot{x}$ and the independent variable t. The fourth derivative is thus always quadratic with respect to the third derivative. Hence, by § 116, we have this result:

In the xt-plane (or, more generally, in any plane perpendicular to the plane xy in which the motion actually takes place), the ∞^1 curves having any element of curvature in common are such that the locus of the centers C' of their osculating conics (constructed at the common point) is a conic passing through the common point in the direction of the common tangent.

As indicated above, the ∞^4 curves in the xy-plane, that is, the trajectories, do not usually enjoy this simple property. Even in the case where the time enters only through the mass, the locus of the centers of the osculating conics may be of any degree of complication. Its shape depends on the law of variation of the mass. Only for the special laws stated at the bottom of page 112, together with certain combinations of them, is the equation of the trajectories of the quadratic type.

138. It is possible to obtain additional general properties of the xt-system, describing how the locus conic, corresponding to a curvature element, changes when the element changes. For the coefficients A, B, C determining the position of the conic have the following forms: A does not involve $\dot{x}$, B is linear and integral in $\dot{x}$, C is quadratic and integral in $\dot{x}$. Hence these results:

If the curvature element is varied, at the given point O, in such a way that the second derivative $\ddot{x}$ is constant, so that only $\dot{x}$ varies, the center C'' of the corresponding locus conic describes a new conic.

At the same time a certain two-to-one correspondence arises between the initial direction of the element and the direction of the line joining O to the center C''.

139. A clearer picture is perhaps obtained by changing the notation to correspond with the usual x, y, z notation for rectangular coordinates in space. It is then desirable to lay off the time on the x-axis, since this is the independent variable. The actual motion then takes place in the yz-plane, and the differential equations of motion are

$$\frac{d^2y}{dx^2} = \varphi(x, y, z), \qquad \frac{d^2z}{dx^2} = \psi(x, y, z).$$

The curves in space x, y, z are then the space-time curves. Their projections on the yz-plane are the trajectories (whose explicit properties have not been derived). Their projections on the xy-plane (or on the xz-plane, or on any plane parallel to the z-axis) are curves whose properties have just been stated (§§ 137, 138). The differential equation in the xy-plane is

$$y^{IV} = \alpha y'''^2 + (\beta_1 + \beta_2 y')y''' + (\gamma_1 + \gamma_2 y' + \gamma_3 y'^2),$$

where the coefficients involve only x, y, and y'.

140. We have not attempted a complete direct characterization of the systems of curves arising in any one of the coordinate planes. Such a characterization has however been given (§ 136) for the system of ∞^4 space-time curves. Indirectly this really solves all the problems. A system of curves in the plane can be regarded as trajectories of a force depending on time and position if and only if the curves can be obtained by orthogonal projection from some system of ∞^4 curves in space having the properties (1) and (2) of § 136. If, furthermore, the space system is invariant under translation perpendicular to the given plane, the plane system, then consisting of only ∞^3 curves, belongs to a positional field.

141. For any force depending on time and position

$$\ddot{x} = \varphi(x, y, t), \qquad \ddot{y} = \psi(x, y, t),$$

the number of space-time curves is always ∞^4. When we project these on the xy-plane, to obtain the trajectories, the number is usually ∞^4. The number reduces to ∞^3 if the force is positional but does not vanish; in the latter case the trajectories are merely the ∞^2 straight lines.

In the xt-plane the usual number of curves is ∞^4. The only exception arises when the function φ is free from the variable y. In this case the xt-curves all satisfy the equation of second order $\ddot{x} = \varphi(x, t)$ and therefore their number is only ∞^2. Similar statements hold of course for the yt-plane.

Consider, as a single example, gravity, taken as uniform and acting in the vertical xy-plane. The equations of motion are

$$\ddot{x} = 0, \quad \ddot{y} = g.$$

The xyt-curves are

$$x = at + b, \quad y = \tfrac{1}{2}gt^2 + ct + d,$$

a certain family of ∞^4 parabolas in space. The xt-curves are ∞^2 straight lines. The yt-curves are ∞^2 parabolas. The xy-curves (that is, the trajectories) are ∞^3 parabolas

$$y = \alpha x^2 + \beta x^2 + \gamma.$$

It is to be observed that if the gravity constant g is changed, the new problem, while giving the same trajectories, gives a distinct family of xyt-curves. If g takes all possible values, the totality of space-time curves obtained is formed of ∞^5 parabolas (namely, those whose axes are parallel to the t-axis). These curves, in accordance with the general statement made in the footnote on page 114, are the trajectories of a positional field in space, the generating force being constant and acting in the t-direction.

All the results can be extended so as to apply to the four-dimensional space-time curves depicting motion in ordinary space.